AN INTRODUCTION TO

X-Ray Physics, Optics, and Applications

AN INTRODUCTION TO

X-Ray Physics, Optics,
and Applications

CAROLYN A. MacDONALD

PRINCETON UNIVERSITY PRESS
Princeton and Oxford

Published by Princeton University Press, 41 William Street,
Princeton, New Jersey 08540
In the United Kingdom: Princeton University Press,
6 Oxford Street, Woodstock, Oxfordshire OX20 1TR

press.princeton.edu

Jacket images (top, left to right): Fig. 1. Courtesy of Michael
Flynn, University of Michigan. Fig. 2. From *Journal of Analytical
Atomic Spectrometry* (2011) / The Royal Society of Chemistry.
Fig. 3. From *Proceedings of the 46th Annual Denver X-ray
Conference* (1997) / Courtesy of Scott Rohrbach. (Bottom)
Fig. 4 © American and Science Engineering, Inc.

ISBN 978-0-691-13965-4

Library of Congress Control Number 2017935618

British Library Cataloging-in-Publication Data is available

This book has been composed in Minion Pro and
Gotham Narrow

Printed on acid-free paper. ∞

Printed in the United States of America

10 9 8 7 6 5 4 3 2

In memory of my father *Harry Edward MacDonald*

and my mentor *Walter Maxwell Gibson*

Contents

This text focuses on the physics of x-ray generation and of x-ray interaction with materials. It particularly emphasizes the common physics necessary to understand diverse phenomena and applications. For example, knowledge of characteristic line emission is necessary to explain the spectrum from a mammography system or to employ fluorescence analysis. Alternatively, understanding the process of broadband emission from a radiography tube provides the basis for assessing quantitative limits in chemical analysis in an electron microscope. Photoelectric absorption gives rise to patient dose in mammography, absorption spectroscopy, and imaging detectors. Surprisingly, given that the physics has been well known for a century, there has been a recent dramatic increase in the availability, variety, and performance of x-ray sources, optics, and detectors, and examples are provided throughout the chapters.

The text is organized around a few key physical concepts. Chapter subtopics, examples, and end-of-chapter problems include applications drawn from medicine, astronomy, and materials analysis. Detailed solutions to all problems are provided in the appendix; these have significant extra information which should be considered part of the text.

The material is suitable for a semester or year-long introductory course in x-ray physics, optics, analysis, or imaging. It is intended for advanced undergraduate or graduate students in physics or related sciences who have a basic familiarity with electromagnetism and quantum mechanics but no specialized knowledge of x-ray or optical physics. It is also intended for researchers in related fields who wish to gain knowledge of x-ray application areas. Each chapter is kept fairly short and includes examples and end-of-chapter problems. A bibliography is presented at the end of each chapter for additional reading to cover topics in more depth. The basic physics is concentrated in chapters 4–7, 9, 11, and 13, and these could be used as the core of a one-semester course. X-ray optics are explored in chapters 12 and 15.

ACKNOWLEDGMENTS

Very many people have contributed to this book, including a host of colleagues and researchers. I am grateful to a generation of students who have commented on numerous versions of the class notes which eventually mutated into this book, and to two dozen doctoral students whose graphs and data appear within. I am particularly indebted to Ariel Caticha for theoretical discussions, Robert Schmitz and Laila Hassan for extensive proofreading and commentary, and Barbara Liguori for painstaking copyediting.

Mostly, I am enormously thankful for the patience and tolerance of friends and family and the loving encouragement of my mother, Betty Rutter MacDonald, and especially my husband, Norman Ross Stewart, Jr.

Constants

Symbol	Units	Description	Section
c	m/s	speed of light	1.2
e^-		electron	
e		2.718 . . .	
h	J·s	Planck's constant	1.2
$\hbar$	J·s	Planck's constant/2π	1.2
i		$\sqrt{-1}$	
k_B	J/K	Boltzmann's constant	3.1, 3.5, 8.1
m		metastable	2.1
n		neutron	2.2
p		shell name, proton	4.2, 2.2
q_e	Coul	magnitude of the electron charge	2.2, 3.2
r_e	m	classical electron radius	9.3
s		shell name	4.2
A		Einstein A coefficient	8.1
B		Einstein B coefficient	8.1
C_A		Auger constant	4.5
C_K	Js^2	Kramers' constant	6.4
C_{char}		characteristic emission rate constant	4.4
C_η	1/volt	efficiency constant	6.5
H		Hamiltonian, quantum operators	4.3, 9.1
K		shell name	4.3
L		shell name	4.3
M		shell name	4.3
M_e	kg	e^- rest mass	1.2
N_A		Avogadro's number	9.1
O_{char}		Exponent for characteristic emission	4.4
U_{eo}	eV	electron rest mass energy	1.2
U_R	eV	Rydberg energy	3.1, 4.2
α		fine-structure constant, or transition name	7.3, 4.3
β		transition name	4.3
β ray		electron	2.2
γ ray		gamma ray	2.1

Symbol	Units	Description	Section
ε_o	$Coul^2/N \cdot m^2$	vacuum permittivity	3.1
μ_o	N/A^2	vacuum permeability	6.1
σ_o	m^2	absorption cross section constant	9.1
σ_s	m^2	Thompson scatter cross section	11.6
π		$3.14159 \ldots$	
Φ_b	J	Electron binding energy	4.4
		Vectors	
$\vec{g}$	$1/m$	reciprocal lattice vector	13.5
$\vec{r}$	m	position	6.2
$\vec{t}$	m	translation vector (in crystal)	13.5
$\hat{x}$		unit vector	
$\hat{y}$		unit vector	
$\hat{z}$		unit vector	
$\dot{G}$	$1/m$	reciprocal translation vector	13.5
		Variable modifications	
$[X]$		units of variable	5.1
$<X>$		time average of variable	5.7
$\bar{X}$		average of variable	5.1
Δ		change in, range of variable	3.5
X_o		initial, vacuum, central	8.4

Variables (lowercase italic)

a	m/s^2	acceleration	6.1
b_A	$\#/s/mrad^2/m^2$ in 0.1% bw	brilliance	5.3
d	m	aperture or plane spacing	5.8, 13.6, 15.1
f		atomic scattering factor	9.3, 11.2
f_1		real part of the atomic scattering factor	9.3
$-f_2$		imaginary part of atomic scattering factor	9.3
f'		Rayleigh scattering factor	11.2
f_o		Thompson scattering factor	11.2
g		radial distribution function	9.6
h		integer (as in (hkl) plane)	13.6
j		total angular momentum quantum number, or generic integer	4.2
k		integer (as in (hkl) plane)	13.6
l		integer (as in (hkl) plane)	13.6
ℓ		orbital quantum number	4.3
m_l		magnetic quantum number	4.3
m_s		spin quantum number	4.3
m_j		quantum number	4.3
m		integer, quantum number	4.2
n		index of refraction	9.3, 11.1

Symbol	Units	Description	Section
p	kg·m/s	photon momentum	1.2
p_e	kg·m/s	e^- momentum	10.1
q	Coul	charge	3.3
r	m	observer location, basis atom locations	3.5, 6.2, 13.10
$\boldsymbol{r}$		reflection coefficient	11.9
s		spin, spring constant	4.2, 11.4
$sdev$		standard deviation	2.5
t	s	time	2.1
$t_{1/2}$	s	half-life	2.1
t_e	s	time in electron frame	7.5
t_{ph}	s	time in x-ray frame	7.5
$\boldsymbol{t}$		transmission coefficient	14.11
u	m	lattice vector	13.4
v	m/s	velocity	3.4, 6.1
w	m	variable in width direction, width	5.8, 12.2.4
x	m	distance variable, generally perpendicular to beam	2.3
Δx	m	blur or spacing	2.3, 5.5
y	m	distance along detector plane	5.8
Δy	m	blur, fringe spacing	5.5, 5.8
z	m	distance from source, in beam direction	2.3

Variables (uppercase italic)

Symbol	Units	Description	Section
A	m^2	area	3.3, 4.7, 5.1
A_s	m^2	area of source	5.3
A_D	m^2	detector area	5.1
B	tesla	magnetic field	3.2, 6.1, 7.1
$Bkgnd$	counts	background	4.7
$Bndwth$		frequency or energy bandwidth	5.3
C	F	capacitance	2.4
C_Y	N/m^2	Young's modulus	13.12.2
C_v		Poisson's ratio	15.3.3
$\boldsymbol{C}$		contrast	9.8.1, 10.5
D	Coul/m^2	electric displacement	11.1
$Dose$	J/kg	dose	9.8.2
E	N/Coul	electric field	3.3, 5.7, 6.2
F	N	force	3.3, 7.1, 11.1
$\boldsymbol{F}$		structure factor	13.10
G	1/m	magnitude of the reciprocal lattice vector	13.5
I	J/s/m^2	intensity	5.1, 5.7
I_Ω	J/s/steradian	(solid angle) angular intensity	5.1
I_χ	J/s/mrad^2	(linear angle) angular intensity	5.1
I_U	J/s/m^2/eV	intensity spectrum	6.3
J	Coul/s	current	4.4, 6.1

Symbol	Units	Description	Section
J_A	Coul/s/m^2	current density	6.1
K_e	J	electron kinetic energy	1.2
K_{insert}		wiggler or undulator insertion parameter	7.6
L	m	length in beam direction, object-to-detector distance	2.3, 5.4
L_c	m	longitudinal coherence length	5.9
M	kg	mass	1.2
M_{atom}	g/atom	mass per atom	
M_M	g/mole	Molar mass	4.7, 9.1
MDL	ppm	minimum detection limit	4.7
N		number of photons	2.5, 4.7
ΔN		variation	2.5
N_{atom}		number of atoms	2.1, 9.1
$\boldsymbol{N}$		noise	4.7
O_{char}		exponent in characteristic emission	4.4
P_{elec}	J/s	electrical power	6.5
P	J/s	x-ray power	5.1, 6.5
P_U	J/s/eV	power spectrum	6.6
$\boldsymbol{P}$		probability	2.5, 9.6, 13.3
Q	1/m	momentum transfer, scattering vector	11.9, 13.1
R	m	radius, location of atoms	7.1, 13.2, 15.2
$\boldsymbol{R}$		reflectivity	11.9
S		signal	4.7, 9.8.1
SNR		signal-to-noise ratio	4.7, 9.8.1
SF		scatter fraction	10.4
Sen		sensitivity	2.3
$\boldsymbol{T}$		transmission	9.1, 10.6, 12.2.4
T	kelvin	temperature	3.1
U	eV	photon energy	1.2
U_b	eV	binding energy	4.4
V	m^3	volume	3.3, 13.5
W	m	width	9.9
W_D	m	detector width	2.3
W_a	m	aperture width	2.3
W_s	m	source width	5.5
X		concentration (atomic fraction)	4.7
Y	m	distance along aperture plane	5.8
Y_C	m	transverse coherence length	5.8
Z		atomic number	2.1

Variables (Greek)

Symbol	Units	Description	Section
β		Im $\{n\}$	9.3
γ		Lorentz factor	7.2
Γ	photons/s	rate	2.1
Γ_e	electrons/s	electron arrival rate	4.4
δ		refractive index decrement	11.1
ε	Coul^2/N·m^2	permittivity	
ζ		screening constant	4.2
η		efficiency	6.2
θ	radian	incidence angle, angle relative to electron or beam direction	7.1, 10.1, 11.8
θ_B	radian	Bragg angle	13.7
Θ	radian	global divergence	5.4
κ	1/m	x-ray wavevector	1.2
λ	m	x-ray wavelength	1.2
Λ	m	path length	5.7, 13.2, 15.2
Λ_D	m	Debye length	3.4
Λ_S	m	correlation length	11.11
μ_{ab}	1/m	absorption coefficient	9.1
μ_{ab}^{-1}	m	absorption length	9.1
μ_s	1/m	scatter coefficient	11.6
μ_{tot}	1/m	attenuation coefficient	9.1
ν	Hz	x-ray frequency	1.2
ξ	radian	local divergence	5.5
Π	Coul/m^2	polarization field	11.1
ρ	g/cm^3	mass density	3.3, 9.1
ρ_e	#/m^3	e$^-$ density	3.3
ρ_q	Coul/m^3	charge density	3.3
ρ_{atom}	#/m^3	atom density	9.1
Σ		sum	
σ	1/m^2	cross section	9.1
ς	N/m	damping constant	11.4
τ	s	period	5.7
τ_c	s	coherence time	4.6, 5.9
ϕ	radian	phase angle	
φ	radian	azimuthal or other angle, e.g., in Compton scatter	10.1
Φ	volt	voltage	4.4
χ	radian	angle	5.1
ψ		quantum state	4.4
Ψ	photons/m^2/s	photon intensity	5.2
Ψ_χ	photons/mrad^2/s	photon angular intensity	5.3

Symbol	Units	Description	Section
$\Psi_{\chi Bndwth}$	#/s/mrad² in 0.1% bw	brightness	5.3
ω	rad/s	angular frequency	3.3, 5.7
ω_p	rad/s	plasma frequency	3.3, 11.1
Ω	steradian	solid angle	5.1

PART I
FOUNDATIONS

1

INTRODUCTION

1.1 The discovery

Near the end of the nineteenth century, Röntgen was experimenting with cathode ray tubes—vacuum tubes similar to old-fashioned computer monitors. After placing a metal target in the electron beam, he noticed that phosphor behind a wood screen glowed and nearby photographic plates became exposed even though protected from light. Röntgen realized that these effects must be due to some unknown, "x" radiation, and he quickly began to investigate, placing various objects in the beam (including his wife's hand—the famous image is shown in Figure 1-1).

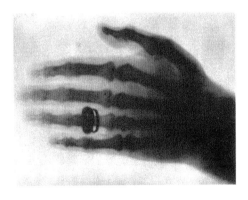

Figure 1-1. An x-ray image of Frau Röntgen's hand, from On a new kind of rays, *Nature* 53 (January 23, 1896): 274–76.

This commonly told story of the discovery of x rays is a classic tale of serendipity. Imagine it—a laboratory in Würtzburg, Germany, cluttered with all the latest scientific apparatus of 1895: vacuum tubes, photographic plates, jars and sheets of phosphors and metals, and an excited scientist randomly applying high voltages. Suddenly he notices the phosphor on the other side of the room is glowing, and William Conrad Röntgen is on his way to receiving the first Nobel Prize in Physics. Shortly into his investigations, he happens to see the outline of the bones in his wife's hand exposed on a photographic plate, and the field of diagnostic imaging is born.

As with most great advances, the real story is a bit more deliberate. Röntgen, along with several of his contemporaries, including Tesla and Hertz, was actively engaged in research on the emissions from cathode ray tubes. The inventor of his tube, Crookes, had previously seen shadows on photographic plates, and may have suggested that Röntgen investigate them. However, Röntgen did quite quickly realize the significance of his observations and rapidly began identifying many of the characteristics of x rays—for example, the dependence of penetration on density and the lack of significant refraction—as well as pioneering some of their applications.

Today, x rays are important not only for medical imaging and baggage inspection but also for astronomical observations, for materials analysis, for structure determination of viruses and pharmaceuticals, for fluorescence analysis in manufacturing quality control, and for fraud detection in art. An increasing interest in x-ray astronomy was one of the major forces behind the development of x-ray optics in the latter half of the last century. Mirror systems similar to those developed for astronomy also proved useful for synchrotron beamlines. Just as x-ray tubes were an accidental offshoot of cathode ray research, synchrotron x-ray sources were originally parasitic to particle physics: the synchrotron radiation was an unwanted consequence of accelerating the particles, because it removes energy from the particle beam. The subsequent development of specialized synchrotron sources with increasing brightness and numbers of beamlines led to creation of a whole new array of x-ray tools and a consequent demand for an increasing array of optics.

The rapid development of x-ray optics also has been symbiotic with the development of detectors and compact sources. Detectors developed for particle physics, medicine, and crystallography have found application across the different fields. Similarly, the increasing capabilities of x-ray systems have stimulated the development of new science, with ever-growing requirements for intensity, coherence, and spatial and energy resolution. X-ray diffraction and fluorescence were early tools during the rapid development of materials science after World War II. More recently advancements have been made to meet the demands of shrinking feature sizes and allowed defect concentrations in semiconductors. The use of x-ray diffraction, especially the development of dedicated synchrotron beamlines, has also been stimulated by the growing demands for rapid protein crystallography in biophysics and pharmaceutical development.

1.2 What is an x ray?

Despite Röntgen's early identification of his unknown, "x" rays as longitudinal vibrations of the ether (this was just nine years after the Michelson-Morley experiment), x rays proved to be simply light waves, electromagnetic radiation, with very short wavelengths. The definition of the wavelength range considered to be in the x-ray regime differs somewhat among fields and applications, but is typically between 0.1 and 10 Å (0.01 to 1 nm). Longer wavelengths are considered to be in the range of extreme ultraviolet (EUV), and shorter wavelengths are generally considered to be in the gamma ray regime (although some fields make a distinction that "gamma ray" refers only to the products of nuclear reactions).

The usual relation holds between wavelength λ and wavenumber κ (the magnitude of the wavevector),

$$\lambda = \frac{2\pi}{\kappa}.$$

(1-1)

Quantum mechanics gives us the relationships between wavevector and momentum,

$$\vec{p} = \hbar \vec{\kappa} \Rightarrow \lambda = \frac{h}{p},$$

(1-2)

where h is Planck's constant and, as usual,

$$\hbar = \frac{h}{2\pi}. \tag{1-3}$$

Relativity gives us the relationship between momentum and energy U,

$$U^2 = (Mc^2)^2 + p^2c^2 \Rightarrow p = \frac{\sqrt{U^2 - (Mc^2)^2}}{c} = \frac{U}{c}, \tag{1-4}$$

since, as photons are massless, $M = 0$. Thus, wavelength and energy are related by

$$U = \frac{hc}{\lambda}, \tag{1-5}$$

where c is the speed of light. Expressing h in units of eV · s, and c in units of Å/s gives the useful result that

$$hc \approx 12.4 \text{ keV} \cdot \text{Å.} \tag{1-6}$$

Thus, the wavelength range from 10 to 0.1 Å corresponds to 1.2 to 124 keV in photon energy. For comparison, a visible light photon with a wavelength of 0.5 μm corresponds to a photon energy of 2.5 eV, or 2.5×10^{-3} keV. Quantum mechanics also gives us the relationship between photon energy and frequency v,

$$U = hv, \tag{1-7}$$

which gives us the expected relationship between wavelength and frequency,

$$\lambda = \frac{hc}{U} = \frac{c}{v}. \tag{1-8}$$

EXAMPLE 1-1

a) What are the wavelength and frequency of the 8 keV x rays frequently used for protein crystallography experiments?

From equation 1-5,

$$\lambda = \frac{hc}{U} \approx \frac{12.4 \text{ keV} \cdot \text{Å}}{8 \text{ keV}} \approx 1.55 \text{ Å.}$$

From equation 1-8,

$$v = \frac{c}{\lambda} \approx \frac{3 \times 10^{18} \text{ Å/s}}{1.55 \text{ Å}} \approx 1.9 \times 10^{18} \text{ Hz} \approx 1.9 \text{ exahertz} \approx 1.9 \text{ EHz.}$$

b) For comparison, what is the wavelength of an electron with a kinetic energy of 8 keV?

The difference between the photon and electron wavelengths arises in applying equation 1-4, because the electron is not massless. The kinetic energy is the difference between the total energy U and the rest mass energy, which for small (nonrelativistic) momentum is

$$K_e = U - M_e c^2 = \sqrt{\left(M_e c^2\right)^2 + p^2 c^2} - M_e c^2$$

$$= M_e c^2 \sqrt{1 + \frac{p^2 c^2}{\left(M_e c^2\right)^2}} - M_e c^2 \approx M_e c^2 \left(1 + \frac{1}{2}\frac{p^2 c^2}{\left(M_e c^2\right)^2}\right) - M_e c^2 = \frac{p^2}{2M_e}$$

which is the expected, classical result. Then,

$$\lambda = \frac{h}{p} = \frac{h}{\sqrt{2M_e(K_e)}} \approx \frac{(6.6 \times 10^{-34} \text{ J} \cdot \text{s})}{\sqrt{2(9.1 \times 10^{-31} \text{ kg})(8 \times 10^3 \text{ eV})\left(\dfrac{1.6 \times 10^{-19} \text{ J}}{1 \text{ eV}}\right)}}$$

$$\approx 1.4 \times 10^{-11} \frac{\text{kg}\dfrac{\text{m}^2}{\text{s}^2}\text{s}}{\sqrt{\text{kg}^2 \dfrac{\text{m}^2}{\text{s}^2}}} \approx 1.4 \times 10^{-11} \text{ m} \approx 0.14 \text{ Å}.$$

The electron wavelength is more than a factor of 10 smaller than that of the x ray with the same kinetic energy.

1.3 What makes x rays useful?

The wavelength of x rays is in the angstrom range, similar to the spacing of atoms in a crystal. Thus, the arrays of atoms in a crystal can act as a diffraction grating for x rays. The 1914 Nobel Prize in Physics was awarded to Laue for the first demonstration of diffraction of x rays by a crystal. The 1915 prize went to William Henry Bragg and William Lawrence Bragg for the development of the theory that allows for association of crystal structure with the diffraction pattern. X-ray crystallography is routinely used today for applications such as verifying the crystal quality of films grown on silicon wafers, detecting stress in airplane engines, and determining the structure of proteins to understand their function in cancer growth. Diffraction is also used as a way to control the direction or wavelength of x rays used in a particular experiment, just as gratings are used for visible light. The 1936 Nobel Prize in Chemistry was awarded to Peter Debye for, among other things, development of the theory of diffraction from powders and liquids.

FIGURE 1-2. Baggage x-ray imaging, Gemini Dual-Energy system. The color images are produced by comparing absorption at two different x-ray photon energies. Copyright 2016 American Science and Engineering, Inc.

Short-wavelength, high-energy photons are not easily absorbed—their high energy and momentum makes them difficult to stop. This means that x rays easily pass through materials such as human tissue for radiography or luggage for baggage inspection, as shown in Figure 1-2, or the dark paper Röntgen had used to protect

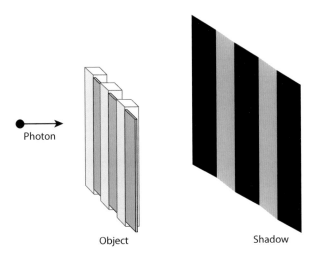

Photon

Object   Shadow

FIGURE 1-3. Making a shadow image.

his photographic plate. Absorption increases with the electron density of the material but is lower for higher-energy photons.

By way of analogy, consider an object with thick and thin regions like that of Figure 1-3. If the object was made with alternating painted plywood and tissue paper, you could map out the areas of tissue paper by throwing balls at the object and letting them mark the wall behind the object when they passed through. If the object was constructed of thick wood and bricks, you would need higher-momentum projectiles, perhaps bullets, to make a shadow. However, bullets would do a poor job of making a shadow image of the tissue paper area, because they would pass through the plywood as well. Thus, high-energy ("hard") x rays are used for inspecting vehicles and steel cargo containers, as shown in Figure 1-4. Since almost all the hard x rays would pass through a thinner object or one with a lower atomic number, creating very little shadow, lower-energy ("softer") x rays must be used to diagnose a broken hand.

Because x rays barely interact with materials, their index of refraction in any material is only slightly different from unity. This results in sharp shadows for radiography, because the rays are hardly refracted, but means it is very difficult to make refractive optics such as the lenses normally used for visible light. The penetrating nature of

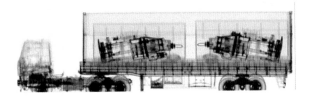

FIGURE 1-4. High-energy x-ray images of a cargo truck, OmniView Dual-Energy Transmission system. Copyright 2016 American Science and Engineering, Inc.

x rays also makes it difficult to construct optics such as Fresnel zone plates, or even pinholes for pinhole cameras, since the masking material must be thick compared with the dimensions of the apertures.

The energy of an x-ray photon, in the kiloelectronvolt range, is very much larger than the sub-electronvolt range typical for valence electron transitions in materials. Hence, x-ray properties are relatively insensitive to chemical state, unlike the changes in color or opacity that can easily be induced for visible light. However, x-ray energies are similar to ionization energies for *core* electrons and thus can be used to probe for characteristic atomic transitions. X-ray absorption spectroscopy and x-ray fluorescence are extremely important for elemental analysis. The 1924 Nobel Prize in Physics was awarded to Siegbahn for developing the field of x-ray spectroscopy.

1.4 The layout of the text

Any x-ray application or experiment requires an x-ray source and some material for the x-ray to interact with, including, in most cases, a detector. The next section of the book discusses methods of generating x rays. While some sources are naturally occurring—radioactive materials, black holes—the most common technique for generating x rays in the laboratory is by accelerating electrons, which generates a continuum (bremsstrahlung or synchrotron radiation) and characteristic emission lines (the same lines used for fluorescence analysis). X rays can also be emitted by blackbody radiation from very hot plasmas such as the sun, or those created by very intense lasers. The mechanisms for x-ray interactions with matter (including x-ray detection) are discussed in part III. These include absorption, scattering, refraction, reflection, and diffraction. Applications and optics enabled by these interactions are included in this section. The solutions to end-of-chapter problems are given in the appendix.

1.5 The elusive hyphen

Just as the definition of an x ray varies between applications, so does its hyphenation and capitalization, and you will encounter several styles. Grammatically, "x" is a modifier, like "optical," so no hyphen is required. When the noun string is used as an adjective, as in x-ray beam, the hyphen is necessary. For example, when a child says "I am three years old," you refer to him or to her as a "three-year-old child." Some journals are very strict about removing extraneous hyphens. In other journals, it is the practice to always use the hyphen and/or to capitalize the x.

Problems

SECTION 1.2

1. Planck's constant, h, is 6.6×10^{-34} J · s. 1 eV is the energy associated with an electron charge, $q_e \approx 1.6 \times 10^{-19}$ Coul, in a potential of 1 V. The speed of light is approximately 3×10^8 m/s. Verify equation 1-6.

2. What is the wavelength of 30 keV x rays?

Further reading

General references for x-ray topics

Jens Als-Nielsen and Des McMorrow, **Elements of Modern X-ray Physics**, John Wiley & Sons, 2001.

Eric Lifshin, **X-ray Characterization of Materials**, John Wiley & Sons, 1999.

A. G. Michette and C. J. Buckley, **X-ray Science and Technology**, Institute of Physics Publishing, 1993.

Alan Michette and Sławka Pfauntsch, **X-Rays: The First Hundred Years**, John Wiley & Sons, 1996.

E. Spiller, **Soft X-ray Optics**, SPIE Press, 1994.

David Attwood and A. Sakdinawat, **X-Rays and Extreme Ultraviolet Radiation: Principles and Applications**, Cambridge University Press, 2016.

Relativity

D. Halliday, R. Resnick, and J. Walker, **Fundamentals of Physics**, 10th ed., John Wiley & Sons, 2013, chapter 37.

Historical references

Arthur Stanton, Wilhelm Conrad Röntgen on a new kind of rays: Translation of a paper read before the Würtzburg Physical and Medical Society, *Nature* 53 (1895): 274–76.

New York Times, February 16, 1896, Nature of the X Rays.

2

A CASE STUDY: NUCLEAR MEDICINE

This chapter provides an introduction and motivation for upcoming chapters by taking a brief look at the techniques required for an example application, radioactive imaging, also known as nuclear medicine. Radioactive decay is used as a source of x rays and so can be considered a method of x-ray generation, the topic of the next section. In addition, a discussion of nuclear medicine necessarily requires an introduction to many of the of the topics of part III, on x-ray physics, including x-ray scatter, absorption, detection, imaging, and statistical noise.

2.1 Metastable emitters and half-life

Radioactive materials are commonly used as x-ray sources for nuclear medicine, for some portable x-ray fluorescence systems, and for tagging particular proteins for DNA analysis, as shown in Figure 2-1. In nuclear medicine, as shown in Figure 2-2, radioactive substances are injected into, inhaled, or swallowed by patients to be used as x-ray sources in diagnostic imaging. Because the radioactive material must circulate to the area of interest, nuclear medicine is generally considered to be *functional* imaging, dependent on blood flow or other anatomical function, whereas normal radiographic imaging is said to be *morphological*, providing an image of the inherent structure. For

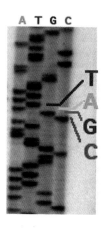

FIGURE 2-1. Gel electrophoresis image made by autoradiography of DNA segments tagged with radioactive markers. The molecules' drift to the bottom of the gel is impeded by the long stringy molecules of the gel. Larger molecules are more impeded, so that the vertical axis in the picture is proportional to molecular weight. The segments in the column marked A are from a DNA solution mixed with a deoxynucleotide that cleaves the DNA at all the adenine protein sites (T is for thymine, C for cytosine, G for guanine). The DNA sequence can then be read as a letter sequence from top to bottom. Copyright J. W. Schmidt.

example, the image in Figure 2-2 does not display the bones of the feet in the manner that Figure 1-1 shows the bones of the hand but, rather, shows blood pooling from a subtle fracture, which resulted in a higher concentration of the radioactive isotope.

One of the most commonly used isotopes in medical imaging is ^{99m}Tc. The superscript 99 is the number of nucleons, the sum of the number of protons and neutrons. Because technetium has an atomic number of 43, that is, 43 protons, it can also be written as $^{99m}_{43}$Tc. In general, the symbols for elemental isotopes are written $^{Z+N_n}_{Z}X_{N_n}$, where X is a one- or two-letter elemental symbol; Z is the atomic number, which is the number of protons; and N_n is the number of neutrons, so that $N_n + Z$ gives the total number of nucleons. For example, $^{235}_{92}U_{43}$ is a uranium isotope with 92 protons and 143 neutrons, for a total of 235 nucleons. This notation is redundant, because Z defines the element; writing U means the same as writing $_{92}$U. Similarly, the symbol ^{235}U (read "U 235") represents the same isotope as $^{235}_{92}U_{143}$.

The m in ^{99m}Tc stands for *metastable*. (A metastable nuclear state is one in which the nucleus is not in the lowest energy state. It is unstable but has a relatively long lifetime.) The metastable technetium decays to a stable state, giving off a 140 keV photon, which is detected and used to map the distribution of radiation in the patient. The decay reaction is

$$^{99m}_{43}Tc \rightarrow \,^{99}_{43}Tc + \gamma, \tag{2-1}$$

where the γ in the reaction indicates the emitted photon. X-ray photons emitted from nuclear reactions are often referred to as gamma rays, although sometimes the term *gamma ray* is reserved for photons with energies of more than 1 MeV.

The isotope ^{99m}Tc has a half-life of 6 hours. *Half-life $t_{1/2}$* is the time required for half the material to decay. The number of atoms N_{atom} remaining after a time t is related to the amount of starting material N_o by

$$N_{atom} = N_o e^{\frac{-t\ln(2)}{t_{1/2}}}. \tag{2-2}$$

EXAMPLE 2-1

The isotope ^{14}C has a half-life of 5730 years. It is produced by cosmic ray bombardment in the upper atmosphere, so that the fraction of carbon in the atmosphere and living things that breathe the atmosphere is one ^{14}C atom per 10^{12} stable atoms. A sample of a well-preserved plant was found to have a concentration of $X = 3 \times 10^6$ atoms of ^{14}C per mole of carbon. How long ago did the plant die?

The total number of carbon-14 atoms in the plant depends on its size, $N_o = X_o N_{total}$, and so is unknown, but we will assume the total number of atoms remains unchanged in the specimen (or at least that atoms were lost with the same isotope ratio as in the rest of the plant). In that case,

$$N_{atom} = N_o e^{-\frac{\ln(2)}{t_{1/2}}t}$$

$$\Rightarrow t = -\frac{t_{1/2}}{\ln(2)}\ln\left(\frac{N}{N_o}\right) = -\frac{5730 \text{ yr}}{\ln(2)}\ln\left(\frac{3\times10^6 \frac{^{14}C}{\text{mole}}}{6\times10^{11}\frac{^{14}C}{\text{mole}}}\right) \approx 100,000 \text{ yr.}$$

The rate of decay, or activity, is given by

$$\Gamma = \left|\frac{dN}{dt}\right| = N\frac{\ln(2)}{t_{1/2}}. \tag{2-3}$$

The activity of a radioactive source is often specified in curies, Ci, where 1 Ci is the amount of material required to produce 3.7×10^{10} decays/s. One decay per second is also known as 1 becquerel, 1 Bq.

EXAMPLE 2-2

a) The isotope ^{125}I has a half-life of 60 days. How many atoms of ^{125}I are required to have an activity of 1 mCi?
b) What mass of pure ^{125}I is required?
c) What will the activity be 30 days later?

a) From equation 2-3,

$$\Gamma = N\frac{\ln(2)}{t_{1/2}} \Rightarrow N = \Gamma\frac{t_{1/2}}{\ln(2)} \approx 10^{-3}\text{Ci}\frac{3.7\times10^{10}\frac{1}{s}}{\text{Ci}}\frac{60 \text{ days}}{\ln(2)}\frac{24 \text{ hr}}{\text{day}}\frac{3600 \text{ s}}{\text{hr}}$$

$$\approx 2.8\times10^{14}.$$

b) Now, we need the mass of that number of atoms, which can be computed from the molar mass M_M,

$$M = \frac{N}{N_A}M_M \approx \frac{2.8\times10^{14} \text{ atoms}}{6\times10^{23}\frac{\text{atoms}}{\text{mole}}}\left(125\frac{g}{\text{mole}}\right) \approx 5.7\times10^{-8} \text{ g} \approx 0.06\,\mu g.$$

c) $\Gamma = N\left[\dfrac{\ln(2)}{t_{1/2}}\right] = \left\{N_0\left[\dfrac{\ln(2)}{t_{1/2}}\right]\right\}e^{-\frac{\ln(2)}{t_{1/2}}t} = \{\Gamma_0\}e^{-\frac{\ln(2)}{t_{1/2}}t}$

$\approx [1\,\text{mCi}]\,e^{-\ln(2)\frac{30\,\text{days}}{60\,\text{days}}} \approx 0.7\,\text{mCi}.$

2.2 A brief introduction to nuclear decay

The radioactive technetium used in nuclear medicine is produced by nuclear β (beta) decay.* Nuclear forces can be described by analogy with atomic forces. An atom is held together by electrostatic attraction between the electrons and the positive nucleus. Residual electrostatic attraction, between electrons of one atom and the nucleus of another, hold atoms together in molecules and solids. Similarly, the quarks in a proton are held together by the strong force. Residual forces between the nucleons hold the nucleus together. Because the strong force must overcome the electrostatic repulsion of the protons, nuclei with more than one proton must contain neutrons. As the total charge increases, the number of necessary neutrons per proton also increases. The ratio rises from 1 for helium to 1.6 for heavy elements. Elements with the same atomic number but different numbers of neutrons are called *isotopes*. Isotopes with more neutrons than the ratio required for stability are called "neutron-rich," and isotopes with fewer neutrons than required are "neutron-poor."

For example, ^{14}C has six protons and eight neutrons, giving a neutron-to-proton ratio of 4/3, which is more than the stable ratio for lightweight elements, so ^{14}C is neutron-rich. The isotope reduces its number of neutrons by β decay, in which a neutron decays to a proton and an electron.

A neutron, n, is made up of three quarks, one "up," with +2/3 electron charge, $2/3q_e$, and two "down" quarks, each with 1/3 electron charge, $-1/3q_e$. Under the influence of the weak force, one of the neutron's "down" quarks changes "flavor" and becomes an "up" quark. The former neutron now has a positive charge and becomes a proton, p. This process can be written

$$_0^1 n \rightarrow {}_1^1 p + {}_{-1}^0 e^-.\tag{2-4}$$

The electron is necessarily created to conserve charge, and the emitted electron is referred to as a β ray (or cathode ray—recall that Röntgen was investigating these when he discovered x rays). The electron, e$^-$, is written, oddly, as $_{-1}^0 e^-$ here simply to emphasize that charge and nucleon number are conserved in all reactions.†

β decay for ^{14}C is written as

$$_6^{14}C \rightarrow {}_7^{14}N + {}_{-1}^0 e^-.\tag{2-5}$$

---

*This is an optional section. The details here are not necessary for the rest of the chapter.

†Lepton number is also preserved, so that the suppressed details include the emission of a short-lived W$^-$ boson, which decays into an electron and an antineutrino to retain a net zero lepton number. The neutrino is difficult to detect and so is neglected in the equation.

The reaction which produces $^{99\mathrm{m}}$Tc is

$$^{99}_{42}\mathrm{Mo} \rightarrow {}^{99\mathrm{m}}_{43}\mathrm{Tc} + {}^{0}_{-1}\mathrm{e}^{-}. \tag{2-6}$$

The isotope ^{99}Mo has a half-life of 2.8 days, so it is more easily shipped to a clinic than the short-lived $^{99\mathrm{m}}$Tc. It is produced by fission in a nuclear reactor by bombarding ^{235}U with energetic neutrons,

$$^{235}_{92}\mathrm{U} + \left({}^{1}_{0}\mathrm{n}\right) \rightarrow {}^{134}_{50}\mathrm{Sn} + {}^{99}_{42}\mathrm{Mo} + 3\left({}^{1}_{0}\mathrm{n}\right) + \gamma. \tag{2-7}$$

^{235}U has a neutron-to-proton ratio that is slightly less than the 1.6 required for stability for heavy elements. The deficiency of neutrons is resolved by decay into the two daughter products, Sn and Mo, which, because they have fewer protons, require fewer neutrons. The newly created ^{99}Mo ends up with too many neutrons for the lower atomic number and undergoes β decay, as seen in reaction 2-6. The energetic neutrons emitted from the fission are the source for further fission events if there is sufficient uranium so that the neutrons are not absorbed by other materials first.

Heavy neutron-poor isotopes can capture an electron because the electron cloud is attracted to the large positively charged nucleus. Radioactive iodine-125 decays by electron capture,

$$^{125}_{53}\mathrm{I} + \mathrm{e}^{-} \rightarrow {}^{125}_{52}\mathrm{Te} + \gamma. \tag{2-8}$$

In this case a proton and an electron became a neutron, the reverse of reaction 2-4, adding to the number of neutrons to stabilize the isotope. A photon is directly emitted in this reaction. Additional photons are emitted as the tellurium falls to its ground state, so the photon emission rate is not simply one per decay.

Just as radioactive daughter products are formed in fission reaction 2-7 by bombarding uranium with fast neutrons, unstable radioactive materials can also be produced by bombarding targets with high-velocity charged particles. Some large hospitals have linear accelerators for this purpose. As an aside, very large accelerators can also be used to produce x rays directly, because, as will be seen in chapter 7, accelerating charges give off electromagnetic radiation. Some accelerators built especially for this purpose are called *synchrotrons*.

Radioactive materials are used as x-ray sources for x-ray fluorescence, discussed in chapter 4, and for nuclear medicine.

2.3 Nuclear medicine

In nuclear medicine, the radioactive element is chemically attached to a radiopharmaceutical and administered to the patient. If the pharmaceutical selectively collects in a tumor or other area of interest, it will produce a *hot spot* in the patient. The goal of nuclear medicine is to accurately image the distribution of radioactivity in the patient, which is different from the goal of radiography with an external source, which is to produce a map of transmission, as shown in Figure 2-3, as well as in Figures 1-1, 1-2, and 1-4. The problem of imaging internal radiation is illustrated in Figure 2-4, where only a blur is produced on the film, because every spot within the tumor is a source of

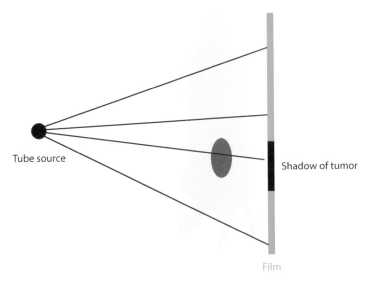

FIGURE 2-3. Radiography of a patient—in this case a mouse—with an external source. A shadow of the tumor appears on the film (or other imaging detector).

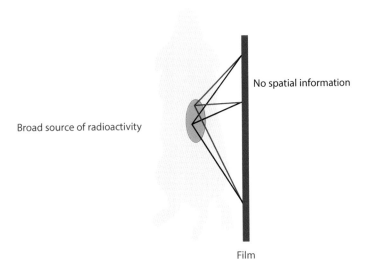

FIGURE 2-4. Radiography of a mouse using internal radiation gives only a blur of overlapping x rays on the film. The red and black rays hit the film at the same points, but come from different parts of the mouse.

overlapping *isotropic* radiation, that is, radiation traveling outward in all directions. This problem can be remedied in one of two ways. In autoradiography, shown in Figure 2-1, the gel is placed very close to the film or other x-ray detector, so there is no space for the blur to develop. For thicker objects with internal sources, the detector cannot be placed at the source of the radiation. Instead, the image is taken through a pinhole camera, as shown in Figure 2-5. The pinhole aperture, which can be either a

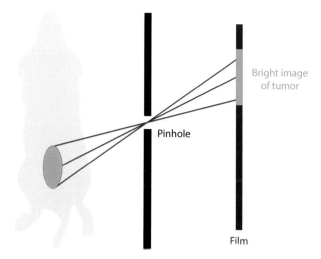

FIGURE 2-5. Image of tumor using a pinhole camera. Only the areas containing radiation show up on the film. Note that the image is reversed (upside down). Scatter has not yet been taken into account.

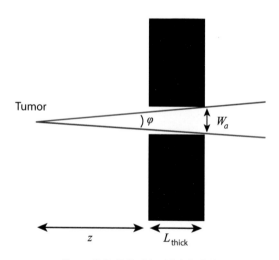

FIGURE 2-6. Cylindrical "pinhole."

cylinder, as shown in Figure 2-6, or designed to accept a fixed angle χ, as in Figure 2-7, rejects most of the isotropic radiation from each point on the tumor, allowing only a small cone of radiation to pass. (A pinhole with a large χ can still have a small diameter W_a and so still accept rays from only a small range of angles φ from a single emission point on the tumor.) A small pinhole diameter provides good spatial resolution but results in low sensitivity—that is, only a small fraction of the radioactive decays present in the patient can be detected. In addition, the pinhole thickness, L_{thick}, must be large enough so that there is enough surrounding material to absorb the x rays, which makes such pinholes challenging to manufacture.

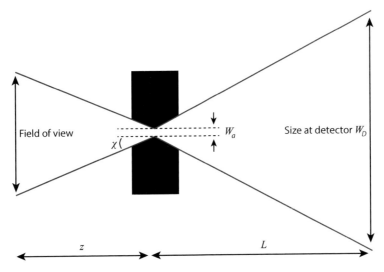

FIGURE 2-7. Pinhole collimator for nuclear imaging.

EXAMPLE 2-3

a) If the pinhole in Figure 2-6 or Figure 2-7 has an aperture diameter of $W_a = 1$ mm and is placed $z = 100$ mm from the tumor, for what range of angles can an x ray be emitted by a point on the tumor and still pass through the hole?
b) What fraction of the emitted radiation is accepted by the pinhole?
c) What activity, in microcuries, is required to produce 1 photon/s through the pinhole, assuming each decay creates one photon?

  a) The geometry is sketched in Figure 2-6. Assuming L_{thick} is small, the acceptance angle φ is given by

$$\tan\left(\frac{\varphi}{2}\right) = \frac{W_a/2}{z} \Rightarrow \varphi \approx 2\tan^{-1}\left(\frac{1/2}{100}\right) \approx 0.6°.$$

  b) The fraction, known as the sensitivity Sen is the ratio of the area of the pinhole to the area of the imaginary sphere centered at the source. This sphere, with a radius equal to the pinhole distance, represents the area equally irradiated by the point source, so that

$$Sen = \frac{\pi(W_a/2)^2}{4\pi(z)^2} = \frac{W_a^2}{16(z)^2} \approx \frac{(1\,\text{mm})^2}{16(100\,\text{mm})^2} \approx 6 \times 10^{-6}.$$

  c) The rate through the aperture, Γ_a, is the sensitivity times the decay rate (of the material within the field of view of the pinhole,

$$\Gamma_a = \Gamma Sen \Rightarrow \Gamma = \frac{\Gamma_a}{Sen} \approx \frac{1\,\dfrac{\text{photon}}{\text{s}}}{6 \times 10^{-6}} \approx 1.6 \times 10^5\,\frac{\text{photons}}{\text{s}}$$

$$= \left(1.6 \times 10^5\,\frac{\text{photons}}{\text{s}}\right)\frac{1\,\text{Ci}}{3.7 \times 10^{10}\,\dfrac{\text{photons}}{\text{s}}} \approx 4 \times 10^{-6}\,\text{Ci} = 4\,\mu\text{Ci}.$$

The sensitivity depends on the pinhole diameter W_a and so could be increased by using a larger pinhole. However, that would increase the uncertainty in tumor location—the resolution—as shown in Figure 2-8. The resolution could be improved without changing the sensitivity by increasing the detector distance L, but that would reduce the field of view, as shown in Figure 2-7, unless the detector size is also increased.

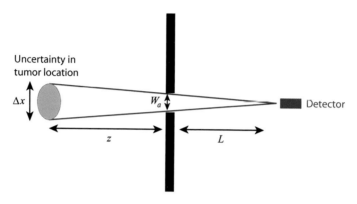

FIGURE 2-8. Resolution of a pinhole.

EXAMPLE 2-4

a) If the pinhole in Figure 2-6 or Figure 2-7 is 1 mm in diameter and is placed 100 mm from the tumor, what is the minimum resolution of the system if the detector is 200 mm from the pinhole? Assume the pixels are very tiny and so do not affect the resolution. b) If the detector is 50 mm in diameter, what is the field of view, the maximum size of the mouse that can be imaged?

a) The geometry is sketched in Figure 2-8. The uncertainty in emission location for a point on the detector, Δx, is given by

$$\Delta x = \frac{z+L}{L}W_a = \frac{100\,\text{mm} + 200\,\text{mm}}{200\,\text{mm}}(1\,\text{mm}) = 1.5\,\text{mm}$$

b) The geometry is sketched in Figure 2-7. The field of view, FOV, is

$$\text{FOV} \approx z\tan(2\chi) \approx z\frac{2\dfrac{W_D}{2}}{L} = 100\,\text{mm}\,\frac{50\,\text{mm}}{200\,\text{mm}} = 25\,\text{mm},$$

where W_D is the size of the detector.

The resolution in example 2-4 was calculated for a detector with tiny pixels. Real nuclear medicine detectors have large pixels, which greatly degrades the resolution further. Usually, the low sensitivity and field of view are improved by adding a two-dimensional array of pinholes, as shown in Figure 2-9.

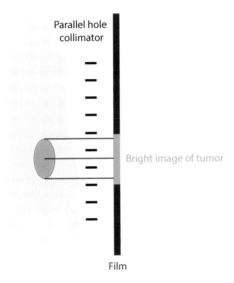

FIGURE 2-9. Image of tumor using a parallel-hole collimator. Only the areas containing radiation show up on the film. Scatter has not yet been taken into account.

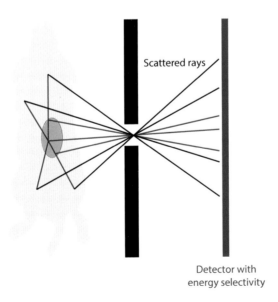

FIGURE 2-10. The rays emitted from the tumor (shown in red) are scattered by other tissue and obscure the image of tumor unless an energy-sensitive detector is employed to remove scattered x rays. The scattered rays are shown in black.

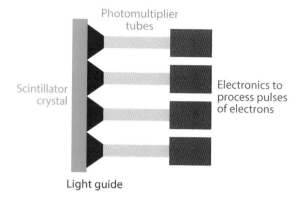

FIGURE 2-11. Energy-sensitive x-ray detector.

2.4 Photon detection and scatter rejection

In practice the images of both Figures 2-5 and 2-9 will be completely masked by the scattering of x rays within the mouse, as shown in Figure 2-10. Each hot spot on the tumor radiates isotropically, not just toward the pinhole but toward every point on the mouse. Most of those x rays will be scattered by the mouse tissue, so that some of the scattered rays, shown in black in Figure 2-10, also pass through the pinhole and completely obscure the tumor image. However, as will be seen in chapter 10, Compton scattering transfers energy from the photon to an electron, resulting in a lower-energy photon. If a detector capable of distinguishing between the original photon energy and the scattered photon energy is employed, the scattered photons can be removed from the image.

One such detector, shown in Figure 2-11, is a combination of a scintillator such as NaI(Tl), which converts x rays to visible light, and a photomultiplier tube, which converts light to electrons. The conversion of x rays to light occurs when an x-ray beam excites electrons in the NaI. The excited electrons share energy with other electrons in the crystal, raising a number of them into the conduction band, where they can travel and drop into the unoccupied "trap" states created by the small concentration of Tl dopant added to the NaI crystal. The electrons then fall from the metastable trap state to the ground state, emitting visible blue-violet photons. The number of electrons

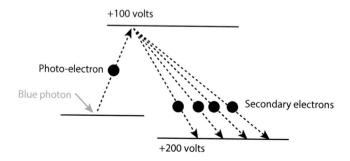

FIGURE 2-12. Geometry for photomultiplier tube showing first stage of gain.

moved into trap states depends on the amount of energy carried by the x-ray photon. About 13% of the absorbed energy is reemitted as light. X-ray absorption, scintillators, and phosphors are discussed in more detail in chapter 9.

EXAMPLE 2-5

How many light photons with wavelength 0.41 μm are created from a single 23 keV x-ray photon absorbed in a NaI scintillator?

$$U_{blue\text{-}violet} = \frac{hc}{\lambda} \approx \frac{12.4 \, keV \, \text{Å}}{0.41 \times 10^{-6} m} \left(\frac{10^3 \, eV}{keV} \right) \left(\frac{10^{-10} m}{\text{Å}} \right) \approx 3 \, eV.$$

$$U_{avail} = (0.13) U_{x \, ray} \approx (0.13)(23 \, keV) \approx 3 \, keV.$$

$$N \approx \frac{\left(\dfrac{3 \, keV}{x\text{-ray photon}} \right)}{\left(\dfrac{3 \, eV}{blue \, photon} \right)} = 1000 \, \frac{visible \, photons}{x \, ray}.$$

Some of the light emitted isotropically by the NaI is incident on a photomultiplier tube (PMT). The first stage of the tube is a metal plate in vacuum, from which electrons are ejected by the light photons via the photoelectric effect. Since only a small number of electrons are created from a single x-ray photon, the charge must be amplified to be detectable. The electrons are accelerated by a potential of about 100 V into a second plate. Each of the original electrons is then energetic enough to knock off multiple electrons from this second plate, which are then accelerated toward a third plate, as shown in Figure 2-12. After several stages of multiplication, tens of millions of electrons are produced from a single x-ray photon, producing a pulse of charge that is collected on a capacitor and turned into a voltage pulse. The number of electrons (and thus the voltage) depends on the number of visible-light photons, which depends on the original x-ray photon energy. Low-energy x-ray photons can be rejected electronically by counting only those pulses above a certain voltage threshold. This provides a way of rejecting scatter in the geometry of Figure 2-10. Since the NaI emits the visible light isotropically, the light created by a single x-ray photon can be detected on multiple PMT detectors in the array shown in Figure 2-11. The location of the original x-ray photon is calculated by weighting the intensity of the light signals from the neighboring PMT tubes. This allows the array to have finer spatial resolution than the size of the PMT tube. The combination of the collimator, scintillator, and photomultiplier tube is called a *gamma camera*.

EXAMPLE 2-6

---

a) If 1% of the blue-violet light photons created by a 23 keV x-ray photon (from example 2-5) land on a PMT tube with a gain of 10^6, how many electrons are created?

b) If these are collected with 2% efficiency onto a 1 pF capacitor, what is the height of the resultant voltage peak?

c) What is the conversion factor, that is, the ratio of voltage in the electron signal to kiloelectronvolts of photon energy?

a) If we assume 1 photoelectron per blue-violet photon, then

$$N_{elec} \approx (1000 \text{ blue-light photons}) \left(\frac{1 \text{ electron}}{\text{photon}} \right)(0.01)(10^6) \approx 10^7 \text{ electrons}.$$

b) $q = C\Phi \Rightarrow \Phi = \dfrac{q}{C} \approx \dfrac{0.02(10^7 \text{ electrons})}{1 \times 10^{-12} \text{ farad}} \left(\dfrac{1.6 \times 10^{-19} \text{ Coul}}{\text{electron}} \right) \approx 0.032 \text{ V}.$

c) $\dfrac{0.032 \text{ V}}{23 \text{ keV}} \approx 1.4 \times 10^{-3} \dfrac{\text{V}}{\text{keV}}.$

So, for example, a signal pulse with a height of 14 mV would indicate that a 10 keV photon was detected.

Alternatively, rather than using energy sensitivity to reject Compton scatter, it may be possible, with very small hole parallel collimators, such as the polycapillary optics discussed in section 12.2.4, to provide some scatter rejection by simply increasing the number of pixels. This method is effective only if the radiation from the object to be imaged is concentrated into a very small region.

2.5 Photon statistics

For most visible light applications, the number of photons is so large, and the energy of an individual photon is so small, that the effect of the quantization of light into photons is insignificant. The reverse is true for many x-ray applications. In fact, as in the preceding example in nuclear medicine, many x-ray applications rely on detecting and counting individual photons. As a result, counting statistics are an important contribution to *noise* or uncertainty in measurement results. In general, if the mean result from a measurement is N, Poisson counting statistics give a standard deviation of $\sqrt{N}$. For example, if the mean is 100, the standard deviation is 10, and two-thirds of the results will be between 90 and 110; 95% of the results will be within two standard deviations, in the range 80–120. A similar effect is responsible for *quantum noise* in radiography. If an area is irradiated uniformly, so that each pixel receives, on average, N photons, the standard deviation, the variation between pixels, is also $\sqrt{N}$.

Any real variation in intensity (for example, due to a tumor) must be greater than the noise to be visible. Random processes give rise to Poisson statistics. Other processes give rise to Gaussian and Lorentz profiles, for example, the line shapes discussed in section 4.5.

Poisson statistics arise from the binomial distribution. If the probability of a photon occurring in a time interval Δt is $\Gamma \Delta t$, and one waits for m time intervals, $t = m\Delta t$, then the probability of getting N photons in the time t is

$$P(N) = \frac{m! [\Gamma \Delta t]^N (1-[\Gamma \Delta t])^{m-N}}{N! (m-N)!}. \tag{2-9}$$

Since the probability of getting a photon N times is $[\Gamma \Delta t]^N$, not getting any photons in $m - N$ intervals gives $(1 - [\Gamma \Delta t])^{m-N}$, and the N intervals with an event can be arranged $N!$ ways, while those with no detected photon can be arranged $(m - N)!$ ways. This distribution has a mean

$$\overline{N} = m[\Gamma \Delta t] \tag{2-10}$$

and a standard deviation

$$stdev = \sqrt{m[\Gamma \Delta t](1-[\Gamma \Delta t])}. \tag{2-11}$$

If $\Gamma \Delta t$ is small (this can always be the case if the time interval Δt is chosen to be small), then using Stirling's formula, $\ln(x!) \approx x \ln x - x$, the probability becomes

$$P(N) \approx \frac{\overline{N}^N e^{-\overline{N}}}{N!}, \tag{2-12}$$

and

$$stdev \approx \sqrt{m[\Gamma \Delta t]} = \sqrt{\overline{N}}. \tag{2-13}$$

The noise $\boldsymbol{N}$ is defined as the variation in the measured signal,

$$\boldsymbol{N} = \Delta N = \sqrt{N}. \tag{2-14}$$

EXAMPLE 2-7

A radioactive source produces a count rate per pixel in the image of the hot spot of 3 counts/s. How many seconds must the image be recorded to keep the quantum noise to less than 5%?

$$\Delta N = \sqrt{N} \Rightarrow \frac{\Delta N}{N} = \frac{\sqrt{N}}{N} = \frac{1}{\sqrt{N}} \Rightarrow N = \left(\frac{1}{\Delta N/N}\right)^2 = \left(\frac{1}{0.05}\right)^2 = 400.$$

To obtain 400 counts the time required is $t = \dfrac{400\,\text{counts}}{3\dfrac{\text{counts}}{\text{s}}} \approx 133\,\text{s}.$

EXAMPLE 2-8

Measurements of two different samples give count rates of 396 counts/s and 414 counts/s, respectively. Are these results consistent with the two samples having the same activity?

$\sqrt{396} \approx 20;\ \sqrt{414} \approx 20 \Rightarrow 396 = 396 \pm 20;\ 414 = 414 \pm 20.$ The first result falls between 376 and 416, the second between 394 and 434. The numbers overlap and are consistent. One-third of the time two measurements will differ by two standard deviations, so even in that case, one could not rule out consistency from a single set of measurements.

2.6 SPECT

The images produced from the radiography or nuclear medicine image geometries of Figures 2-3 and 2-10 are two-dimensional. *Computed tomography*, CT, is a technique that produces a three-dimensional representation of an object from a series of two-dimensional images taken at different angles, as diagramed in Figure 2-13. For applications such as nuclear medicine, in which photons are emitted from the object itself and are counted individually, the technique is called *single-photon emission computed tomography*, or SPECT.*

To provide a less complex example than three-dimensional SPECT, consider the "patient" in Figure 2-13 to be two-dimensional, and the detector to be a one-dimensional array. The goal is then to create a two-dimensional array representing the rate of photon emission from different points within the patient by recording a series of one-dimensional "images." A common mathematical method for performing this operation is simple back projection. For example, say that the detector in position A for some fixed counting period collects 10 photons in the top pixel from the upper hot spot and 5 photons in the fourth pixel from the lower spot. The "real" patient could then be represented as a two-dimensional matrix,

$$\text{Source} = \begin{pmatrix} 0 & 10 & 0 & 0 & 0 \\ 0 & 0 & 0 & 0 & 0 \\ 0 & 0 & 0 & 0 & 0 \\ 0 & 0 & 0 & 5 & 0 \\ 0 & 0 & 0 & 0 & 0 \end{pmatrix}. \tag{2-15}$$

---

*PET is positron emission tomography, which is similar to SPECT but takes into account the fact that the interaction of a positron with an electron generates two identical photons traveling in opposite directions. In PET coincidence timing is used instead of discrimination based on the photon energy to remove the scatter background.

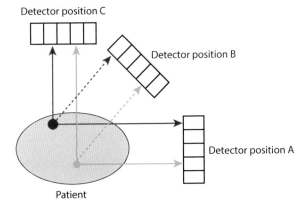

FIGURE 2-13. SPECT of an object with two hot spots, taken by placing the detector consecutively at A, B, and C. The detector assembly includes a parallel-hole collimator, as shown in Figure 2-9, so that only rays perpendicular to the pixel array are accepted.

The data from the single measurement can be represented as a column array,

$$\text{Detector}_A = \begin{pmatrix} 10 \\ 0 \\ 0 \\ 5 \\ 0 \end{pmatrix}. \tag{2-16}$$

There is not enough information from a single image to make a two-dimensional map of the distribution of hot spots. All that is known is the vertical location of the spots. So as a first step in creating the matrix, the data are projected perpendicular to the detector, in this case along the rows,

$$\text{Projection}_A = \begin{pmatrix} 10 & 10 & 10 & 10 & 10 \\ 0 & 0 & 0 & 0 & 0 \\ 0 & 0 & 0 & 0 & 0 \\ 5 & 5 & 5 & 5 & 5 \\ 0 & 0 & 0 & 0 & 0 \end{pmatrix}. \tag{2-17}$$

The object to the right of the array is a grayscale representation. The data obtained when the detector is at position C can be represented as a row array,

$$\text{Detector}_C = (0 \quad 10 \quad 0 \quad 5 \quad 0), \tag{2-18}$$

which can be back projected down the columns to produce a second patient array,

$$\text{Projection}_C = \begin{pmatrix} 0 & 10 & 0 & 5 & 0 \\ 0 & 10 & 0 & 5 & 0 \\ 0 & 10 & 0 & 5 & 0 \\ 0 & 10 & 0 & 5 & 0 \\ 0 & 10 & 0 & 5 & 0 \end{pmatrix}. \tag{2-19}$$

Summing the two arrays gives

$$\text{Projection}_A + \text{Projection}_C = \begin{pmatrix} 10 & 20 & 10 & 15 & 10 \\ 0 & 10 & 0 & 5 & 0 \\ 0 & 10 & 0 & 5 & 0 \\ 5 & 15 & 5 & 10 & 5 \\ 0 & 10 & 0 & 5 & 0 \end{pmatrix}. \qquad (2\text{-}20)$$

The array position (1,2) corresponds to the hot spot in the upper left of Figure 2-13, and array position (4,4) corresponds to the hot spot at lower right. There are *streak artifacts* (nonzero values where there is actually no radiation) in the columns and rows adjacent to the hot spots.

Determining what the data would look like from position B requires specifying the size of the pixels and hot spots and the center of rotation of the detector module position. Hot spots that overlap two pixels would cause a fraction of their intensity to be collected in each pixel. Assuming the hot spots are small and align exactly with the pixels as shown, the array holds (10 0 0 5 0). The back projection is first done in the rotated coordinates of Figure 2-14, and then the coordinate of each unrotated pixel must be computed from the rotated data, by converting the x, z coordinate using the relations

$$x'_j = x\cos\phi_j - z\sin\phi_j$$
$$z'_j = x\sin\phi_j + z\cos\phi_j \qquad (2\text{-}21)$$

and interpolating as necessary, because the pixel indices are no longer integers. The result for this case at 45° is

$$\text{Projection}_B \approx \begin{pmatrix} 1.72 & 8.79 & 4.14 & 0.00 & 0.00 \\ 8.79 & 4.14 & 0.00 & 0.00 & 3.54 \\ 4.14 & 0.00 & 0.00 & 3.54 & 2.93 \\ 0.00 & 0.00 & 3.54 & 2.93 & 0.00 \\ 0.00 & 3.54 & 2.93 & 0.00 & 0.00 \end{pmatrix}. \qquad (2\text{-}22)$$

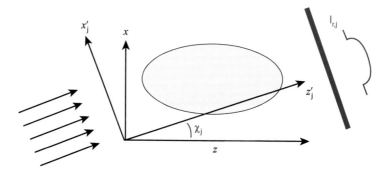

FIGURE 2-14. Coordinates for the jth projection, rotated an angle χ_j from the original axis in a radiographic image.

Adding that array to the sum $\text{Projection}_A + \text{Projection}_C$ yields

$$
\text{Projection}_{\text{Sum}} =
\begin{pmatrix}
11.7 & 28.8 & 14.1 & 15.0 & 10.0 \\
8.8 & 14.1 & 0.0 & 5.0 & 3.5 \\
4.1 & 10.0 & 0.0 & 8.5 & 2.9 \\
5.0 & 15.0 & 8.5 & 12.9 & 5.0 \\
0.0 & 13.5 & 2.9 & 5.0 & 0.0
\end{pmatrix}.
\tag{2-23}
$$

With each additional detector position (projection), the array (after dividing by the number of measurements) becomes a somewhat more accurate two-dimensional representation of the actual activity of the source described by equation 2-15. For a better final array, more images are required. Figure 2-15 is an example of a full SPECT image. The displayed views are two-dimensional cuts through a three-dimensional array of source intensity computed from a series of images. In practice, the problem of determining the three-dimensional distribution of radiation from the planar images is complicated by noise, the unknown number of sources, hot spots outside the region of interest, and the like. The reconstruction is often done after application of Fourier or other spatial frequency transforms and noise-reducing filters. CT for conventional radiography is discussed in section 9.10.

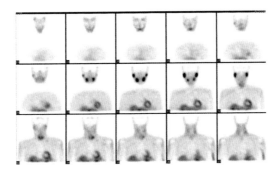

FIGURE 2-15. Series of reconstructed views of SPECT image of primary hyperparathyroidism (overactive parathyroid gland in neck). Salivary and thyroid glands and heart muscle are also highlighted.

Problems

SECTION 2.1

1. Verify using equation 2-2 that after a time $t = t_{1/2}$ half of the original material remains.
2. How many atoms of ^{99m}Tc are required to produce an activity of 1 Ci? What mass of pure ^{99m}Tc is required?
3. Assuming one photon per decay, what is the emission rate from a 69-year-old radioactive source that has a half-life of 10 years and an original strength of 2 mCi?
4. A sample of freshly produced ^{125}I, with a half-life of 60 days, was originally measured to have an activity of 1000 counts/s. It now gives 400 counts/s in the same geometry. How old is the sample?

5. To compute the activity, in millicuries, of a 100 kg person, due to the normal concentration of 10^{-3} ppb (10^{-3} parts per billion, or 1 part per trillion, a concentration of 10^{-12} g/kg) of ^{14}C with a half-life of 5730 years, we will assume people are the same concentration as Lucite, $C_5H_8O_2$.
   a) Compute the number of moles of Lucite in 100 kg. b) Estimate the number of atoms of carbon in a 100 kg person c) Estimate the number of atoms of ^{14}C in a 100 kg person. d) Compute the activity of the person.

SECTION 2.3

6. An 8 mm diameter cylindrical pinhole is placed 100 mm from a source. What is the sensitivity of the pinhole?

7. If the cylindrical hole in Figure 2-6 is 1 mm in diameter and is placed 200 mm from the tumor, what fraction of the emitted radiation is intercepted by the pinhole? What activity, in millicuries, is required to produce 1000 photons per second through the pinhole (assuming one photon per decay)?

SECTION 2.4

8. A 10 keV x-ray photon is captured by a NaI(Tl) scintillator. How many blue-violet photons are emitted? If 1% of the blue-violet photons land on a PMT tube, and each creates a single photoelectron, how many electrons are there at the entrance to the PMT? If the PMT has a gain of 10^6, how many electrons exit the PMT? If these are collected with 2% efficiency onto a 1 pF capacitor, what is the total charge on the capacitor? What is the height of the resultant voltage peak? What is the conversion factor (ratio of the electronic signal in volts to photon energy in kiloelectronvolts)?

SECTION 2.5

9. A radioactive source produces a count rate of 10 counts/s. How many seconds must the count be recorded to keep the quantum noise to less than 1%?

10. A pinhole with a sensitivity of 10^{-6} is to be used to image a tumor with three hot spots, with activity of 1, 2, and 3 µCi, respectively. How long a count time is required to distinguish the intensities?

SECTION 2.6

11. An area detector placed in one location records an intensity of (0 4 0 2 0)

µCi on its five pixels. A second detector, placed at 90°, records $\begin{pmatrix} 0 \\ 3 \\ 0 \\ 3 \\ 0 \end{pmatrix}$ You don't

know how many sources there are. Find at least three possible distributions of integer source values (three possible 5×5 arrays) that fit the data. *Hint:* The first row and column of the 5×5 array must be all zeros for the first pixel of the detector to read zero in the two orientations. One possible arrangement includes the three values 1, 2, and 3.

Further reading

General references for medical imaging

Jerrold T. Bushberg, J. Anthony Seibert, Edwin M. Leidholdt Jr, and John M. Boone, **The Essential Physics of Medical Imaging**, 3rd ed., Lippincott Williams & Wilkins, 2012.

S. Webb, **The Physics of Medical Imaging**, IOP Press, 1988.

William R. Hendee and E. Russell Ritenour, **Medical Imaging Physics**, 4th ed., Wiley-Liss, 2002.

PART II
X-RAY GENERATION

3

THERMAL SOURCES AND PLASMAS

3.1 Blackbody radiation

Hot things glow. Hotter things glow brighter and at higher frequencies. Really hot things give off x rays. Statistical mechanics tells us that the intensity of radiation at a frequency v from an ideal "black" body at temperature T is*

$$I_v dv = \frac{8\pi}{(hc)^2} \frac{(hv)^3}{e^{\left(\frac{hv}{k_B T}\right)} - 1} dv, \qquad (3\text{-}1)$$

where h is Planck's constant, k_B is Boltzmann's constant, and c is the speed of light. The frequency at which the peak occurs is proportional to temperature T, and the total emitted power is proportional to T^4. The distribution is plotted in Figure 3-1 with photon energy, $U = hv$, substituted for frequency. As can be seen from the graph, doubling the temperature doubles the peak frequency and increases the total emission by a factor of 16. For the emission peak to be at photon energies of 1 keV, the temperature must be about 4 million kelvin. At that temperature, $T = 4$ MK, the approximate mean thermal kinetic energy, $k_B T$, of an electron in a material is about 345 eV (with a broad distribution out to kilovolt energies). Since valence electron ionization energies range from 4 to 25 eV, all the valence electrons have energies exceeding their ionization energy, and so the atom is at least partially ionized, and the material is a *plasma*, a collection of ions and free electrons, rather than a collection of atoms. (Atoms with high atomic number are only partially ionized, because core ionization energies can be several thousands of electronvolts, so that when the valence electrons are stripped off the atom, an ionic core consisting of the nucleus and bound core electrons is left behind.) While plasmas are somewhat unfamiliar compared with solid-state materials, the mechanism of light emission from plasmas is similar to that from cooler materials: energetic electrons make transitions to lower energy states, giving off the

*Sections 3.1 and 3.2 describe hot plasma sources (including astronomical objects) but are optional for a study focusing on tube sources or synchrotron radiation. Section 3.3 provides a derivation of plasma frequency, which is useful in developing the index of refraction in chapter 11. The energy levels described in section 3.1 and the screening described in section 3.5 are similar to the computation of energy levels in section 4.2 but are not essential to that development.

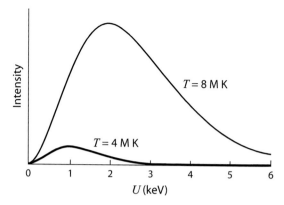

FIGURE 3-1. Blackbody intensity I_U as a function of photon energy U for temperatures of 4 million and 8 million K.

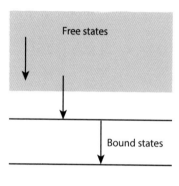

FIGURE 3-2. Electronic structure of very hot, highly ionized materials. The bound states are core electronic levels in an atom from which most of the outer electrons have been stripped off.

excess energy as photons. The electronic energy levels consist of a continuum of states for the free electrons, as well as bound states for the electrons in the ions, as shown in Figure 3-2. The arrow at the left in the figure represents a free electron transition, which can have a continuous distribution of energies. This is similar to the bremsstrahlung emission that will be discussed in chapter 6. It is the dominant mechanism for high-Z elements (those with high atomic number). The arrow at right represents transitions between bound states. These transitions have discrete energies, similar to the characteristic atomic emission, which will be discussed in chapter 4, except that for plasmas the relevant energies are for states of the highly ionized atoms.

The energy levels of the bound states in the plasma can be approximated as those of hydrogenic (hydrogen-like) atoms, those with a single electron. Those states are described in terms of their principal quantum number, m,[*]

$$U_m = -U_R \frac{Z^2}{m^2},\qquad\text{(3-2)}$$

where Z is the atomic number, the number of protons in the nucleus, and U_R is the Rydberg energy,

$$U_R = \frac{M_e q_e^4}{8\varepsilon_o^2 h^2} \approx 13.6 \text{ eV},\qquad\text{(3-3)}$$

*Note that m is an integer. The variable n is reserved for index of refraction, and M is used for mass.

where M_e is the rest mass of the electron, q_e is its charge, and ε_o is the vacuum permittivity, $\varepsilon_o \approx 8.85 \times 10^{-12} \dfrac{Coul^2}{N \cdot m^2}$.

EXAMPLE 3-1: EMISSION LINE WAVELENGTH

a) What are the approximate $m = 1$ and $m = 2$ energy levels for highly ionized carbon?

The atomic number for carbon is $Z = 6$, which gives

$$U_1 = -U_R \frac{Z^2}{m^2} \approx -(13.6 \text{ eV}) \frac{6^2}{1^2} \approx -490 \text{ eV}.$$

$$U_2 \approx -(13.6 \text{ eV}) \frac{6^2}{2^2} \approx -122 \text{ eV}.$$

b) What is the wavelength for the Lyman α emission from carbon plasmas, caused by an electron dropping from the $m = 2$ to the $m = 1$ level?

The photon carries away the energy difference between the levels,

$$U = U_2 - U_1 \approx -123 \text{ eV} - (-491 \text{ eV}) \approx 367 \text{ eV}.$$

$$\lambda = \frac{hc}{U} \approx \frac{12.4 \times 10^3 \text{ eV} \cdot \text{Å}}{367 \text{ eV}} \approx 34 \text{ Å}.$$

Emission from plasmas is discussed further in chapter 8. Transitions can also occur between free and bound states, as depicted by the middle arrow; this is known as recombination radiation.

3.2 Generation of very hot plasmas

The very hot temperatures required for blackbody emission of x rays exist naturally in astronomical bodies such as stars, including our sun, as shown in Figure 3-3. X-ray astronomy allows for the direct observation of very energetic material in these bodies, and spectroscopy on the bound-to-bound emissions reveals information about composition as well as temperature. Additional astronomical images appear in chapter 12. Very hot plasmas also exist in terrestrial fusion reactors, and x-ray emission is used as a diagnostic tool for fusion plasmas such as those produced by the laser system shown in Figure 3-4.

Hot plasmas are also created especially to generate x rays. In these x-ray sources, the plasma is created as a short-lived burst, for example by a short-pulse laser

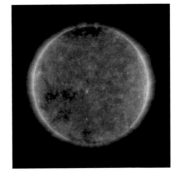

FIGURE 3-3. Image of the sun taken from the NASA SOHO orbiting telescope using 171 Å radiation. Courtesy of SOHO/EIT consortium. SOHO is a project of international cooperation between ESA and NASA.

Figure 3-4. Laser beams at the National Ignition Facility, used to study fusion as a potential energy source. Photo by Damien Jemison/LLNL.

beam focused onto a metal or gaseous target. Laser light can be focused to a diameter on the order of the wavelength, for example, approximately 0.5 μm for green light. Green light would penetrate on the order of 40 nm into a metal target, so the volume irradiated for a typical laser spot 18 μm in diameter is about 10^{-17} m³. If the pulse is short enough that no appreciable thermal diffusion or atomic motion can take place, all the energy is deposited in this volume. For example, if the pulse duration is 1 ps, then evaporated atoms traveling at 1000 m/s will move only about 1 nm and will not escape the hot spot during irradiation. Thus a fairly moderate 10 mJ laser pulse creates an energy density of 10^{15} J/m³. A 10 mJ laser pulse with a 1 ps pulse duration carries an instantaneous power of 10^{10} W and a power density of 10^{27} W/m³, which is quite capable of generating a plasma and x-ray emission. Laser-generated x-ray sources are used for x-ray lithography in the manufacture of specialized computer circuitry.

Figure 3-5. A pinch plasma source, with the "load" of crossed wires in the shape of an X. Photo courtesy of Cornell Plasma Studies Lab; apparatus used by Isaac Blesener.

Plasmas can also be created by a short pulse of very high current through a wire. In a laboratory system such as that shown in Figure 3-5, a current of approximately 100,000 A is shunted through the wire by rapidly shorting a capacitor bank. The wire quickly gets so hot that it evaporates and is ionized, creating a plasma. The plasma is then compressed into a small volume by the magnetic field created by the current

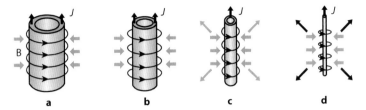

FIGURE 3-6. a. The current J (traveling upward in the sketches) creates a magnetic field B. The short pulse of current is carried largely by the outer edges of the wire, owing to skin-depth effects. The material becomes hot and ionizes. b. The B field creates an inward force of $qv \times B$ on the electrons and ions in the plasma. c. The hot, high-density plasma begins radiating. d. Radiative cooling (loss of energy due to the emitted radiation) allows further collapse of the plasma. From V. Kantsyrev, Pinch Plasma Sources, chap. 57, *Handbook of Optics*, 3rd ed., vol. 5, McGraw-Hill, 2010.

pulse itself, as shown in Figure 3-6. A nanosecond pulse of intense soft (low photon energy) x-ray emission is generated.

3.3 Plasma frequency

Plasmas are characterized by their temperature and density or, alternatively, by their plasma frequency and screening length, which can be computed from the density and temperature. The plasma frequency, ω_p, is the natural frequency of oscillation. As will be seen in chapter 11, the response of a material (not just a plasma) to the oscillating electric field of an x-ray beam can be described in terms of the plasma frequency.

A plasma is modeled as a collection of free electrons and positive ions, as shown in Figure 3-7. The electric field, E, on the cylindrical container walls can be computed from Gauss's law,

$$\varepsilon_0 \int \vec{E} \cdot \overrightarrow{dA} = \int (q_e \rho_+ - q_e \rho_e) dV = q_{enclosed}, \tag{3-4}$$

where the area integral is over the surface of the container, the second integral is over its volume, ε_0 is the vacuum permittivity, ρ_+ is the positive charge density, ρ_e is the electron density, q_e is the magnitude of the charge on an electron, and $q_{enclosed}$ is the total enclosed charge. Normally, because $\rho_+ = \rho_e$, the total charge $q_{enclosed}$ is zero, and the field is also zero. This is the static case, analogous to a mass hanging at rest on a spring. To determine the natural frequency of oscillation of the mass, you can pull down on the mass and then release it. Similarly, one can imagine displacing the electrons in a plasma and then releasing them. A brief application of an electric field can cause the electrons to move a distance x away from the positive charges, as shown in Figure 3-8. The total charge and the field on the container walls is still zero,

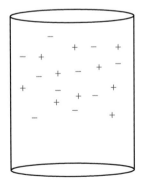

FIGURE 3-7. A plasma can be modeled as a collection of free charges.

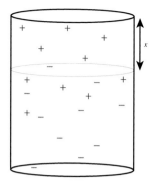

but the field on the (imaginary) circular plate in the middle is not. Reapplying Gauss's law to the top section of the cylinder gives an expression for the electric field,

$$\varepsilon_o \int \vec{E} \cdot \vec{dA} = \varepsilon_o EA = \int (q_e \rho_+ - 0)\, dV \qquad (3\text{-}5)$$

$$= q_e \rho_+ \, xA \Rightarrow E = \frac{x q_e \rho_+}{\varepsilon_o}.$$

The field created by temporarily separating the positive and negative charges in the container exerts a force on the charges that causes them to accelerate back toward each other,

$$F = -q_e E = -q_e \frac{x q_e \rho_+}{\varepsilon_o} = -q_e \frac{x q_e \rho_e}{\varepsilon_o}, \qquad (3\text{-}6)$$

where, for convenience, the positive charge density ρ_+ is written as equal to the initial electron charge density ρ_e. Because the mass of a positive ion is several thousand times the mass of an electron, the positive ions remain essentially stationary while the electrons undergo a motion described by

$$F = M_e \frac{d^2 x}{dt^2} = -q_e \frac{x q_e \rho_e}{\varepsilon_o} \Rightarrow \frac{d^2 x}{dt^2} + \frac{q_e^2 \rho_e}{\varepsilon_o M_e} x = 0. \qquad (3\text{-}7)$$

The solution to this differential equation is oscillatory motion of the charges,

$$x = x_o \cos(\omega_p t), \qquad (3\text{-}8)$$

with an angular frequency equal to the plasma frequency,

$$\omega_p = \sqrt{\frac{q_e^2 \rho_e}{\varepsilon_o M_e}}. \qquad (3\text{-}9)$$

This characteristic frequency of the plasma depends only on universal constants and the electron density ρ_e. It is characteristic of the plasma in the same way that the resonant frequency is characteristic of a mass on a spring. The amplitude of the response of a mass on the spring to a driving force at a given frequency depends on the system's resonant frequency (for example, whether the frequency is near, far above, or far below the resonance). In the same way, the optical properties, including the x-ray optical properties, of a material depend on its plasma frequency. Just like a mass on a spring does not respond appreciably to oscillations much above its resonant frequency, a plasma can barely respond to oscillating electric fields at frequencies very much higher than the plasma frequency. This means it is nearly transparent at those frequencies, and neither absorbs or reflects the radiation. For example, the plasma formed by cosmic radiation in the upper atmosphere has a plasma frequency between that of AM and FM radio signals. FM transmissions pass through it, while the lower-frequency AM transmissions are reflected back to earth. For that reason, you can receive AM radio signals even if you are far away, around the curve of the earth from the line of sight of the transmitter.

EXAMPLE 3-2: PLASMA FREQUENCY

---

a) What is the plasma frequency of copper at a temperature high enough that it has been fully ionized, assuming (unrealistically) that the density is unchanged from that at room temperature?

To apply equation 3-9, we need the charge density, the number of electrons per unit volume, in copper. Copper is atomic number $Z=29$, so it has 29 electrons per atom. Since the copper has been fully ionized, all the electrons are participating in the plasma. It has a molecular mass of 63.5, that is a mass of $M_M = 63.5 \times 10^{-3}$ kg/mole. The specific density of copper is 8.96, so its mass density is $\rho = 8.96 \times 10^3$ kg/m³. Therefore the electron density is

$$\rho_e = ZN_A \frac{\rho}{M_M} \approx \left(29 \frac{\text{electrons}}{\text{atom}} \right) \left(6.02 \times 10^{23} \frac{\text{atoms}}{\text{mole}} \right) \left(\frac{8.96 \times 10^3 \frac{\text{kg}}{\text{m}^3}}{63.5 \times 10^{-3} \frac{\text{kg}}{\text{mole}}} \right)$$

$$\approx 2.46 \times 10^{30} \frac{\text{electrons}}{\text{m}^3}.$$

Inserting this result into equation 3-9 gives

$$\omega_p = \sqrt{\frac{\rho_e q_e^2}{\varepsilon_0 M_e}} \approx \sqrt{\frac{\left(2.46 \times 10^{30} \frac{\text{electrons}}{\text{m}^3} \right) \left(1.6 \times 10^{-19} \frac{\text{Coul}}{\text{electron}} \right)^2}{\left(8.85 \times 10^{-12} \frac{\text{Coul}^2}{\text{N} \cdot \text{m}^2} \right) (9.1 \times 10^{-31} \text{kg})}}$$

$$\approx 8.8 \times 10^{16} \frac{\text{rad}}{\text{s}}.$$

For comparison, expressed in units of energy,

$$U_p = \hbar \omega_p \approx (1.05 \times 10^{-34} \text{J} \cdot \text{s}) \left(8.8 \times 10^{16} \frac{\text{rad}}{\text{s}} \right) \left(\frac{1 \text{eV}}{1.6 \times 10^{-19} \text{J}} \right) \approx 58 \text{eV}.$$

This energy is very much smaller than x-ray photon energies but larger than visible photon energies (which are about 2 eV). In analogy with radio transmission, this means that copper should be opaque to visible light and relatively transparent to x rays.

b) What is the plasma frequency of the ionosphere if the electron density is 6×10^{12} electrons/m³?

In this case, the plasma frequency is $\omega_p = 1.4 \times 10^8$ rad/s. Note that from equation 3-8, ω_p is an angular frequency. The frequency $\nu = \omega_p/2\pi$ is 22 MHz.

The plasma frequency even for solids is very much less than the frequency of x-ray radiation, so the electrons in most materials do not oscillate strongly in response to x-ray radiation. Thus, x rays generally pass straight through solids without being appreciably reflected or deflected. Optical properties, such as refraction of x rays, are discussed in more detail in chapter 11.

3.4 Debye length

As noted earlier, plasmas can be described in terms of their density and temperature or in terms of their plasma frequency and their screening length, also known as the *Debye length* (after the same Peter Debye who, as noted in section 1.3, received a Nobel Prize partly for work in x-ray diffraction). The Debye length depends on both density and temperature. A convenient, if somewhat artificial, way of bringing temperature into the discussion is to consider thermal motion. In thermal equilibrium at very high temperatures electrons will have very large thermal velocities. The linear velocity can be estimated from the thermal energy,

$$\frac{3}{2}k_B T \approx \frac{1}{2}M_e |v|^2 = \frac{1}{2}M_e (v_x^2 + v_y^2 + v_z^2) \approx \frac{3}{2}M_e v_z^2 \Rightarrow v_z \approx \sqrt{\frac{k_B T}{M_e}}. \tag{3-10}$$

The distance an electron moves in one dimension during a time t equal to the period of a plasma frequency oscillation is

$$z \approx v_z t \approx \sqrt{\frac{k_B T}{M_e}} \frac{2\pi}{\omega_p} = \sqrt{\frac{k_B T}{M_e}} \frac{2\pi}{\sqrt{\frac{\rho_e q_e^2}{\varepsilon_o M_e}}} = 2\pi \sqrt{\frac{\varepsilon_o k_B T}{\rho_e q_e^2}}. \tag{3-11}$$

The quantity under the square root is the Debye length,

$$\Lambda_D = \sqrt{\frac{\varepsilon_o k_B T}{\rho_e q_e^2}}. \tag{3-12}$$

Note that this is not the amplitude x_o of the plasma oscillation, since the plasma oscillation is not connected to the thermal velocity. Rather, the Debye length here resulted from combining the time from the plasma oscillation with the velocity from the random thermal motions.

EXAMPLE 3-3: DEBYE LENGTH

What is the Debye length of the ionosphere at 10,000 K?

From equation 3-12,

$$\Lambda_D = \sqrt{\frac{\varepsilon_o k_B T}{\rho_e q_e^2}} \approx \sqrt{\frac{\left(8.8\times10^{-12}\,\frac{Coul^2}{N\cdot m^2}\right)\left(1.38\times10^{-23}\,\frac{J/electron}{K}\right)(10{,}000\,K)}{\left(6\times10^{12}\,\frac{electrons}{m^3}\right)\left(1.6\times10^{-19}\,\frac{Coul}{electron}\right)^2}}$$

$$\approx 2.8 \text{ mm}.$$

3.5 Screening and the Debye length

The Debye length more directly describes the screening of a point charge by a plasma. Screening also becomes important in computing electronic energy levels, and hence x-ray emission energies, in chapter 4.

Without a plasma, the voltage Φ_{bare} at a distance r from a bare, unscreened, point charge q is

$$\Phi_{bare} = \frac{q}{4\pi\varepsilon_o r}. \qquad \textbf{Point charge} \qquad (3\text{-}13)$$

For the neutral plasma, before the charge q is introduced, the total charge and field are zero, so the voltage is a constant, Φ_o, which, since the absolute potential is arbitrary, we will take as our reference, $\Phi_o = 0$. In response to the introduction of the charge q, the electrons in the plasma move and redistribute their charge density, creating a non-zero voltage distribution. To give a rough outline of the computation of the screened voltage distribution, assume that the electron density at a distance r changes from the uniform ρ_e to $\rho_e'(r) = \rho_e + \Delta\rho_e(r)$. As a result of the redistribution of charge, the voltage at r will change to $\Phi(r)$. The change in voltage gives each electron a change in energy of $\Delta U = -q_e\Phi$. In thermal equilibrium, the new electron density and the energy should be related by Boltzmann statistics; that is, the number of charges at a location r, $\rho_e'(r)$, should depend on the extra energy required to place them at r, $\Delta U(r)$,

$$\rho_e' = \rho_e\left(e^{-\frac{\Delta U}{k_B T}} \right). \qquad (3\text{-}14)$$

The requirement that the disturbed plasma be in local thermal equilibrium introduces temperature into the problem. Substituting $\Delta U = -q_e\Phi$, and assuming that the change in energy is small yields

$$\rho_e' = \rho_e\left(e^{-\frac{\Delta U}{k_B T}} \right) = \rho_e\left(e^{\frac{q_e\Phi}{k_B T}} \right) \approx \rho_e\left(1 + \frac{q_e\Phi}{k_B T} \right)$$

$$\Rightarrow \Delta\rho_e = \rho_e' - \rho_e \approx \rho_e\frac{q_e\Phi}{k_B T}. \qquad (3\text{-}15)$$

The resultant voltage Φ and the charge distribution are related by the differential form of Gauss's law (which was given in integral form in equation 3-4)

$$\nabla^2\Phi = \frac{1}{\varepsilon_o}\rho_q(r), \qquad (3\text{-}16)$$

where $\rho_q(r)$ is the net charge density distribution. Since the plasma was initially neutral, the net charge density is

$$\rho_q(r) = -q_e\Delta\rho_e. \qquad (3\text{-}17)$$

Hence the voltage obeys

$$\nabla^2\Phi = \frac{1}{r^2}\frac{\partial}{\partial r}\left(r^2\frac{\partial\Phi}{\partial r} \right) = -\frac{1}{\varepsilon_o}q_e\Delta\rho_e = -\frac{\rho_e q_e^2}{\varepsilon_o k_B T}\Phi = -\frac{\Phi}{\Lambda_D^2}. \qquad (3\text{-}18)$$

Solving the differential equation gives

$$\Phi = \frac{q}{4\pi\varepsilon_o r} e^{\frac{-r}{\Lambda_D}} = \Phi_{bare}\, e^{\frac{-r}{\Lambda_D}}. \tag{3-19}$$

The voltage due to the bare charge, when screened by the plasma, falls off exponentially, much more rapidly than the usual dependence on $1/r$. At distances much larger than the Debye screening length, the voltage due to the point charge q is nearly zero.

3.6 Fluctuations and the Debye length

One can also describe the Debye length in terms of the expected size of thermal fluctuations in the charge density of the plasma. If the plasma has an electron density fluctuation of $\Delta\rho_e \sim \rho_e$ over a region of size R, then the charge contained in a sphere of radius R, assuming the positive charge remains constant, is

$$q = -q_e\rho_e \frac{4}{3}\pi R^3. \tag{3-20}$$

The resultant fluctuation in potential and hence energy at the surface of the spherical fluctuation is

$$\Delta U = -q_e\Phi = -q_e \frac{q}{4\pi\varepsilon_o R} = \frac{q_e^2\left(\frac{4}{3}\right)\pi R^3\rho_e}{4\pi\varepsilon_o R} = \frac{q_e^2 R^2\rho_e}{3\varepsilon_o}. \tag{3-21}$$

If this energy is to be supplied by thermal fluctuations, then it should not be much more than the thermal energy, $\Delta U < k_B T$, so that

$$\Delta U < k_B T \Rightarrow \frac{q_e^2 R^2\rho_e}{3\varepsilon_o} < k_B T \Rightarrow R < \sqrt{3}\sqrt{\frac{\varepsilon_o k_B T}{q_e^2\rho_e}} = \sqrt{3}\Lambda_D. \tag{3-22}$$

Thermal fluctuations in the charge density within the plasma will be limited to a scale of approximately a Debye length.

Problems

SECTION 3.1

1. Verify the value of U_R in equation 3-3.
2. Show that the ratio of radiation intensity at two high temperatures is approximately $\dfrac{I(T_1)}{I(T_2)} \approx e^{\frac{U}{k_B}\left(\frac{1}{T_2}-\frac{1}{T_1}\right)}$ and then estimate the ratio of intensity of 1 keV radiation at 500,000 K to that at a temperature of "only" 100,000 K.
3. Compute the energy of the (visible light) hydrogen Balmer α line, associated with the transition of an electron from the $m = 3$ to the $m = 2$ levels in hydrogen.

SECTION 3.2

4. a) Estimate the velocity of nickel atoms at 1000 K by setting $k_B T \approx M_{Nickel} v^2$, then use that result to estimate the distance a hot nickel atom could travel during a 1 ps laser pulse.

   b) Calculate the approximate power density due to a 2 mJ laser pulse of 1 ps duration focused to a diameter of 0.5 μm that is absorbed within a depth of 50 nm of a nickel surface.

SECTIONS 3.3–3.5

5. A gas discharge in a vacuum is used to create a plasma. The chamber is filled with He, then evacuated to a density of 10^{-9} kg/m³. The spark has a temperature of about 1 million K. Calculate the electron density, the plasma frequency, and the Debye length.

6. A plasma has a plasma frequency of 1 GHz and a Debye length of 60 μm. The volume is reduced by a factor of 4, increasing the electron density by the same factor. The temperature is unchanged.

   a) What is the new plasma frequency?
   b) What is the new screening length?

7. Show that the energy corresponding to the plasma frequency of fully ionized SiO_2 with a density of 2.2 g/cm³ is 30 eV.

Further reading

A. G. Michette and C. J. Buckley, **X-ray Science and Technology**, Institute of Physics Publishing, 1993.

Alan Michette, Laser-Generated Plasmas, chap. 56 in M. Bass, C. DeCusatis, J. Enoch, V. Lakshminarayanan, G. Li, C. MacDonald, V. N. Mahajan, and E. Van Stryland, eds., **Handbook of Optics,** 3rd ed., vol. 5, **Atmospheric Optics, Modulators, Fiber Optics, X-Ray and Neutron Optics,** McGraw-Hill, 2010.

Victor Kantsyrev, Pinch Plasma Sources, chap. 57 in M. Bass, C. DeCusatis, J. Enoch, V. Lakshminarayanan, G. Li, C. MacDonald, V. N. Mahajan, and E. Van Stryland, eds., **Handbook of Optics,** 3rd ed., vol. 5, McGraw-Hill, 2010.

4

CHARACTERISTIC RADIATION, X-RAY TUBES, AND X-RAY FLUORESCENCE SPECTROSCOPY

The discussion of x-ray tubes is divided into three sections. Chapter 4 includes narrowband characteristic radiation, chapter 5 discusses some tube design issues as well as properties of importance for comparing the applicability of sources for various uses, and chapter 6 continues with the origin of broadband bremsstrahlung radiation.

4.1 Introduction

The most common x-ray source for a wide variety of applications is an x-ray tube similar to that used by Röntgen. A simplified schematic is shown in Figure 4-1. Most cathodes work by thermionic emission: the wire gets so hot from the current flowing through it that the thermal energy of the electrons becomes greater than their binding energy, and they can escape the cathode. More recently, field emission cathodes have been developed. Such cathodes have very fine tips, for example, carbon nanotubes, so that the electric fields in the vicinity of the tip are high enough to strip off the electrons even from a cold cathode. With either cathode, a high voltage, typically

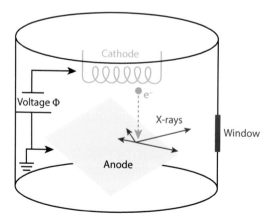

Figure 4-1. Schematic of an x-ray tube. Electrons emitted from the cathode in a vacuum vessel are accelerated toward a metal anode by an accelerating voltage, usually 10,000–200,000 V.

between 10 kV and 200 kV, then accelerates the electrons removed from the cathode toward the anode, where they strike a metal target (which is usually the same as the anode).

In unipolar sources, such as the one shown in Figure 4-2, the anode is grounded, and the cathode is kept at a large negative voltage. This keeps the high voltages away from the anode, which allows the x-ray window (and the point of user contact) to be placed close to the x-ray emission point. In bipolar circuits, the anode is usually held at half the full potential between the cathode and anode; the cathode is then held at the same negative potential. By splitting the potential, smaller high-voltage standoffs and lighter electrical cables can be used, which increases the portability of such sources. Since portability is important in many nondestructive testing applications such as x-ray fluorescence, industrial x-ray tubes are frequently bipolar.

FIGURE 4-2. A small x-ray source sitting on its power supply. The beryllium x-ray window is visible on the copper heat sink at the front of the source. Photo courtesy of Oxford Instruments.

However the circuit is arranged, the kinetic energy of the electrons striking the anode is

$$K_e = q_e \Phi,$$ (4-1)

where q_e is the electron charge, and Φ is the applied voltage.* If the incoming electrons have sufficient energy, they can knock out deep core electrons in the atoms of the metal anode, as shown in Figure 4-3. The resulting core vacancy is unstable and is quickly filled by an outer electron dropping down from a valence level. The excess energy released when the valence electron moves to the lower energy level can be carried away from the atom by a photon. For deep core vacancies, the photon energy can be in the x-ray range. Because the photon energy is the difference in energy between two atomic levels, the photon energy is unique to the element, and the emission is called *characteristic radiation*. Characteristic radiation can be used to identify the composition of materials, as will be discussed in section 4.7. After the characteristic emission process, in which free electrons are emitted, the atom is left as an ion, with too few electrons. Eventually, the free electrons lose energy and are recaptured by some of the ionized atoms, so that the material remains electrically neutral.

4.2 Core atomic levels

In chapter 3, equation 3-2 gave the energy levels of highly ionized atoms, those with a single electron. For atoms with many electrons, the core electrons partially shield the valence electrons from the nucleus, so the mth energy level becomes

$$U_{e,m} = -U_R' \frac{(Z-\zeta)^2}{m^2},$$ (4-2)

*The variable V is reserved for volume.

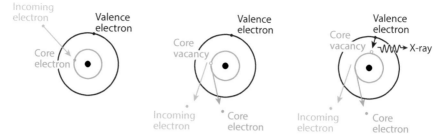

FIGURE 4-3. Emission of characteristic x rays from an electron-impact source. The incoming electron (left sketch) knocks out a core electron (center sketch). The resulting core vacancy is filled by a valence electron (right), giving off excess energy in the form of an x-ray photon.

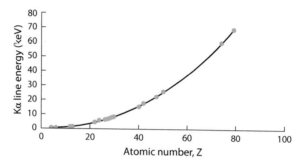

FIGURE 4-4. Energy of the x-ray emission line versus atomic number (points) and quadratic fit (curve). Data from http://wwwastro.msfc.nasa.gov/xraycal/linewidths.html.

where ζ is a screening factor (the screening is similar to that discussed in section 3.5), and m is the principal quantum number. The Rydberg energy is very slightly modified to

$$U'_R = U_R \frac{M_A}{M_e + M_A},\qquad (4\text{-}3)$$

where M_A is the atomic mass, and U_R is, as noted in equation 3-3, approximately 13.6 eV. The ionization energy for an electron in a particular level m_1 is given by the difference between the minimum final energy and the energy of the m_1 level,

$$U_{ionization} = U_{continuum} - U_{e,m_1} = 0 - U_{e,m_1} = -U_{e,m_1} = U'_R \frac{(Z-\zeta)^2}{m_1^2}.\qquad (4\text{-}4)$$

For an electron in the m_1 level to be knocked out by an incoming electron, as shown in Figure 4-3, the incoming electron must have a kinetic energy greater than the m_1 level binding energy. In general, the binding energy is roughly equal to the ionization energy, although relaxation effects can be important for high-Z elements such as tungsten. When the core vacancy is filled by an electron falling from the m_2 level, as in

Figure 4-3, the emitted photon will carry the difference in energy between the two levels,

$$U_{photon} = \Delta U_e = U_R' \left[-\frac{(Z-\zeta_2)^2}{m_2^2} + \frac{(Z-\zeta_1)^2}{m_1^2} \right]. \quad \textbf{Moseley's law} \quad (4\text{-}5)$$

The relation between ΔU and Z is known as *Moseley's law*. Comparison of characteristic emission for the same emission line (same m_1 and m_2) from different atoms reveals that the photon energy is quadratic in atomic number Z, as shown in Figure 4-4.

EXAMPLE 4-1: ENERGY LEVELS

---

a) What are the approximate $m=1$, $m=2$, and $m=3$ energy levels in copper, assuming screening factors $\zeta_1 = 3.2$, $\zeta_2 = 12$, and $\zeta_3 = 21$?

Noting that the atomic number for copper is $Z=29$, substituting into equation 4-2 we get

$$U_1 = -U_R' \frac{(Z-\zeta)^2}{m^2} \approx -(13.6 \text{ eV}) \frac{(29-3.2)^2}{1^2} \approx -9000 \text{ eV} = -9 \text{ keV};$$

$$U_2 \approx (13.6 \text{ eV}) \frac{(29-12)^2}{2^2} \approx -1 \text{ keV; and}$$

$$U_3 \approx -(13.6 \text{ eV}) \frac{(29-21)^2}{3^2} \approx -0.1 \text{ keV}.$$

b) What is the emission line energy due to electrons falling from the $m=2$ to the $m=1$ level in copper?

The photon energy is the energy lost by the electron in falling from the $m=2$ to the $m=1$ level, $U_{photon} = \Delta U_e = (U_2 - U_1) \approx (-1 \text{ keV}) - (-9 \text{ keV}) \approx 8 \text{ keV}$.

c) What is the ionization energy for the $m=1$ electron for copper?

The ionization energy is the energy required to move an electron from the $m=1$ level to the continuum, outside the atom. An electron in the continuum by definition has potential energy $U_e = 0$, so the ionization energy is $\Delta U \approx 0 - (-9 \text{ keV}) \approx 9 \text{ keV}$.

d) What tube voltage is required to see the 8 keV characteristic emission of copper?

The emission results from an electron falling into an $m=1$ hole, so enough energy is required to ionize the $m=1$ level, or 9 keV. Since the energy of an electron in a voltage Φ is $K_e = q_e \Phi$, a voltage of 9 kV = 9000 V is the minimum required.

e) What is the energy of the characteristic emission from copper if the tube voltage is 30 kV?

It is still 8 keV. The excess kinetic energy of the incoming electron does not change the characteristic energy of the photon.

4.3 Characteristic spectra

The energy of the characteristic emission given in equation 4-5 depends on the principal quantum numbers, m_1 and m_2, of the initial and final electron energy levels. The lines in the emission spectrum are named according to the quantum numbers of those levels, with a few historic twists. First, the level with principal quantum number $m = 1$

TABLE 4-1. Energy levels for copper

m	Name	Energy, keV	Orbital sublevel	ℓ	m_ℓ	$m_s = \pm s$	$j = \ell \pm s$	$m_j = m_\ell + m_s$	Number of electrons (total = 29)
1	K	−8.98	1s	0	0	±1/2	1/2	±1/2	2
2	L_1	−1.10	2s	0	0	±1/2	1/2	±1/2	2
2	L_2	−0.95	$2p_{1/2}$	1	−1	1/2	1/2	−1/2	2
					1	−1/2		1/2	
2	L_3	−0.93	$2p_{3/2}$	1	−1	−1/2	3/2	−3/2	4
					0	±1/2		±1/2	
					1	1/2		3/2	
3	M_1	−0.13	3s	0		±1/2	1/2	±1/2	2
3	M_2	−0.09	$3p_{1/2}$	1	−1/2		1/2	±1/2	2
3	M_3	−0.08	$3p_{3/2}$	1		1/2	3/2	−3/2, −1/2, 1/2, 3/2	4
3	M_4	−0.01	$3d_{3/2}$	2	−2	1/2	3/2	−3/2	4
					−1	1/2		−1/2	
					1	−1/2		1/2	
					2	−1/2		3/2	
3	M_5	−0.01	$3d_{5/2}$	2	−2	−1/2	5/2	−5/2	6
					−1	−1/2		−3/2	
					0	−1/2		−1/2	
					0	1/2		1/2	
					1	1/2		3/2	
					2	1/2		5/2	
4	N_1	−0.01	4s	0		±1/2	1/2	±1/2	1 (partially filled)

Source: NIST website, http://physics.nist.gov/PhysRefData/XrayTrans/Html/search.html and LBL website, henke.lbl.gov/optical_constants.
Note: The energies are rounded to the nearest 0.01 keV. m is the principal quantum number. ℓ is the orbital quantum number, m_ℓ is the magnetic quantum number, s is the spin, m_s is the spin quantum number, and j is the total angular momentum quantum number (from $\vec{J} = \vec{L} + \vec{s}$). ℓ can take on integer values from 0 to $m−1$, m_ℓ from $−\ell$ to ℓ, and m_j from $−j$ to j. Energies are relative to the continuum. The energy is determined primarily by m, as given in equation 4-2. For a single value of m, levels with the lower j have the lower energy.

is called the K shell, the one with $m=2$ is called L, $m=3$ is M, and so forth. If the core electron vacancy is in the K shell, the photon emitted when an electron falls into that hole is called K emission. If the electron falls from a level that is one quantum number above the hole, it is called an α line, if it falls from a level with principal quantum number $m+2$, it is a β line. Hence an electron falling from an $m=3$ level to one with $m=1$ emits a Kβ photon, and one falling from $m=3$ to $m=2$ emits an Lα photon (note that α and β are not subscripts in this notation).

The energy levels, and hence the emission lines, are further split according to the orbital quantum number ℓ and total angular momentum j.* The spin of the electron is $s=1/2$. The available values of ℓ for the mth level run from 0 to $m-1$. The energy then depends on whether the z components of the spin and orbital angular momentum are parallel or antiparallel, which is characterized by $j=\ell\pm s$. Lower j levels are antiparallel and have lower energies. Table 4-1 gives the filled levels for copper (atomic number 29).

The details of computing transition probabilities are outside the topic of this work and are described in any quantum physics text (see the further reading for this chapter). However, it is useful to know that the angular momentum quantum numbers also give rise to transition rules for emission of a photon. Fermi's "golden rule" of quantum mechanics states that the probability of the atom making the transition from the initial state ψ_i to the final state ψ_f is proportional to the square of the matrix element of the interaction potential,

$$H_{if}=\left\langle\psi_f\,|\,H\,|\,\psi_i\right\rangle=\int\psi_f H\psi_i d^3r. \tag{4-6}$$

For the case of a transition that emits a photon (which has spin $s=1$), the operator H is represented by the electric dipole operator $q_e\vec{r}$, so $H_{if}=\left\langle\psi_f\,|\,q_e\vec{r}\,|\,\psi_i\right\rangle=\int\psi_f q_e\vec{r}\psi_i d^3r$. For the integral, and hence the probability of the transition, to be nonzero it is necessary for

$$\Delta m>0,\,\Delta\ell=\pm1,\quad\text{and}\quad\Delta j=0\text{ or }\pm1.\qquad\textbf{Transition rules}\qquad(4\text{-}7)$$

EXAMPLE 4-2: EMISSION LINES

---

a) If a 1s electron is knocked out of a copper atom, what transitions are possible, and what are the names of the emission lines and their energies? Refer to table 4-1 for the energy levels of copper.

K$\alpha_{1,2}$: Because the core hole is a 1s, that is, $m=1$, level, called a K level, this will be a K emission. If the electron filling the hole comes from an $m=2$ level, it will be a Kα emission, since α denotes $\Delta m=1$. The hole cannot be filled from the first level up, 2s (called L$_1$) because both the initial and final states have $\ell=0$, so that $\Delta\ell=0$, in violation of the transition rules if a photon is to be emitted.

---

*The details of the quantum analysis here and in the following three short sections are not essential to the rest of the development if the material is unfamiliar. The results, however, are useful.

If the electron comes instead from one of the 2p levels, then $\Delta \ell = 1$, which is allowed.

In addition, $\Delta j = 0$ for the L_2, $j = 1/2$ level, and $\Delta j = 1$ for the L_3, $j = 3/2$ level, so both are allowed. The transition from the L_2 level is called KL2 or $K\alpha_2$. It has energy (8.98 keV) − (0.95 keV) = 8.03 keV. The transition from the L3 level, called $K\alpha_1$, has energy (8.98 keV) − (0.93 keV) = 8.05 keV. Together the two emission lines are referred to as the $K\alpha$ doublet. The KL3 line is called $K\alpha_1$ because it is the more intense. Because there are twice as many electrons available in the L_3 level with $j = 3/2$ as in the L_2 level with $j = 1/2$, twice as many can fall into the core hole, and the $K\alpha_1$ line is twice as intense as the $K\alpha_2$ line.

Kβ: If the electron comes from one of the $m = 3$ levels, it will be a Kβ emission, since $\Delta m = 2$. Again the 3s level is not possible because that would give $\Delta \ell = 0$. Similarly, the 3d levels would give $\Delta \ell = 2$. Thus the emission would come from the M_2 and M_3 levels, both with energy near (8.98 keV) − (0.08 keV) = 8.9 keV. In atoms with 4p electrons, there will be additional emission near this same energy from the N_2 and N_3 levels. Historically (but confusingly), these emissions also are often called Kβ lines.

4.4 Emission rates and intensity

As noted in table 4-1, the number of electrons in a filled level depends on the number of available states. Since magnetic quantum number can range from $-j$ to j, there are $2j + 1$ available states for a level with a fixed m and j. As seen in example 4-2, this gives twice as many electrons in the upper state available to fall into the core hole for $K\alpha_1$ emission as are available in the lower state to create the $K\alpha_2$ emission. In general, then, the relative emission ratio for a multiplet (multiple emission lines with the same starting m and the same final level, such as the $K\alpha$ doublet) is given by

$$\frac{\Gamma_1}{\Gamma_2} = \frac{2j_1 + 1}{2j_2 + 1} \Rightarrow \frac{\Gamma_{K\alpha_1}}{\Gamma_{K\alpha_2}} = \frac{2\left(\frac{3}{2}\right) + 1}{2\left(\frac{1}{2}\right) + 1} = \frac{4}{2} = 2. \quad \textbf{Multiplet intensity ratio} \quad (4\text{-}8)$$

Measured doublet emission profiles are shown in chapter 13.

The absolute emission rate Γ, the number of photons emitted per second, depends on the number of incident electrons with sufficient energy to knock out the core electron. It is found empirically to rise with increasing electron kinetic energy $q_e\Phi$ as

$$\Gamma = C_{char} \frac{J}{q_e} \left(\frac{q_e\Phi - q_e\Phi_b}{q_e\Phi_b} \right)^{O_{char}}, \quad (4\text{-}9)$$

where $q_e\Phi_b$ is the binding energy for the core electron, Φ is the tube voltage, and J is the tube current, so that J/q_e is the rate at which electrons impact the anode. The exponent O_{char} is around 1.7 at low tube voltages. At very high tube voltages, the electrons penetrate deeper into the target, so that more x rays are absorbed before escaping the target material, and O_{char} decreases to about 1.2. When the tube voltage is twice the binding voltage, $\Phi = 2\Phi_b$,

$$\Gamma_{\Phi=2\Phi_b} = C_{char} \frac{J}{q_e} \left(\frac{q_e 2\Phi_b - q_e \Phi_b}{q_e \Phi_b} \right)^{O_{char}} = C_{char} \frac{J}{q_e}, \tag{4-10}$$

so that the coefficient C_{char} gives the number of characteristic photons emitted per incident electron when the tube voltage is twice the ionization voltage; typically, C_{char} is on the order of 1–4 photons per 10,000 incident electrons. It is lower for high-Z anodes. The low emission rate means that most of the electrical power is not converted to x rays but contributes simply to heating the metal anode. Anode design and heat flux issues are discussed in chapter 5.

EXAMPLE 4-3: INTENSITY

---

a) If the efficiency of characteristic photon production is 4×10^{-4} when the tube voltage is twice the ionization voltage, what is the Kα emission rate from a copper source at 30 kV and 150 W? Assume $O_{char} = 1.6$, typical for this energy.

First, to find C_{char}: $\dfrac{\Gamma}{J/q_e} = 4 \times 10^{-4} \dfrac{photons}{electron} = C_{char} \left(\dfrac{2\Phi_b - \Phi_b}{\Phi_b} \right)^{1.6} = C_{char}.$

The current is $J = \dfrac{P_{elec}}{\Phi} \approx \dfrac{150\ W}{30\ kV} \approx 5\ mA.$

Then,

$$\Gamma = C_{char} \frac{J}{q_e} \left(\frac{\Phi - \Phi_b}{\Phi_b} \right)^{1.6}$$

$$\approx \left(4 \times 10^{-4} \frac{photons}{electron} \right) \left(\frac{5 \times 10^{-3} \dfrac{Coul}{s}}{1.6 \times 10^{-19} \dfrac{Coul}{electron}} \right) \left(\frac{30\ kV - 9\ kV}{9\ kV} \right)^{1.6}$$

$$\approx 4.8 \times 10^{13} \frac{photons}{s}.$$

b) What is the photon intensity onto a sample 100 mm from the source?

The source radiates isotropically into a sphere of surface area $4\pi z^2$, so the photon intensity or the emission density is

$$\Gamma_A = \frac{\Gamma}{A} \approx \frac{4.8 \times 10^{13} \dfrac{photons}{s}}{4\pi (100\ mm)^2} \approx 3.9 \times 10^8 \frac{photons}{mm^2 \cdot s}.$$

c) What is the emitted power?

The copper characteristic photons have energy $U = 8$ keV, so the power is

$$P = \Gamma U \approx \left(4.8 \times 10^{13} \, \frac{\text{photons}}{\text{s}} \right) \left(\frac{8 \times 10^3 \, \text{eV}}{\text{photon}} \right) \left(\frac{1.6 \times 10^{-19} \, \text{J}}{1 \, \text{eV}} \right) \approx 0.06 \, \text{W}.$$

Note that this value is much less than the electrical power of 150 W.

d) What is the intensity onto a sample 100 mm from the source?

The source radiates isotropically into a sphere of surface area $4\pi z^2$, so the intensity is

$$I = \frac{P}{A} \approx \frac{0.06 \, \text{W}}{4\pi (0.1 \, \text{m})^2} \approx 0.49 \, \frac{\text{W}}{\text{m}^2}.$$

Power, intensity, emission rate, and their relationship are discussed at greater length in chapter 5.

4.5 Auger emission

The bombardment of atoms by high-energy electrons (or other particles) creates a core vacancy and can result in characteristic emission. However, the final image in the sequence in Figure 4-3 is not the only possibility. Alternatively, the excess energy can be used to ionize a second valence electron. This process, in which the incoming electron knocks out a core electron and causes emission of a third electron, is called *Auger emission*. The kinetic energy of the emitted Auger electron is also characteristic of the atom and can be used to identify the material (but only if the material is kept under vacuum so that the electron is not absorbed before detection).

The Auger emission probability is roughly independent of the atomic number Z. However, the probability for x-ray emission grows with increasing Z. As noted for equation 4-6, the transition probability is proportional to the square of the appropriate Hamiltonian matrix element. In equation 4-2, the energy is quadratic in Z, so the probability, which is proportional to its square, rises as Z^4. Thus, the fraction of x-ray versus Auger electron emission is roughly

$$\text{Fraction of transitions that yields x-ray emission} \approx \left| \frac{Z^4}{C_A + Z^4} \right|, \qquad (4\text{-}11)$$

where C_A is a constant arising from the Auger emission that is approximately 10^6 if the core vacancy is in the lowest electron level, $m = 1$. $C_A \sim 10^8$ for the next level, $m = 2$. The Z^4 dependence means Auger emission predominates for carbon (the fraction of x-ray emission is low), but x-ray emission predominates for tungsten, for which the fraction is near unity.

4.6 Line widths

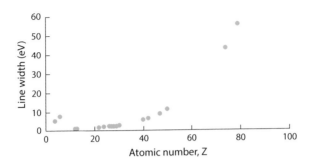

FIGURE 4-5. Line width for the Kα_1 line, as a function of atomic number. Data from http://wwwastro.msfc.nasa.gov/xraycal/linewidths.html.

All atomic energy levels to which an electron can be excited have a finite lifetime τ, which means there is a Heisenberg uncertainty in their energy (the ground state is assumed to be fixed at zero energy). As a result of this uncertainty in energy, the levels given in table 4-1 have some energy width. Thus, the characteristic emission lines also have some line width Δv. The line width is inversely related to the lifetime τ of the upper level by

$$\Delta U = h\Delta v \sim \frac{h}{\tau}. \tag{4-12}$$

Because the lifetimes are long, the line widths are quite small. For example the energy width of either of the Cu Kα lines near 8000 eV is only 2.5 eV.

Because emission rates increase with atomic number Z, as given in equation 4-11, the lifetimes decrease, and the line widths increase rapidly with atomic number, as shown in Figure 4-5. The line width for W Kα is about 60 eV (which, while much larger than for copper, is still very small compared with x-ray energies).

More formally, the time evolution of a quantum state is given by Schrödinger's equation. A state with fixed energy $U_0 = hv_0$ can be written as

$$\psi = \psi_0 (e^{2\pi i\, v_o t}). \tag{4-13}$$

A state with lifetime τ decays exponentially with time as

$$\psi(t) = \psi_0 \left(e^{-\frac{t}{\tau}} \right)(e^{2\pi i v_o t}). \tag{4-14}$$

The frequency distribution is given by the Fourier transform of the time dependence, so that the line shape of the intensity is given by the square of the Fourier transform,

$$I(v) = \left| \Im\{\psi(t)\} \right|^2 = I(v_o)\frac{1}{1 + \left(2\pi(v - v_o)\tau\right)^2}. \tag{4-15}$$

This $I(v)$ is a Lorentzian distribution. Broadening effects, such as Doppler broadening due to thermal motion of the emitting atoms, tend to cause Gaussian distributions or

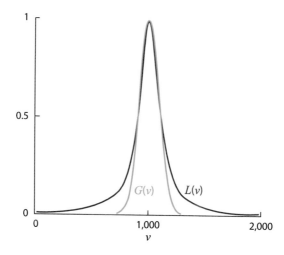

F$_{IGURE}$ 4-6. A Lorentzian $L(v)$ and a Gaussian $G(v)$ with the same full width at half maximum and peak location.

mixed Gaussian/Lorentzian distributions. A comparison of a Gaussian and a Lorentzian distribution is shown in Figure 4-6.

EXAMPLE 4-4: LINE WIDTH

What is the full width at half maximum (FWHM) of the Lorentzian distribution?

At half maximum,

$$I(v_{1/2}) = \frac{1}{2}I(v_o) \Rightarrow 1 + \left(2\pi(v - v_o)\tau\right)^2 = 2 \Rightarrow v_{1/2} = v_o \pm \frac{1}{2\pi\tau}.$$

The full width at half maximum is thus $\Delta v_{FWHM} = 2(v_{1/2} - v_o) = 1/\pi\tau$.

EXAMPLE 4-5: LINE WIDTH

What is the approximate lifetime of the upper level of a transition with a line width of 4 eV?

From the preceding example, if the distribution is Lorentzian, and the given width is the FWHM, then

$$\tau \sim \frac{1}{\pi\Delta v} = \frac{h}{\pi\Delta(hv)} = \frac{h}{\pi\Delta U} \approx \frac{1}{\pi}\frac{4\times10^{-15}\text{eV}\cdot\text{s}}{4\,\text{eV}} \approx 3\times10^{-16}\,\text{s} \approx 0.3\,\text{fs},$$

where fs is the abbreviation for femtosecond.

This is a "long" lifetime relative only to much shorter lifetimes, not to human timescales.

4.7 X-ray fluorescence

Characteristic emission occurs not just in x-ray tubes but whenever core electrons are ejected from atoms by bombardment of high-energy particles, whether they are electrons, ions, or higher-energy x-ray photons. The ejection of electrons by photons is known as the photoelectric effect, as described in chapter 2 and discussed in more detail in chapter 9. The ejection by high-energy particles is known as *particle-induced x-ray emission*, or PIXE. High-energy electrons, for example in electron microscopes, can also be used, in which case the analysis is called *energy-dispersive x-ray spectroscopy* or EDS.

In general, the emission of the resulting characteristic radiation is called *fluorescence*. The detection of the x rays to assess composition is known as *x-ray fluorescence analysis*, XRF. Typically, a sample is bombarded with a high-energy x-ray beam (from a tube,

Figure 4-7. Handheld x-ray fluorescence analyzer. Photo courtesy of Bruker Corp.

synchrotron, or radioactive source), and the resulting characteristic fluorescence spectra are collected and analyzed. XRF is commonly used, for example, to detect impurities in silicon wafers, lead in paint and toys, and a variety of heavy metals in biological samples, and for forensic analysis. A handheld analyzer is shown in Figure 4-7. Spatially resolved XRF is used to map interconnects in computer chips and paint irregularities in fraudulent art masterpieces, as shown in Figure 4-8. An example of an XRF map is shown in Figure 4-9.

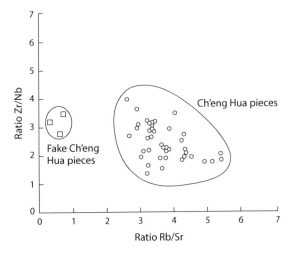

Figure 4-8. XRF analysis of real (circles) and fake (squares) Ming dynasty porcelain. From V. Mazo-Gray and M. Alvarez, X-ray fluorescence analysis of Ming Dynasty porcelain rescued from a Spanish shipwreck, *Archaeometry* 34 (1992): 37–42.

F<small>IGURE</small> 4-9. Elemental map obtained on Vincent van Gogh's *Patch of Grass* showing a hidden portrait of a woman from the XRF map of antimony in the various pigments. From Alfeld et al., *Journal of Analytical Atomic Spectrometry* 26 (2011): 899. Reproduced by permission of The Royal Society of Chemistry.

A typical spectrum, the result of irradiating a thin-film sample with x rays from a Mo anode x-ray tube, is shown in Figure 4-10. Solid-state detectors, like the system described in section 2.4, produce a voltage pulse from each photon that is proportional to the photon energy. The spectrum in Figure 4-10 is obtained by recording the voltage of each pulse and using a known conversion factor to photon energy. Because the line width of characteristic emission is quite narrow, the apparent energy width of the peaks is entirely instrumental, due to the energy resolution of the detector. Detectors are also discussed in more detail in chapter 9.

Better energy resolution can be obtained by diffracting the fluorescence emission from a monochromator crystal. Crystal optics, including monochromators, are discussed in chapter 15. Fluorescence analysis with monochromator crystals is known as *wavelength-dispersive* spectroscopy, for example, WDXRF or WDEDS.

A plot such as the spectrum in Figure 4-10 is obtained by dividing the total energy range into a number of bins, each of which is centered on an energy U and has an energy width ΔU_{bin}. The displayed height is then N_U, which is the number of detected photons with energies between $U - (\Delta U_{bin}/2)$ and $U + (\Delta U_{bin}/2)$. N_U has units of number of photons per bin and is measured for bins of a particular energy width for a particular duration.

In a typical spectrum, such as that in Figure 4-10, in addition to the fluorescence peaks there is considerable background, largely due to scattering of the incident radiation. The x-ray tube emission used to irradiate the sample includes not only the

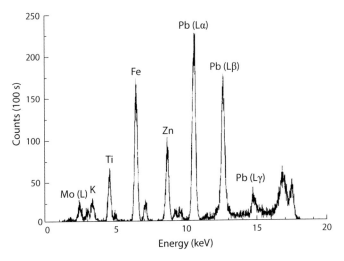

FIGURE 4-10. Spectrum of a thin-film sample taken with a Mo anode x-ray tube at 40 kV with a Si-Li detector. Each labeled peak is at a characteristic energy for one of the elements in the thin film. The peaks just to the right of the Ti, Fe, and Zn Kα peaks are the Ti, Fe, and Zn Kβ peaks. The two peaks at the highest energy are due to the Kα and Kβ emission from the Mo x-ray tube, scattered off the sample. From Gao et al., *Applied Physics Letters* 71 (1997): 3441. Reproduced by permission of AIP Publishing.

characteristic radiation but, as will be seen in chapter 6, also a continuous background from bremsstrahlung radiation. This background is white radiation; that is, it contributes photons at all energies. Thus the detected spectrum includes not only the fluorescence peaks but also the characteristic and white radiation from the tube, scattered from the sample. As a result, XRF analysis depends on the ability to detect fluorescence peaks above the background of scattered radiation.

The number of fluorescence photons is taken as the total integrated counts under the spectral peak minus the integrated background counts,

$$N_S = N - N_{Bkgnd}, \qquad (4\text{-}16)$$

as shown in Figure 4-11. The total number of counts under the spectral peak is the sum over the width of the peak of the number of counts in each bin,

$$N = \sum_{U_m=U_1}^{U_m=U_2} N_{U_m}. \qquad (4\text{-}17)$$

The number of bins in the peak is

$$N_{bins} = \frac{U_2 - U_1}{\Delta U_{bin}}. \qquad (4\text{-}18)$$

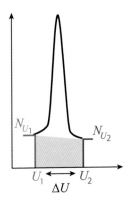

FIGURE 4-11. The background under an XRF peak is the gray area.

The integrated background counts are typically found by fitting to the intensity outside the peak, as shown in Figures 4-11 and 4-12.

$$N_{Bkgnd} \approx \left(\frac{N_{U_1} + N_{U_2}}{2} \right) N_{bins}. \tag{4-19}$$

EXAMPLE 4-6: FLUORESCENCE

A (fake) example of a fluorescence peak is shown in Figure 4-12. The total count time was 100 s, and the bin size is 100 eV. The total number of counts between 6 and 9 keV in that time interval was 62987. The number in the bin centered at 6 keV is 1562, and at 9 keV is 1879.

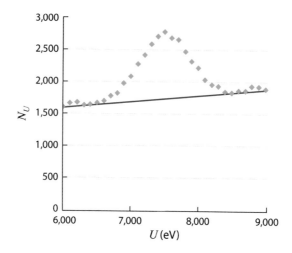

FIGURE 4-12. An example XRF peak, with a linear fit to the background.

a) What is the approximate energy of the fluorescence characteristic photon? b) What is the number of fluorescence photons emitted per second?

a) The peak is centered at about 7.5 keV. Simple observation of the graph reveals that the actual peak could be one bin on either side of the apparent maximum, so the characteristic energy is about 7.5 ± 0.1 keV.

b) The total emission is given. To find the signal in the fluorescence peak we need the background. The number of bins in the whole range is

$$N_{bins} = \frac{U_2 - U_1}{\Delta U_{bin}} = \frac{9000 \text{ eV} - 6000 \text{ eV}}{100 \text{ eV}} = 30.$$

The background is thus

$$N_{Bkgnd} = \frac{N_{U_1} + N_{U_2}}{2} N_{bins} = \left(\frac{1562 + 1879}{2} \right)(30) = 51615.$$

The number of fluorescence photons is then $N_s = N - N_{Bkgnd} = 62897 - 51615 = 11372$. Finally, the emission rate is

$$\Gamma_S = \frac{N_S}{t} = \frac{11372}{100 \text{ s}} \approx 114 \text{ per second.}$$

In part a, a better estimate for the peak energy and for the uncertainty in peak location could be found by fitting a function to the signal after subtracting the background. In Figure 4-13, the values at 6000 and 9000 eV were used to estimate a straight-line fit to the background, $N_{Bkgnd} = C_1 U + C_2$, which was then subtracted from the total count data. Two Gaussians with different parameters are shown compared with the data. The uncertainty in peak location is found to be about 20 eV. The peak might be modeled with a Gaussian, a Lorentzian, or a combination of spectral shapes to give the best fit.

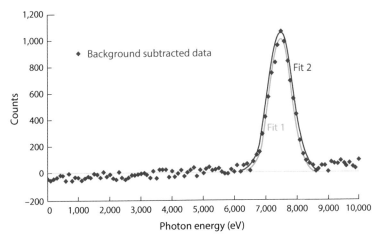

FIGURE 4-13. Background subtracted peak, showing Gaussian fits to the peak, with central values of 7500 and 7480 eV.

Clearly, the numerical values for both the signal and the background depend on the cutoff values U_1 and U_2 chosen to describe the edges of the peak. In the example, the values were chosen to extend beyond the edges of the peak for clarity, but that exaggerated the background. It is typical to choose U_1 and U_2 at the point at which the signal has fallen to about half its peak value. The uncertainty in the starting and ending energy values contributes to an uncertainty in the signal.

A larger problem is deciding whether a small peak is real or is due to random statistical fluctuations, called *noise*. A fluorescence signal

$$S = N_s \tag{4-20}$$

is said to be detectable if the signal is twice the noise **N**.

$$S_{min} = 2\mathbf{N}.$$

(4-21)

The signal-to-noise ratio is defined as

$$SNR = \frac{S}{\mathbf{N}}.$$

(4-22)

Thus for fluorescence analysis, the minimum usable signal-to-noise ratio SNR is considered to be 2,

$$SNR_{min} = 2.$$

(4-23)

The noise is defined as the variation in the measured signal. As discussed in section 2.5, the variation in the total number of counts is

$$\Delta N = \sqrt{N}.$$

(4-24)

The variation in the number of background counts is

$$\Delta N_{Bkgnd} = \sqrt{N_{Bkgnd}}.$$

(4-25)

From equation 4-16, the signal is taken as the difference between the measured total counts and the measured background. The variance in both the total and the background contributes to the uncertainty in the calculated signal. The two uncertainties are independent, so they add in quadrature; that is,

$$\Delta N_S^2 = (\Delta N)^2 + (\Delta N_{Bkgnd})^2.$$

(4-26)

Since the minimum signal S_{min} is small, the variation in the total integrated counts will be similar to the background variation, $\Delta N \approx \Delta N_{Bkgnd}$, so that the noise for a small signal is

$$\mathbf{N} = \Delta N_s = \sqrt{(\Delta N_{Bkgnd})^2 + (\Delta N)^2} \approx \sqrt{2}(\Delta N_{Bkgnd}) = \sqrt{2}\sqrt{N_{Bkgnd}}.$$

(4-27)

Thus, the minimum detectable signal is

$$S_{min} = 2\mathbf{N} = 2\sqrt{2}\sqrt{N_{Bkgnd}}.$$

(4-28)

If a sample with atomic concentration X of the element associated with the particular fluorescence peak gives a signal of N_S counts, then a sample with half that concentration, $X' = X/2$, will give $N_S' = N_S/2$ counts, or, more generally,

$$N_S' = \frac{X'}{X} N_S.$$

(4-29)

Changing the concentration does not change the background level very much, since the background is due to scattering of the incident beam by the bulk of the sample. The lowest detectable concentration of the element, X', is the *minimum detectable limit* (MDL), $X' = MDL$, when

$$N_S' = S_{min} = 2\sqrt{2}\sqrt{N_{Bkgnd}} = \frac{MDL}{X}N_S$$

$$\Rightarrow MDL = 2\sqrt{2}\frac{\sqrt{N_{Bkgnd}}}{N_S}X \approx 3\frac{\sqrt{N_{Bkgnd}}}{N_S}X, \qquad (4\text{-}30)$$

where N_S is the signal when the concentration is X. The MDL is not a property of the element, or even of the sample, but is used as a quality factor to compare measurement systems. The system with the lower MDL is deemed to be better at detecting small quantities of the desired element.

Clearly, the MDL depends on the time used for the measurement, as the background and total counts are both proportional to time, so that

$$MDL \approx 3\frac{\sqrt{\Gamma_{Bkgnd}\,t}}{\Gamma_s\,t}X = 3\frac{\sqrt{\Gamma_{Bkgnd}}}{\Gamma_s}X\frac{1}{\sqrt{t}}. \qquad (4\text{-}31)$$

A longer count time gives a lower detection limit and hence allows the detection of smaller concentrations. If not otherwise specified, a count time of 100 s is standard, although 1000 s is also used.

EXAMPLE 4-7: MDL

---

a) If there is an average background count rate of 10 photons per second, what is the minimum fluorescence count rate that will give a signal-to-noise ratio of 2 for an experiment lasting 100 seconds?

$$S_{min} = 2N \approx 3\sqrt{N_{Bkgnd}} = 3\sqrt{\left(10\frac{1}{s}\right)(100\ s)} \approx 95 \Rightarrow \Gamma_{min} \approx \frac{95}{100\ s} = 0.95\frac{1}{s}.$$

For an experiment lasting 1000 seconds?

$$S_{min} \approx 3\sqrt{\left(10\frac{1}{s}\right)(1000\ s)} \approx 300 \Rightarrow \Gamma_{min} = \frac{300}{1000\,s} = 0.3\frac{1}{s}.$$

b) If a fluorescence signal count rate of 5 photons/s is achieved with a sample of concentration 10% copper, what is the MDL for a 100 second experiment?

Since our minimum count rate is almost 1/s, and we have 5/s, we expect we could have around 1/5 the original concentration, or $X \sim 2\%$. Working through the formula,

$$MDL = \frac{3X\sqrt{N_{Bkgnd}}}{N_s} \approx \frac{3(0.1)\sqrt{\left(10\frac{1}{s}\right)(100\ s)}}{\left(5\frac{1}{s}\right)(100\ s)} \approx 0.019 \approx 2\%.$$

EXAMPLE 4-8: LEAD IN TOYS

A fluorescence peak is measured for a 100% lead sample for 10 s with 0.1 keV energy bins. The peak is 1 keV wide and has a total of 4.0×10^5 counts above the background. Bins outside the peak energy contain an average of 400 counts per bin.

a) What is the total background under the peak?

$$N_{bins} \approx \left(\frac{1 \, keV}{0.1 \, keV/bin} \right) = 10.$$

$$N_{Bkgnd} \approx \left(400 \, \frac{photons}{bin} \right) (10 \, bins) = 4000.$$

b) What is the signal (the total minus background)?

$$N_S = N - N_{Bkgnd} \approx 4.04 \times 10^5 - 4000 \approx 4 \times 10^5.$$

c) What is the variability in the background due to Poisson statistics?

$$\Delta N_{Bkgnd} \approx \sqrt{N_{Bkgnd}} \approx \sqrt{4000} \approx 63.$$

d) What is the variability in the total counts due to Poisson statistics?

$$\Delta N \approx \sqrt{N} = \sqrt{4.04 \times 10^5} \approx 636.$$

e) What is the signal-to-noise ratio?

$$\mathbf{N} = \Delta N_s = \sqrt{(\Delta N_{Bkgnd})^2 + (\Delta N)^2} \approx \sqrt{63^2 + 636^2} \approx 639.$$

$$SNR = \frac{N_S}{\mathbf{N}} \approx \frac{4 \times 10^5}{639} \approx 626.$$

Note that the signal here is not small, so that the approximation of equation 4-27 is not valid.

A toy manufacturer wants to know whether a 100 ppm concentration could be detected in 100 s with this system.

f) What would background be for a 100 s measurement?

From the 10 s measurement,

$$\Gamma_{Bkgnd} = \frac{4000 \, counts}{10 \, s} = 400 \frac{counts}{s},$$

so that

$$N_{Bkgnd} = \left(400 \, \frac{counts}{s} \right) (100 \, s) = 4 \times 10^4.$$

g) What would the signal be for a 100 ppm concentration in 100 s?

$$N_S' = \frac{X'}{X} N_S S = \left(\frac{(100/10^6)}{1} \right) \left(4 \times 10^5 \frac{\text{photons}}{10 \text{ s}} \right) (100 \text{ s}) = 400.$$

h) What is the signal-to-noise ratio for the 100 ppm sample in 100 s?

The total is $N = N_{Bkgnd} + N_S = 4 \times 10^4 + 400 = 4.04 \times 10^4$. In this case, the total count rate is dominated by the background, so equation 4-27 is valid,

$$\boldsymbol{N} = \Delta N_s = \sqrt{(\Delta N_{Bkgnd})^2 + (\Delta N)^2} \approx \sqrt{2} \sqrt{N_{Bkgnd}} \approx 284$$

$$\Rightarrow SNR = \frac{N_S}{\boldsymbol{N}} \approx \frac{400}{284} \approx 1.4.$$

This is slightly below the minimum detectable limit, which requires an SNR of 2. The MDL is

$$MDL \approx 3 \frac{\sqrt{N_{bkgnd}}}{N_s} X = 3 \frac{200}{400} 100 \text{ ppm} = 150 \text{ ppm}.$$

A lead concentration of 100 ppm is not reliably detectable in 100 s with this system.

The MDL scales with count time. It scales in a similar manner with power in the incident beam, since according to equation 4-9, the emission rate depends on the current of exciting particles (which can be electrons or photons). Therefore a high-power source, everything else being equal, would be expected to have a lower MDL. The MDL also depends on the sample. Bulk samples give more scattering and therefore higher background rates and higher MDL than thin samples. Care should be taken to use similar if not identical samples when comparing measurement systems.

If the sample is not homogeneous, the area of the beam and the sample are also significant. A large beam irradiating a sample with a small area of interest will create excess background. The geometry of the sample, the detector, and the shielding blocking the scattered and direct radiation from entering the detector also affect the MDL, as does the resolution of the detector. A detector with good energy resolution, which gives very narrow peaks, will have less integrated background, as calculated from equation 4-19.

The characteristics of the exciting beam are also significant. For x-ray induced fluorescence, the background signal near the peak will be reduced if the incoming beam does not include the x-ray energies of interest. Beam filtering and monochromatization are discussed in chapters 9 and 15. In general, the cross section for creating a core electron vacancy (the precursor to fluorescence emission) is highest for an excitation beam with a photon energy just above the binding energy for the electron.

EXAMPLE 4-9: COMPARING SYSTEMS

The measured MDLs for some elements within a glass substrate taken with a particular set of measurement systems are compared in table 4-2.

TABLE 4-2. MDL for trace elements in $Na_2B_4O_7$ glass, for 400 s count time and three different sources

Atom/Anode	W, 50 kV, 1 μA	Mo, 50 kV, 1 μA	Mo, 15 kV, 200 μA
Na (Kα=1.04 keV)	38,000 ppm	20,000 ppm	480 ppm
K (3.3 keV)	54 ppm	94 ppm	7.4 ppm

Source: Copyright 1995 from Jenkins et al., Quantitative X-ray Spectrometry. Reproduced by permission of Taylor and Francis Group, LLC, a division of Informa PLC.

a) Which system gives the better MDL, and why?

The MDL is much better for the higher-current, lower-voltage system (the third column). Turning down the voltage allows the current to be increased. The high current gives higher incident photon rate, resulting in higher fluorescent count rates.

b) Is the MDL better or worse for Na or K and why?

The sodium MDL values are worse because many of the low-energy fluorescent photons are absorbed by the sample and not detected.

c) Is the MDL better or worse for the Mo or W anode at the same voltage?

For sodium, the MDL is better for the molybdenum anode than for the tungsten anode, because the molybdenum anode L lines are at 2.3 keV, close to the binding energy of the K shell electrons in sodium. Conversely, the potassium MDL is better for the tungsten anode, because the tungsten L lines in the energy range of 8.3–11.3 keV are slightly above the K shell binding energy for potassium.

Another quality factor for measurement systems is *sensitivity*, which is defined as the ratio of the signal rate to the quantity of sample. Higher sensitivity will not necessarily result in lower MDL, as sensitivity does not take the background into account, but it does provide a measure of whether a reasonable count rate can be obtained if the background can be reduced. Depending on the sample details, the units of the sensitivity may be given as count rate per atom or per area. Using the definition of count rate per number of atoms of the desired element, sensitivity for a sample with concentration X, effective area A, effective thickness L, density ρ, and molar mass M_M is

$$Sen = \frac{\Gamma_s}{XN_{atom}} = \frac{\Gamma_s}{X\left(\dfrac{AL\rho}{M_M}N_A\right)}. \tag{4-32}$$

The effective thickness L is the simple thickness for a thin sample, but it is limited by the absorption depth of the x-ray fluorescence emission for bulk samples, as the emission will not escape from deep in the bulk. The effective area A is the smaller of the sample or incident beam area. The effective area and thickness are important for estimating the measured volume if a quantitative estimate of composition is desired.

Quantitative XRF refers to the computation of absolute concentration from XRF measurements. The simplest case is the determination of the concentration of a known impurity in a known matrix (*matrix* here refers to the rest of the sample or sample substrate). In that case, the XRF signal can be determined for a number of standards of known concentration and a calibration curve created. If measurements are desired over a large range of concentration, the curve will be nonlinear, because the absorption and scatter by the impurity will be different than that for the matrix. There can also be significant error due to sample preparation. For example, if the object is inhomogeneous, the sample may not be representative of the whole. Variation in counting statistics can be managed by altering counting time, and the effect of source fluctuations can be reduced by simultaneously monitoring scatter from the source and using a ratio instead of absolute counts. The most complex case is the analysis of unknown sample elements in an unknown matrix. In this case emission peaks may overlap, and the effects of differences in scattering from different elements are difficult to calculate. These analyses require extensive computation with sophisticated software. One standard method is the computation of *influence coefficients,* which describe the effect of the presence of another element on the elemental calibration. Another method is known as *fundamental parameters* computation, which involves using known absorption and scattering cross sections, and source, optic, and detector properties.

Problems

SECTIONS 4.2 AND 4.3

1. X rays are produced by an x-ray tube with a molybdenum anode, with $Z = 42$.

a) What is the approximate energy of the K, L, and M levels, assuming screening constants of 3.2, 12, and 21, respectively?

b) What are the resulting characteristic energies of the Kα, Kβ, and Lα lines?

2. The energy levels of silver (Ag) are listed in table 4-3 with their quantum numbers.

TABLE 4-3.

	E, keV	n	ℓ	j
N$_3$	−0.058	4	1	3/2
N$_2$	−0.06	4	1	1/2
N$_1$	−0.1	4	0	1/2
M$_5$	−0.368	3	2	5/2
M$_4$	−0.37	3	2	3/2
M$_3$	−0.57	3	1	3/2
M$_2$	−0.60	3	1	1/2
M$_1$	−0.7	3	0	1/2
L$_3$	−3.4	2	1	3/2
L$_2$	−3.5	2	1	1/2
L$_1$	−3.8	2	0	1/2
K$_1$	−25.5	1	0	1/2

Source: LBL website, henke.lbl.gov/optical _constants.

a) What are the energies of all allowed Kα lines?
b) What is the minimum x-ray tube voltage required for a silver anode to produce Kα radiation?
c) What are the energies of all allowed Kβ lines?
d) If core electrons are knocked out of silver atoms, what fraction of the emission is x-ray photons (as opposed to Auger electrons; see section 4.5)?
e) X-ray emission from a tungsten anode tube is used to excite Ag fluorescence to determine the concentration of Ag in a sample (see section 4.7). As the tube voltage is reduced from 50 kV to 20 kV the Ag fluorescence count rate drops abruptly to zero. Why?

SECTION 4.4

3. What is the relative emission rate of the two lines in the Mo Kβ doublet?
4. A Mo anode is run with a tube voltage of 25 kV and a current of 1 mA.

a) Assuming that the efficiency coefficient, C_{char}, of Kα of characteristic line emission is 2×10^{-4} x rays per electron and the exponent O_{char} is 1.6, what is the emission rate of Kα x rays in photons per second? The binding energy $q_e \Phi_b$ is 20 keV.
b) Assuming the x rays are emitted isotropically, what is the count rate of Kα x rays for a square detector 10 mm on a side a distance of 150 mm from the source?
c) By what factor does the count rate increase if the tube voltage is doubled to 50 kV, still at 1 mA?

SECTION 4.5

5. What is the x-ray emission probability (as opposed to Auger emission) for an innermost core hole vacancy in carbon? In tungsten?

SECTION 4.6

6. Doppler broadening and other factors tend to create Gaussian line widths, or lines which are mixes of Gaussian and Lorentzian shapes. Find the FWHM, Δv_{FWHM}, when the frequency distribution is Gaussian, $I(v) = F_o e^{-\frac{(v-v_o)^2}{2\sigma^2}}$.

7. Find the Fourier transform of the time dependence of the quantum state $F(v) = \mathfrak{I}\{\psi(t)\} = \int_0^\infty \psi(t) e^{-2\pi i v t} \, dt$ for the decaying exponential of equation 4-14 and show that FF^* is proportional to the Lorentzian distribution of equation 4-15. (The constant of proportionality is resolved by setting the intensity at the peak $I(v)\big|_{v=v_o} = CF(v)F^*(v)\big|_{v=v_o}$ equal to the value at the peak $I(v_o)$.)

8. An energy level has a lifetime with a FWHM of 1 fs. What is the energy width of the emission line created when an electron falls from this level?

SECTION 4.7

9. A fluorescence system has an MDL of 500 ppm for copper with a measurement time of 100 s. The background is 50 counts/s/bin for 20 eV wide bins. The peak is 300 eV wide. What is the total count rate (fluorescence plus background) for a 10% copper sample?

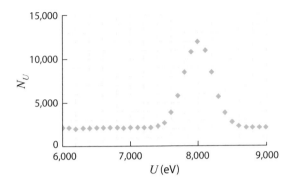

FIGURE 4-14. Example fluorescence peak for a 10% Cu sample.

10. An artificial noise-free fluorescence peak for a 10% copper sample is shown in Figure 4-14. The data plotted in the figure are summed over 100 eV energy bins. Answer the questions by estimating from the graph.

a) What is the background level per bin?
b) What is the FWHM of the signal peak?
c) What is the background if the width is taken as the peak FWHM?
d) What is the signal?
e) What is the expected noise?
f) What is the expected signal-to-noise ratio?
g) With noise, what would be the MDL, in parts per million?

Further reading

Characteristic radiation is described in most of the general references cited in chapter 1.

The following are a few standard texts on modern physics and quantum mechanics:
S. Thornton and A. Rex, **Modern Physics for Scientists and Engineers**, 4th ed., Cengage Learning, 2013.
D. Griffiths, **Introduction to Quantum Mechanics**, 2nd ed., Pearson, 2005.
R. Shankar, **Principles of Quantum Mechanics**, 2nd ed., Plenum Pres, 2011.

The following are a few references on x-ray fluorescence:
Ron Jenkins, **X-Ray Fluorescence Spectrometry**, vol. 152 in **Chemical Analysis: A Series of Monographs on Analytical Chemistry and Its Applications**, 2nd ed., Wiley Interscience, 1999.
E. Lifshin, **X-Ray Characterization of Materials**, Wiley-VCH, 1999.
K. Tsuji, J. Injuk, R. Van Greiken, eds., **X-ray Spectrometry: Recent Technological Advances,** John Wiley & Sons, 2004.
R. Jenkins, R. W. Gould, G. Gedcke, **Quantitative X-Ray Spectrometry**, 2nd ed., Marcel Dekker, 1995.

5

SOURCE INTENSITY, DIVERGENCE, AND COHERENCE

In the previous chapters we discussed x-ray emission from nuclear reactions and plasmas, and the generation of characteristic radiation. In the next few chapters we will discuss bremsstrahlung radiation and the emission from synchrotron sources and x-ray lasers. The usefulness of the x rays generated from any source for a particular application depends on a number of properties of the beam, including the photon energy, emission rate, intensity, brilliance, divergence, and coherence. For example, an imaging application such as the security scan of the tanker truck in Figure 1-4 requires high beam power, while the SPECT example of section 2.6 requires high emission rate. Many imaging applications use detectors that produce a signal proportional to the total energy deposited, integrated over a broad range of x-ray wavelengths, while other detectors, such as those required for x-ray fluorescence or absorption spectroscopy, measure the number of photons in a small range of wavelengths.

5.1 Intensity and angular intensity

In comparing sources for different applications, it is important to consider beam strength in units appropriate to the problem. The emission rate Γ is the number of photons per second. The average energy of N photons is defined as

$$\bar{U} = \frac{\sum_{m=1}^{N} U_m}{N}.$$

(5-1)

For a beam with an average photon energy $\bar{U}$, the power in the x-ray beam is

$$P = \bar{U}\Gamma.$$

(5-2)

The units of emission rate Γ and power P are

$$[\Gamma] = \frac{\text{photons}}{\text{s}} \quad \text{and} \quad [P] = \frac{\text{J}}{\text{s}} = \text{W},$$

(5-3)

where the square brackets are used to mean "units of."

When the application depends on the amount of energy deposited onto a detector or object, an important quantity is intensity,

$$I = \frac{P}{A},$$ (5-4)

where A is the area of the beam. The units of intensity I are watts per meter squared (W/m²),

$$[I] = \frac{[P]}{[A]} = \frac{W}{m^2} = \frac{J}{m^2 s}.$$ (5-5)

At a distance z from an isotropic source, the beam area is

$$A = 4\pi z^2,$$ (5-6)

so that

$$I = \frac{P}{4\pi z^2}.$$ (5-7)

Thus if the x-ray beam is not parallel, the intensity depends on the distance from the source.

The intensity gives the power per area. For example the power deposited onto a detector of diameter W_D at a distance z, as shown in Figure 5-1, is

$$P_D = IA_D = I\pi\left(\frac{W_D}{2}\right)^2.$$ (5-8)

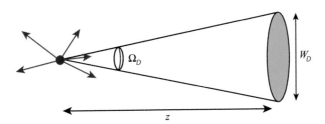

FIGURE 5-1. A detector a distance z from an isotropic source.

EXAMPLE 5-1: INTENSITY

A 1 mm diameter source 100 mm away from a 6 mm diameter detector emits 5×10^{12} photons/s isotropically with a photon energy uniformly distributed from 8 to 12 keV, so that the average energy is 10 keV.

Give the source emission rate and power, the intensity at the detector, and the power deposited onto the detector.

Emission rate: This was given; $\Gamma = 5 \times 10^{12}$ photons/s.

Power in the beam:

$$P = \Gamma \bar{U} \approx \left(5 \times 10^{12} \, \frac{\text{photons}}{\text{s}} \right) \left(10 \times 10^3 \, \frac{\text{eV}}{\text{photon}} \right) \left(1.6 \times 10^{-19} \, \frac{\text{J}}{\text{eV}} \right)$$
$$\approx 8 \times 10^{-3} \, \text{W} \approx 8 \, \text{mW}.$$

Beam area at a distance of 100 mm: $A = 4\pi z^2 \approx 4\pi (0.1 \, \text{m})^2 \approx 0.13 \, \text{m}^2$.

Intensity at 100 mm:

$$I = \frac{P}{A} \approx \frac{8 \times 10^{-3} \, \text{W}}{0.13 \, \text{m}^2} \approx 6.4 \times 10^{-2} \, \frac{\text{W}}{\text{m}^2} \approx 64 \, \frac{\text{mW}}{\text{m}^2}.$$

Area of the detector:

$$A_D = \pi \left(\frac{W_D}{2} \right)^2 \approx \pi \left(\frac{6 \times 10^{-3} \, \text{m}}{2} \right)^2 \approx 2.8 \times 10^{-5} \, \text{m}^2.$$

Power onto the detector at 100 mm:

$$P_D = I A_D \approx \left(6.4 \times 10^{-2} \, \frac{\text{W}}{\text{m}^2} \right) (2.8 \times 10^{-5} \, \text{m}^2) \approx 1.8 \times 10^{-6} \, \text{W} \approx 1.8 \, \mu\text{W}.$$

The computation of beam area in equations 5-6 and 5-7 assumes that the beam is truly isotropic, like an ideal lightbulb. This is rarely true in practice for x-ray sources. For example, aside from more subtle angular dependences, a solid anode inside an x-ray tube like that of Figure 4-1 will actually give off radiation only in the half sphere above the anode, not the full sphere implied by $4\pi z^2$, because half the radiation travels into the anode material and is absorbed. In addition, the beam is generally limited by the x-ray window and other apertures. However, unlike for visible light sources that can be placed inside integrating spheres, it is generally impractical to measure the total power radiated in all directions. The total power is usually computed by measuring the power deposited onto a finite detector and then calculating the intensity from equation 5-8. Computing the total power from equation 5-4 then requires an assumption about the beam area at the detector, which requires knowing the solid angle subtended by the beam. It is common when describing x-ray tubes to simply assume that the beam was initially isotropic, so computations will give the correct answer for the power deposited onto other objects in the beam so long as the same assumption is used. The angular dependence of the beam can then be corrected for more subtle angular anisotropies such as differing absorption lengths through the anode, as will be discussed in section 5.6.

The angular intensity I_Ω is the power per solid angle,

$$I_\Omega = \frac{P}{\Omega_s}, \tag{5-9}$$

where Ω_s is the solid angle subtended by the beam. For an ideal isotropic source, the solid angle is $\Omega_s = 4\pi$ steradians. The angular intensity is then a source property, the same at all distances from the source, so it is not necessary to specify a particular distance, as was required to specify the intensity. The units of angular intensity I_Ω are watts per steradian (W/sr),

$$[I_\Omega] = \frac{[P]}{[\Omega]} = \frac{W}{sr}. \tag{5-10}$$

A detector of area A_D at a distance z subtends a solid angle

$$\Omega_D \approx \frac{A_D}{z^2}, \tag{5-11}$$

so that the power deposited onto the detector is

$$P_D = I_\Omega \Omega_D. \tag{5-12}$$

EXAMPLE 5-2: ANGULAR INTENSITY

For the source of example 5-1, give the angular intensity, the solid angle subtended by the detector, and the power deposited onto the detector.

Angular intensity:

$$I_\Omega = \frac{P}{\Omega_s} \approx \frac{8 \times 10^{-3}\,W}{4\pi\,sr} \approx 6.4 \times 10^{-4}\,\frac{W}{sr}.$$

Angle subtended by the detector:

$$\Omega_D = \frac{\pi\left(\dfrac{W_D}{2}\right)^2}{z^2} \approx \frac{\pi\left(\dfrac{6 \times 10^{-3}\,m}{2}\right)^2}{(10^{-1}\,m)^2} \approx 2.8 \times 10^{-3}\,sr.$$

Power onto the detector:

$$P_D = I_\Omega \Omega_D \approx \left(6.4 \times 10^{-4}\,\frac{W}{sr}\right)(2.8 \times 10^{-3}\,sr) \approx 1.8\,\mu W.$$

The power onto the detector is, of course, the same as in example 5-1.

For more directed beams, it is often useful to express angular intensity in terms of a linear angle, for example, χ_s in Figure 5-2. The solid angle Ω_s subtended by the beam is then

$$\Omega_s = \int_0^{2\pi}\left[\int_0^{\chi_s/2} \sin\chi\,d\chi\right]d\varphi = 2\pi\left(1 - \cos\frac{\chi_s}{2}\right). \tag{5-13}$$

assuming azimuthal symmetry. For an isotropic source, $\chi_s = 2\pi$, so that $\Omega_s = 4\pi$, as before. For a highly directed source, χ_s is small, so that

$$\Omega_s = 2\pi\left(1 - \cos\frac{\chi_s}{2}\right) \approx \pi\left(\frac{\chi_s}{2}\right)^2. \tag{5-14}$$

The angular intensity can be given in units of watts per milliradian squared (W/mrad²),

$$I_\chi = \frac{P}{2\pi\left(1 - \cos\dfrac{\chi_s}{2}\right)} \approx \frac{P}{\pi\left(\dfrac{\chi_s}{2}\right)^2}. \tag{5-15}$$

The angular intensity can be used to find the power deposited onto a detector which subtends a linear angle χ_D, as shown in Figure 5-3,

$$\chi_D = 2\arctan\left(\frac{W_D}{2z}\right) \approx \left(\frac{W_D}{z}\right)\left(\frac{1000\,\text{mrad}}{\text{rad}}\right). \tag{5-16}$$

where the factor of 1000 is used to convert from radians to milliradians. The deposited power on the detector is then

$$P_D = I_\chi \pi\left(\frac{\chi_D}{2}\right)^2, \tag{5-17}$$

assuming χ_D is small.

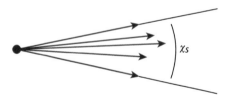

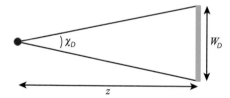

FIGURE 5-2. Linear angle subtended by a source.

FIGURE 5-3. A detector subtends a linear angle χ_D.

EXAMPLE 5-3: ANGULAR INTENSITY IN MILLIRADIANS

For the source of example 5-1, give the angular intensity in W/mrad², the linear angle subtended by the detector, and the power deposited onto the detector.

Angular intensity:

$$I_\chi = \frac{P}{\Omega_s} = \frac{P}{2\pi\left(1 - \cos\dfrac{\chi_s}{2}\right)} \approx \frac{8\times10^{-3}\,\text{W}}{2\pi\left(1 - \cos\dfrac{2\pi}{2}\right)\left(\dfrac{10^3\,\text{mrad}}{\text{rad}}\right)^2}$$

$$\approx 6.4\times10^{-10}\,\frac{\text{W}}{\text{mrad}^2}.$$

Linear angle subtended by the detector:

$$\chi_D \approx \frac{W_D}{z} \approx \left(\frac{6 \times 10^{-3}\,\text{m}}{10^{-1}\,\text{m}}\right)\left(\frac{10^3\,\text{mrad}}{\text{rad}}\right) = 60\,\text{mrad}.$$

Power onto the detector:

$$P_D \approx I_\chi \pi \left(\frac{\chi_D}{2}\right)^2 \approx \left(6.4 \times 10^{-10}\,\frac{\text{W}}{\text{mrad}^2}\right)\left(\pi \left(\frac{60\,\text{mrad}}{2}\right)^2\right) \approx 1.8\,\mu\text{W}.$$

The power onto the detector is again the same as in example 5-1.

5.2 Photon intensity and photon angular intensity

The previous three examples give the power onto, for example, an integrating detector such as film, which records a total energy onto a certain area rather than giving a pulse for each photon. In many cases, as for the SPECT and XRF applications described in chapters 2 and 4, the detector counts photons. In those cases rather than power P, intensity I, and angular intensity I_Ω, the relevant quantities are photon emission rate Γ, photon intensity Ψ, and photon angular intensity Ψ_Ω. The photon intensity is

$$\Psi = \frac{\Gamma}{A}. \tag{5-18}$$

The photon intensity is a measure of the density of photons (at a given distance from the source). The count rate into a detector of diameter W_D at a distance z is

$$\Gamma_D = \Psi A_D = \Psi \pi \left(\frac{W_D}{2}\right)^2. \tag{5-19}$$

EXAMPLE 5-4: PHOTON INTENSITY AND COUNT RATE

For the source of example 5-1, give the photon intensity, count rate into the detector, and power into the detector.

Photon intensity at 100 mm:

$$\Psi = \frac{\Gamma}{A} \approx \frac{5 \times 10^{12}\,\dfrac{\text{photons}}{\text{s}}}{0.13\,\text{m}^2} \approx 4.0 \times 10^{13}\,\frac{\text{photons}}{\text{m}^2 \cdot \text{s}}.$$

Count rate into the detector at 100 mm:

$$\Gamma_D = \Psi A_D \approx \left(4.0 \times 10^{13}\,\frac{\text{photons}}{\text{m}^2 \cdot \text{s}}\right)(2.8 \times 10^{-5}\,\text{m}^2) \approx 1.1 \times 10^9\,\frac{\text{photons}}{\text{s}}.$$

Power into the detector, from equation 5-2:

$$P_D = \Gamma_D \bar{U} \approx \left(1.1 \times 10^9 \, \frac{\text{photons}}{\text{s}}\right)(10 \times 10^3 \, \text{eV})\left(1.6 \times 10^{-19} \, \frac{\text{J}}{\text{eV}}\right) \approx 1.8 \, \mu\text{W}.$$

The power onto the detector is again the same as in example 5-1.

The photon angular intensity Ψ_Ω is the number of photons per second per solid angle Ω_s subtended by the beam

$$\Psi_\Omega = \frac{\Gamma}{\Omega_s}. \tag{5-20}$$

EXAMPLE 5-5: ANGULAR INTENSITY

For the source of example 5-1, give the photon angular intensity and the count rate into the detector.

Photon angular intensity:

$$\Psi_\Omega = \frac{\Gamma}{\Omega_s} \approx \frac{5 \times 10^{12} \, \dfrac{\text{photons}}{\text{s}}}{4\pi \, \text{sr}} \approx 4.0 \times 10^{11} \, \frac{\text{photons}}{\text{s} \cdot \text{sr}}.$$

Count rate into the detector:

$$\Gamma_D = \Psi_\Omega \Omega_D \approx \left(4.0 \times 10^{11} \, \frac{\text{photons}}{\text{s} \cdot \text{sr}}\right)(2.8 \times 10^{-3} \, \text{sr}) \approx 1.1 \times 10^9 \, \frac{\text{photons}}{\text{s}}.$$

The count rate into the detector is, of course, the same as in example 5-4.

As for angular intensity, photon angular intensity is a property of the source and does not depend on the location of the measurement. Photon intensity measures the density of photons in space. Photon angular intensity is a measure of the density of photons in six-dimensional position/momentum space, as it depends on both the spread of photons in space and the spread of directions of travel.

Again, for more directed beams, it is often useful to express angular intensity in terms of a linear angle. The photon angular intensity may then be given in units of photons per milliradian squared (photons/mrad²),

$$\Psi_\chi = \frac{\Gamma}{2\pi\left(1 - \cos\dfrac{\chi_s}{2}\right)} \approx \frac{\Gamma}{\pi\left(\dfrac{\chi_s}{2}\right)^2}, \tag{5-21}$$

where χ_s is the linear angle subtended by the beam, in milliradians.

EXAMPLE 5-6: PHOTON ANGULAR INTENSITY IN MILLIRADIANS

For the source of example 5-1, give the photon linear angular intensity in photons/mrad²s and the count rate into the detector.

Photon linear angular intensity:

$$\Psi_\chi = \frac{\Gamma}{2\pi\left(1-\cos\dfrac{\chi_s}{2}\right)} \approx \frac{5\times10^{12}\,\dfrac{\text{photons}}{\text{s}}}{4\pi\left(\dfrac{10^3\,\text{mrad}}{\text{rad}}\right)^2} \approx 4.0\times10^5\,\frac{\text{photons}}{\text{mrad}^2\text{s}}.$$

Count rate into the detector:

$$\Gamma_D \approx \Psi_\chi\,\pi\left(\frac{\chi_D}{2}\right)^2 \approx \left(4.0\times10^5\,\frac{\text{photons}}{\text{mrad}^2\text{s}}\right)\left(\pi\left(\frac{60\,\text{mrad}}{2}\right)^2\right)$$

$$\approx 1.1\times10^9\,\frac{\text{photons}}{\text{s}}.$$

The count rate into the detector is again the same as in example 5-4.

It should be noted that in some fields it is common to drop the qualifiers and refer to photon angular intensity (or some other quantity) as "intensity." It is important to understand the units being used in a particular application.

5.3 Brightness and brilliance

As we will see in section 5.5, the applicability of the beam in some instances also depends on the source size. This is quantified by dividing by the source area to produce a photon angular intensity density,

$$\Psi_{\chi A} = \frac{\Psi_\chi}{A_s}. \tag{5-22}$$

Its units are

$$[\Psi_{\chi A}] = \frac{\text{photons}}{\text{s}\cdot\text{mrad}^2\cdot(\text{mm}^2\ \text{of source area})}. \tag{5-23}$$

In some texts this quantity is called "brightness" (we use this term for a different quantity) or, in older texts, "spectral brilliance" (and even in some places simply "intensity").

EXAMPLE 5-7: PHOTON ANGULAR INTENSITY DENSITY

For the source of example 5-1, give the photon angular intensity density.

Source area:

$$A_s = \pi \left(\frac{W_s}{2} \right)^2 = \pi \left(\frac{1\,\text{mm}}{2} \right)^2 \approx 0.79\,\text{mm}^2.$$

$$\Psi_{\chi A} = \frac{\Psi_\chi}{A_s} \approx \frac{4.0 \times 10^5 \dfrac{\text{photons}}{\text{mrad}^2 \cdot \text{s}}}{0.79\,\text{mm}^2} \approx 5.1 \times 10^5 \frac{\text{photons}}{\text{mrad}^2 \cdot \text{mm}^2 \cdot \text{s}}.$$

Alternatively, we can express the number of photons in a certain energy range. The term *monochromatic* refers to a single color, or single photon energy. As a consequence of Heisenberg-like uncertainty, as discussed in section 4.6 for line width, no source is perfectly monochromatic. All sources have some energy bandwidth. For spectroscopic applications such as XRF or applications which discriminate against scatter by using the photon energy, such as SPECT, an important quantity is the number of photons within a certain energy range ΔU_{Bndwth} of the desired value U_o.

The photon angular intensity within the given energy bandwidth is

$$\Psi_{\chi\,Bndwth} = (\Psi_\chi)(fraction_{Bndwth}). \tag{5-24}$$

We will call this quantity *brightness*, but again, beware of differing nomenclature. Conventionally, unless otherwise specified, the bandwidth is taken as 0.1% of the central energy, $\Delta U_{Bndwth} = 0.001\,U_o$.

For a broadband uniform spectral distribution of total range ΔU_{tot} the fraction of photons in the given bandwidth is

$$fraction_{Bndwth} = \frac{\Delta U_{Bndwth}}{\Delta U_{tot}}, \qquad \textbf{Broadband} \tag{5-25}$$

so that the brightness is

$$\Psi_{\chi\,Bndwth} = \Psi_\chi \left(\frac{\Delta U_{Bndwth}}{\Delta U_{tot}} \right). \tag{5-26}$$

For a narrow emission line of width $\Delta U_{tot} \ll \Delta U_{Bndwth}$, equation 5-25 is not meaningful because all the photons fall into the energy bandwidth of interest. In that case, the fraction of photons in the given bandwidth is

$$fraction_{Bndwth} = 1, \qquad \textbf{Narrowband} \tag{5-27}$$

so that

$$\Psi_{\chi\,Bndwth} = \Psi_\chi. \tag{5-28}$$

The units of brightness functionally are

$$\left[\Psi_{\chi\,Bndwth}\right] = \frac{\text{photons}}{\text{s} \cdot \text{mrad}^2}, \tag{5-29}$$

but to distinguish it, and to specify the bandwidth range employed, the units are conventionally written as

$$\left[\Psi_{\chi\,Bndwth}\right] = \frac{\text{photons}}{\text{s} \cdot \text{mrad}^2\,(0.1\,\%\ \text{bandwidth})}. \tag{5-30}$$

EXAMPLE 5-8: BRIGHTNESS FOR A BROADBAND SOURCE

For the source of example 5-1, give the brightness at $U_o = 10$ keV and the count rate into a detector energy bin centered at 10 keV with a width of 5 eV.

Brightness: We need the fraction of the photons that are within 0.1 % of the central energy. The bandwidth is $\Delta U_{Bndwth} = 0.1\% \ U_o = 10^{-3}\,(10\ \text{keV}) = 10$ eV. The energy spread of the source was given as $\Delta U_{tot} = 4$ keV $= 4000$ eV, with a uniform (flat) spectrum, so.

$$\Psi_{\chi\,Bndwth} = \Psi_{\chi}\left(\frac{\Delta U_{Bndwth}}{\Delta U_{tot}}\right) \approx \left(3.98 \times 10^5\,\frac{\text{photons}}{\text{mrad}^2 \cdot \text{s}}\right)\frac{10\ \text{eV}}{4 \times 10^3\,\text{eV}}$$

$$\approx 9.9 \times 10^2\,\frac{\text{photons}}{\text{mrad}^2 \cdot \text{s} \cdot (0.1\,\%\ \text{bandwidth})}.$$

Count rate into the detector energy bin:

$$\Gamma_{D_{bin\,o}} = \Psi_{\chi\,Bndwth}\left[\pi\left(\frac{\chi_D}{2}\right)^2\right]\left(\frac{\Delta U_{bin}}{\Delta U_{Bndwth}}\right)$$

$$\approx \left(\frac{9.9 \times 10^2\,\text{photons}}{\text{mrad}^2 \cdot \text{s}}\right)\left[\pi\left(\frac{60\,\text{mrad}}{2}\right)^2\right]\left(\frac{5\,\text{eV}}{10\,\text{eV}}\right)$$

$$\approx 1.4 \times 10^6\,\frac{\text{photons}}{\text{s}}.$$

EXAMPLE 5-9: BRIGHTNESS IN THE CASE OF A NARROW EMISSION LINE

For a source similar to that of example 5-1, except this time emitting a characteristic line with a line width of 3 eV, give the brightness at $U_o = 10$ keV and the count rate into a detector energy bin centered at 10 keV with a width of 5 eV.

Brightness: In this case all the photons are within 0.1% of the central energy, so the fraction $= 1$, and

$$\Psi_{\chi\,Bndwth} = \Psi_{\chi} \approx 3.98 \times 10^5\,\frac{\text{photons}}{\text{mrad}^2 \cdot \text{s} \cdot (0.1\,\%\ \text{bandwidth})}.$$

Similarly, the emission line is completely within the detector energy bin:

$$\Gamma_{D_{bin\,o}} = \Psi_{\chi\,Bndwth}\left[\pi\left(\frac{\chi_D}{2}\right)^2\right] \approx \left(\frac{3.98\times10^5\text{ photons}}{\text{mrad}^2\cdot\text{s}}\right)\left[\pi\left(\frac{60\text{ mrad}}{2}\right)^2\right]$$

$$\approx 1.1\times10^9\,\frac{\text{photons}}{\text{s}}.$$

All the photons hitting the detector from example 5-6 are counted within the single energy bin for this narrowband source.

As noted at the beginning of this section, the source size also affects the useful photon angular intensity. This is expressed in the brilliance b_A,

$$b_A = \frac{\Psi_{\chi\,Bndwth}}{A_s}.\tag{5-31}$$

The functional units of brilliance are

$$[b_A] = \frac{\text{photons}}{\text{s}\cdot\text{mrad}^2\cdot(\text{mm}^2\text{ of source area})},\tag{5-32}$$

but again, the units are conventionally written as

$$[b_A] = \frac{\text{photons}}{\text{s}\cdot\text{mrad}^2\cdot(\text{mm}^2\text{ of source area})\cdot(0.1\%\text{ bandwidth})}.\tag{5-33}$$

EXAMPLE 5-10: BRILLIANCE

For the source of example 5-1, give the brilliance.

$$b_A = \frac{\Psi_{\chi\,Bndwth}}{A_s} \approx \frac{9.9\times10^2\,\dfrac{\text{photons}}{\text{mrad}^2\cdot\text{s}\cdot(0.1\%\text{ bandwidth})}}{0.79\text{ mm}^2}$$

$$\approx 1.3\times10^3\,\frac{\text{photons}}{\text{s}\cdot\text{mrad}^2\cdot\text{mm}^2\cdot(0.1\%\text{ bandwidth})}.$$

Alternatively,

$$b_A = \Psi_{\chi\,A}\left(\frac{\Delta U_{Bndwth}}{\Delta U_{tot}}\right) \approx \left(5.1\times10^5\,\frac{\text{photons}}{\text{mrad}^2\cdot\text{mm}^2\cdot\text{s}}\right)\frac{10\text{ eV}}{4\times10^3\text{ eV}}$$

$$\approx 1.3\times10^3\,\frac{\text{photons}}{\text{mrad}^2\cdot\text{mm}^2\cdot\text{s}\cdot(0.1\%\text{ bandwidth})}.$$

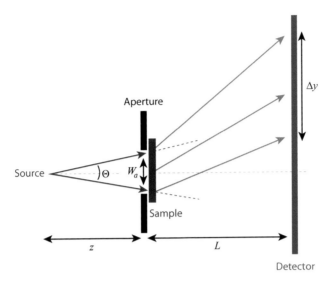

FIGURE 5-4. Diffraction with global divergence.

5.4 Global divergence

As discussed in section 5.1, a conventional isotropic source radiates into a large angle, $\chi = 2\pi$. For any source the angular range of the output beam is called the *global divergence Θ*. Often, an aperture is placed in front of the beam to limit its extent. The global divergence Θ of a beam restricted by an aperture of diameter W_a at a distance $z \gg W_a$ from the source is, as shown in Figure 5-4,

$$\Theta = 2\arctan\left(\frac{W_a/2}{z}\right) \approx \frac{W_a}{z}. \tag{5-34}$$

A perfectly parallel beam has zero global divergence. Because a perfectly parallel beam has an exactly defined (zero) transverse momentum, Heisenberg uncertainty tells us its transverse location cannot be known at all, so it must be an infinite plane wave. Real beams have finite extent, so their location is defined. Thus they must have some uncertainty in transverse momentum, and hence, nonzero divergence.

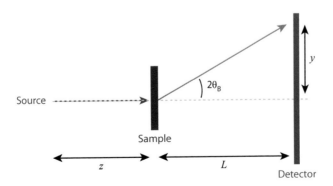

FIGURE 5-5. Diffraction geometry.

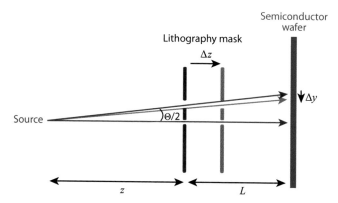

FIGURE 5-6. Proximity lithography showing feature shift Δy as the mask is moved Δz from the red to the green location. The shift is due to the global divergence of the beam.

The global divergence of an x-ray beam has a significant impact on its performance in several applications. For example, as will be seen in chapter 13, Bragg's law, $\lambda = 2d \sin(\theta_B)$, is used in diffraction experiments to determine the plane spacing d of a crystal by determining the diffraction angle $2\theta_B$ of the beam for a given x-ray wavelength λ. The diffraction angle is determined as shown in Figure 5-5 from the measured height y on the detector using

$$\tan(2\theta_B) = \frac{y}{L}. \tag{5-35}$$

If the beam has some global divergence, as shown in Figure 5-4, then there is a blur in y. This leads to some uncertainty in the calculated angle (and therefore in the plane spacing of the crystal).

In addition to causing loss of resolution in diffraction, global divergence can have other effects. For example, it can cause the features to shift in projection lithography. In projection lithography a wafer is exposed through a mask as part of the process of patterning the electrical circuitry onto the wafer. If the mask-to-wafer separation is varied slightly, Δz in Figure 5-6, the shadow on the wafer would remain unchanged if the beam were parallel, but it shifts Δy because of the angle of the beam. The figure is not drawn to scale; in reality L is only about 10–40 μm, but the feature shifts are important in the nanometer range.

5.5 Local divergence

For the beam in Figure 5-4, each part of the sample sees a single well-defined incident angle from the ideal point source. The beam hitting the center of the sample diffracts exactly as in Figure 5-5. Locally, at that one point on the sample, there is no apparent divergence, because the source is infinitely small. In reality, the number of photons emitted per area of the source is finite, so that the source must have some finite size. For example, to avoid melting the anode, the electron beam spot must be spread over a significant area.

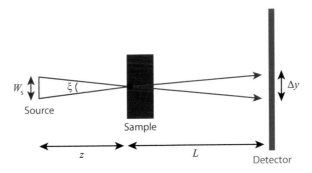

FIGURE 5-7. Geometric blur due to local divergence.

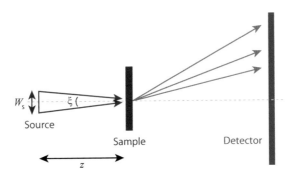

FIGURE 5-8. Local divergence for a diffraction experiment.

As a result, every point in the beam sees a range of beam angles, as shown in Figure 5-7. The local divergence ξ, also called the source divergence, due to a source diameter W_s at a distance z is

$$\xi = 2\arctan\left(\frac{W_s/2}{z}\right) \approx \frac{W_s}{z}. \qquad (5\text{-}36)$$

As seen in Figure 5-8, local divergence also contributes to the uncertainty in angle in a diffraction measurement. The difference between Figure 5-4 and Figure 5-8 is that in Figure 5-4 it is assumed that the source is tiny, but the sample is large, whereas in Figure 5-8 the source is large, but only a tiny part of the sample is illuminated. In reality, both the finite size of the source and the finite size of the sample will affect the measurement.

Another important consequence of local divergence is geometric blur in imaging or lithography, as shown in Figure 5-7. The blur Δy on a detector a distance L from a point in the sample is

$$\Delta y = 2L\tan\left(\frac{\xi}{2}\right) \approx W_s\frac{L}{z} \approx L\xi. \qquad (5\text{-}37)$$

This limits the image resolution of a point object in the sample to the blur Δy.

EXAMPLE 5-11: LOCAL DIVERGENCE

In mammography, the source size is typically about 300 μm, at a distance of about 600 mm from the patient. The detector is placed about 50 mm from the entry side of the breast. What are the local divergence and the blur?

The local divergence is $\xi \approx \dfrac{W_s}{z} \approx \dfrac{0.3\,\text{mm}}{600\,\text{mm}} = 0.5\,\text{mrad}$.

This leads to a blur $\Delta y \approx L\xi = (50\,\text{mm})\,(0.5 \times 10^{-3}) = 25\,\text{μm}$, which is the desired image resolution. The pixel size on the detector is typically also about 25 μm.

The blur caused by local divergence is generally a problem, but it can be beneficial. For example, in the lithography example of Figure 5-6, the edges of the mask can cause diffraction "ringing," rapid variations due to interference effects. A small amount of local divergence will blur out these undesirable effects.

5.6 X-ray tube design

The source size of an x-ray tube is the electron spot size, which must be kept large enough not to melt the target material, as most of the electron energy is deposited as heat. In transmission sources, the electron beam is perpendicular to the target, which is thick enough to stop the electrons but thin enough for the x rays to traverse the target thickness and exit the tube with minimal intensity losses. In reflection sources, the terms *anode* and *target* are used interchangeably, but in transmission targets they are not, since the anode tends to be positioned close to the cathode to provide the greatest electron acceleration, while the target (at ground potential) is farther away, with electron-focusing elements in between. Frequently, the target material also acts as the source x-ray window, enabling focal-spot-to-window distances equal to the thickness of the target, just a few tens of microns. However, transmission source powers are low because of limits on heat dissipation on the thin targets.

More usually, the anode is solid, as shown in Figure 4-1. In that case the power can be increased. Metals with good thermal conductivity and high melting temperatures, such as tungsten and copper, are generally chosen. The required minimum diameter

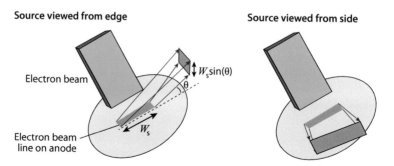

Source viewed from edge Source viewed from side

Electron beam

$W_s \sin(\theta)$

θ

Electron beam line on anode

W_s

FIGURE 5-9. The apparent size of a line source with a length W_s, viewed at an angle θ from the edge, is $W_s \sin \theta$.

increases linearly with the power in the electron beam. (The allowed power is proportional to diameter, not area, because it is more difficult for the heat to flow away from the center of a larger spot.) For sealed-tube copper x-ray sources, the maximum power is typically 5 kW/mm of x-ray beam diameter. To obtain those powers, direct anode cooling is necessary to prevent target melting. In this case, the target material has to be thick enough to separate the cooling medium from the vacuum required for the electron beam operation. Jets of water are frequently sprayed onto the backside of these thick targets to directly cool them. Alternatively, the heat load can be reduced by rotating the target so rapidly that the

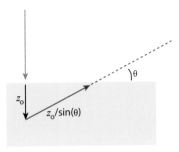

FIGURE 5-10. An electron beam penetrating a distance z_o into an anode, viewed at an angle θ.

electron beam dwell time on the target is too short to thermally overload the target. Maximum powers of roughly 150 kW/mm of x-ray beam diameter can be achieved with rotating anode sources.

To further increase the power, the electron spot can be spread into a line, as shown in Figure 5-9. The x rays are then viewed at a small takeoff angle θ with respect to the target, resulting in what appears to be a nearly square or circular source spot. The apparent brilliance is then found from the apparent source area A'_s,

$$ b'_A = \frac{\Psi_{\chi Bndwth}}{A'_s} = \frac{\Psi_{\chi Bndwth}}{A_s \sin \theta}. \tag{5-38} $$

However, the brightness is reduced because of the additional absorption due to the longer path length through the anode material. If the depth at which the x rays are produced is z_o, then the angled path length is

$$ z = \frac{z_o}{\sin \theta}, \tag{5-39} $$

as shown in Figure 5-10. The output power is reduced by

$$ P = P_o e^{-\frac{z}{\mu^{-1}}} = P_o e^{-\frac{z_o}{\mu^{-1} \sin \theta}}, \tag{5-40} $$

where μ^{-1} is the absorption length, which will be discussed in more detail in chapter 9. Therefore, the apparent brilliance is

$$ b'_A = \frac{\Psi_{\chi Bndwth}}{A'_s} = \frac{\Psi_{\chi Bndwth o} \, e^{-\frac{z_o}{\mu^{-1} \sin \theta}}}{A_s \sin \theta} = b_A \frac{e^{-\frac{z_o}{\mu^{-1} \sin \theta}}}{\sin \theta}, \tag{5-41} $$

which can be written as

$$ b'_A = \frac{\Psi_{\chi Bndwth o} \, \dfrac{e^{-\frac{z_o}{\mu^{-1} \sin \theta}}}{\sin \theta}}{A_s} = \frac{\Psi'_{\chi Bndwth}}{A_s}, \tag{5-42} $$

where $\Psi'_{\chi Bndwth}$ is the *apparent brightness*. This function is plotted in Figure 5-11. The change in brightness with angle is known as the *heel effect* because of the shape of the

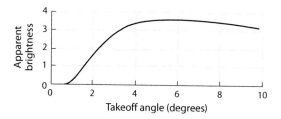

FIGURE 5-11. Apparent source brightness as a function of takeoff angle for an X-ray production depth-to–absorption length ratio z_0/μ^{-1} of 10.

curve. The peak occurs from 6° to 12°, depending on the tube voltage and anode material. Excessively small takeoff angles are more affected by surface roughness, which may increase with target use.

Because the apparent source size depends on the viewing angle, the local divergence and hence the spatial resolution can be different for different parts of the x-ray beam. This effect is especially important for large-area imaging applications such as medical radiography. The image blur, given by equation 5-37, varies across the image as the angle to the source changes.

The discussion of x-ray tubes has been restricted to conventional designs with single electron spots and single cathodes. Alternative designs exist, with multiple cold cathodes that allow rapid switching of the source location, or even *inverse geometry* designs in which the electron beam is rastered over a large area (and the object is usually placed close to the large-area source and far from a tiny detector, the reverse of the usual imaging geometry).

5.7 Coherence

Whatever the design of the source, local divergence affects the spatial coherence of the beam. The quantities used to describe the beam in sections 5.1–5.3 take into account the energy, location, and direction of the photons. However, x rays are also electromagnetic waves and have a phase. Coherence is a description of the phase relationships of the x-ray beam. The spatial coherence length is defined as the transverse distance across the beam over which two parts of the beam have a fixed phase relationship. Consider the light (visible or x-ray) from two sources or two slits falling together on a detector screen. The electric field amplitude from the first source is

$$E_1 = E_o \cos(\kappa r_1 - \omega t), \tag{5-43}$$

where κ is the wavevector, r_1 is the distance from the first source to a point on the detector, and ω is the angular frequency, $\omega = 2\pi\nu$. The electric field can also be written in complex notation,

$$E_1 = \mathrm{Re}\{E_o e^{i(\kappa r_1 - \omega t)}\} = \mathrm{Re}\{\tilde{E}_1 e^{-i\omega t}\}, \tag{5-44}$$

where the complex amplitude is

$$\tilde{E}_1 = E_o e^{i\kappa r_1}. \tag{5-45}$$

The field amplitude from a second source at the same frequency is

$$E_2 = E_o \cos(\kappa r_2 - \omega t + \phi_s) = \text{Re}\{E_o e^{i(\kappa r_2 - \omega t + \phi_s)}\} = \text{Re}\{\tilde{E}_2 e^{-i\omega t}\}, \qquad (5\text{-}46)$$

Where r_2 is the distance from the second source to the same point on the detector, and

$$\tilde{E}_2 = E_o e^{i(\kappa r_2 + \phi_s)}. \qquad (5\text{-}47)$$

The fields are coherent with each other if the phase difference between them, ϕ_s, is a constant and incoherent if ϕ_s is rapidly varying. Two distinct sources, for example, two lightbulbs, or two different lasers with no optical or electrical connection, are always incoherent. Two incoherent sources cannot cause interference.

The intensity of the first source is

$$I_1 = c\varepsilon_o \left\langle \left(E_o \cos(\omega t)\right)^2 \right\rangle, \qquad (5\text{-}48)$$

where c is the speed of light, ε_o is the vacuum dielectric permittivity, and the triangle bracket $\langle \ \rangle$ denotes time average,

$$\left\langle \cos^2(\omega t) \right\rangle = \frac{1}{\tau} \int_0^\tau \cos^2 \omega t \, dt = \frac{1}{\tau} \int_0^\tau \frac{\cos 2\omega t + 1}{2} \, dt = \frac{1}{2}, \qquad (5\text{-}49)$$

where τ is the period, $\tau = 2\pi/\omega$. Thus the intensity is

$$I_1 = \frac{c\varepsilon_o}{2} E_o^2. \qquad (5\text{-}50)$$

Alternatively, using the complex notation,

$$I_1 = \frac{c\varepsilon_o}{2} \left\langle \tilde{E}_1 \tilde{E}_1^* \right\rangle = \frac{c\varepsilon_o}{2} \left\langle E_o^2 \right\rangle = \frac{c\varepsilon_o}{2} E_o^2, \qquad (5\text{-}51)$$

where the asterisk denotes complex conjugate. The intensity of the second source is

$$I_2 = \frac{c\varepsilon_o}{2} E_o^2 = I_1. \qquad (5\text{-}52)$$

The sum field where they overlap on the screen is

$$E = E_1 + E_2 = E_o \cos(kr_1 - \omega t) + E_o \cos(kr_2 - \omega t + \phi), \qquad (5\text{-}53)$$

so that the intensity is

$$\begin{aligned}
I &= c\varepsilon_o \left\langle \left(E_o \cos(\kappa r_1 - \omega t) + E_o \cos(\kappa r_2 - \omega t + \phi_s)\right)^2 \right\rangle \\
&= 2I_1 + 4I_1 \left\langle \cos(\kappa r_1 - \omega t) \cos(\kappa r_2 - \omega t + \phi_s) \right\rangle \\
&= 2I_1 + 2I_1 \left\langle \cos[(\kappa r_2 - \omega t + \phi_s) + (\kappa r_1 - \omega t)] \right. \\
&\qquad \left. + \cos\left[\left((\kappa r_2 - \omega t + \phi_s) - (\kappa r_1 - \omega t)\right)\right] \right\rangle, \\
&= 2I_1 + 2I_1 \left\langle \cos(\kappa \Delta r + \phi_s) \right\rangle \\
&= 2I_1 + 2I_1 \left\langle \cos(\phi_{tot}) \right\rangle \qquad (5\text{-}54)
\end{aligned}$$

where the total phase difference is

$$\phi_{tot} = k\Delta r + \phi_s. \tag{5-55}$$

In complex notation,

$$\tilde{E} = \tilde{E}_1 + \tilde{E}_2 = E_o\, e^{i\kappa r_1} + E_o\, e^{i(\kappa r_2 + \phi)}, \tag{5-56}$$

so that

$$
\begin{aligned}
I &= \frac{c\varepsilon_o}{2}\left\langle \tilde{E}\tilde{E}^*\right\rangle = 2I_1 + I_1\left\langle e^{i\kappa r_1}e^{-i(\kappa r_2 + \phi_s)} + e^{-i\kappa r_1}e^{i(\kappa r_2 + \phi_s)}\right\rangle \\
&= 2I_1 + 2I_1\left\langle\cos(\kappa\Delta r + \phi_s)\right\rangle = 2I_1 + 2I_1\left\langle\cos(\phi_{tot})\right\rangle
\end{aligned} \tag{5-57}
$$

If the two sources are incoherent, then the phase difference between them ϕ_s is rapidly varying on the time scale of the measurement, the time average is zero, and the intensity of the sum is just twice the individual intensities,

$$I = 2I_1. \tag{5-58}$$

If the two sources are coherent with each other, then ϕ_{tot} is a constant, and

$$I = 2I_1 + 2I_1\cos(\phi_{tot}). \tag{5-59}$$

If $\phi_{tot} = 0$ or $2m\pi$, where m is an integer, then there is constructive interference, and the intensity is four times the individual intensities. If $\phi_{tot} = \pi$ or $(2m+1)\pi$, then there is destructive interference, and the intensity is zero.

In general,

$$I = I_1 + I_2 + \sqrt{I_1 I_2}\left|\frac{E_1 \otimes E_2}{E_{1o}\, E_{2o}}\right|\cos(\phi_{tot}), \tag{5-60}$$

where the quantity within the absolute value symbol is the degree of partial coherence, E_{1o} and E_{2o} are the amplitudes of the two fields, and the convolution integral is

$$E_1 \otimes E_2 = \int_{-\infty}^{\infty} E_1(t')E_2(t - t')dt'. \tag{5-61}$$

5.8 Spatial coherence

A classic example of interference is Young's double-slit experiment, sketched in Figure 5-12. The path length Λ_1 for the upper ray from the upper slit at a distance $d/2$ above the centerline to a point at a position y above the centerline on a screen a distance L away is

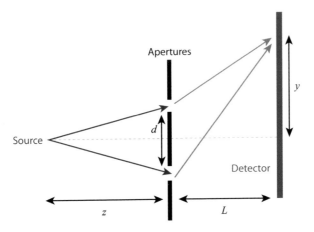

FIGURE 5-12. Young's double slit, with slit separation d.

$$\Lambda_1 = \sqrt{\left(y - \frac{d}{2}\right)^2 + L^2} = \sqrt{y^2 - yd + \frac{d^2}{4} + L^2}$$

$$= L\sqrt{1 + \frac{y^2}{L^2} - \frac{yd}{L^2} + \frac{d^2}{4L^2}} \approx L + \frac{y^2}{2L} - \frac{yd}{2L}, \tag{5-62}$$

where $d \ll y \ll L$. The difference in path lengths between the rays traveling through the upper and lower slits is

$$\Delta\Lambda = \Lambda_2 - \Lambda_1 = \sqrt{\left(y + \frac{d}{2}\right)^2 + L^2} - \sqrt{\left(y - \frac{d}{2}\right)^2 + L^2}$$

$$\approx \left(L + \frac{y^2}{2L} + \frac{yd}{2L}\right) - \left(L + \frac{y^2}{2L} - \frac{yd}{2L}\right) = \frac{yd}{L}. \tag{5-63}$$

Since the two rays originate from the same point source, the inherent phase difference is $\phi_s = 0$. The total phase difference is then the path difference times the wavevector $\kappa = 2\pi/\lambda$,

$$\phi = \frac{2\pi}{\lambda}\Delta\Lambda = \frac{2\pi}{\lambda}\frac{yd}{L} \tag{5-64}$$

Constructive interference, which yields an intensity peak on the screen, occurs when ϕ_{tot} is integral multiples of 2π, or for regular intervals of y along the screen,

$$y_m = m\frac{\lambda L}{d}, \tag{5-65}$$

where m is an integer.

EXAMPLE 5-12: DOUBLE SLIT AND SPATIAL COHERENCE

For an ideal point source, what is the maximum slit separation for seeing fringes (intensity peaks) at least 100 µm apart on a screen 1 m away with x-ray radiation with a 1 Å wavelength?

The separation is. $\Delta y = \dfrac{\lambda L}{d} \Rightarrow d = \dfrac{\lambda L}{\Delta y} = \dfrac{(10^{-10}\,\text{m})(1\,\text{m})}{100 \times 10^{-6}\,\text{m}} = 1\,\mu\text{m}.$

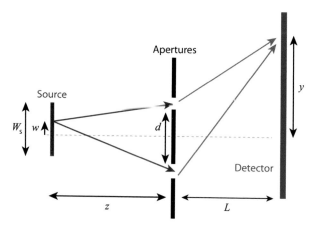

FIGURE 5-13. Young's double slit with a finite source of width W_s.

The two rays, through the upper and lower slits, were fully coherent because they shared an ideal point source. A real source has a finite width W_s, as in Figures 5-8 and 5-13. The finite-width source can be regarded as a collection of independent point sources. Each point source is incoherent with every other, with rapidly varying relative phases, and each point source produces an intensity pattern on the screen. The result of the multiple point sources is the simple intensity sum of each of the individual patterns, as in equation 5-58, since each point source is incoherent with its neighbors. As before, the result of a single point source a distance w above the centerline is the coherent addition of two rays, one through the upper slit and one through the lower slit. The path length for the upper ray is

$$\Lambda_1 = \sqrt{\left(w - \frac{d}{2}\right)^2 + z^2} + \sqrt{\left(y - \frac{d}{2}\right)^2 + L^2} \approx z + \frac{w^2}{2z} - \frac{wd}{2z} + L + \frac{y^2}{2L} - \frac{wd}{2L}, \quad (5\text{-}66)$$

where $w \ll z$, and z is the distance from the source to the slits. The difference in path lengths between the rays traveling through the upper and lower slits is then

$$\Delta\Lambda \approx \frac{wd}{z} + \frac{yd}{L}, \quad (5\text{-}67)$$

so constructive interference occurs when

$$y_m = m\frac{\lambda L}{d} - \frac{wL}{z}. \tag{5-68}$$

Shifting the source an amount w has shifted the peak an amount

$$y_{shift} = \frac{wL}{z}. \tag{5-69}$$

Because one part of the source is incoherent with any other part, the total pattern is the simple sum of intensities. If the shifted intensity maximum overlaps the original intensity minimum, the interference pattern will be washed out to give a uniform intensity, and no interference can be seen. That occurs when

$$y_{shift} = \frac{\Delta y}{2} \Rightarrow \frac{wL}{z} = \frac{\lambda L}{2d} \Rightarrow w_{washout} = \frac{\lambda z}{2d}. \tag{5-70}$$

For an interference pattern to be seen, the source radius must be smaller than this washout distance,

$$\frac{W_s}{2} < w_{washout} \Rightarrow W_s < \frac{\lambda z}{d}. \tag{5-71}$$

For a given source radius, this limits how far apart the slits can be for interference to be observed,

$$W_s < \frac{\lambda z}{d} \Rightarrow d < \frac{\lambda z}{W_s} = \frac{\lambda}{\xi}, \tag{5-72}$$

where ξ is the local divergence, from equation 5-36. The rays going through the two slits produce coherent interference if the slits are closer together than the transverse coherence length, $d < Y_c$, where

$$Y_c = \frac{\lambda}{\xi}. \tag{5-73}$$

EXAMPLE 5-13: DOUBLE SLIT AND SPATIAL COHERENCE

For the slits of example 5-12, if the source is 50 μm in diameter, how far must it be from the slits to have a transverse coherence length larger than the slit separation?

The transverse coherence length is

$$Y_c = \frac{\lambda}{\xi} = \frac{\lambda z}{W_s} \Rightarrow z = \frac{W_s}{\lambda} Y_c \approx \frac{(50 \times 10^{-6}\,\text{m})}{10^{-10}\,\text{m}}(10^{-6}\,\text{m}) \approx 0.5\,\text{m}$$

Even though the wavelength is on an atomic scale, interference effects can be seen for slits spaced apart on a micron scale. However, spatial coherence over larger distances would require greater source distances and hence very much reduced intensities.

5.9 Temporal coherence

Spatial, or transverse coherence, is a measure of how far apart two points of the x-ray beam can be without having random, incoherent phase differences. Temporal coherence is defined as the duration over which a location in the beam maintains a fixed phase relationship. Because the beam is moving with speed c, the coherence time τ_c is related to the longitudinal coherence length L_c by

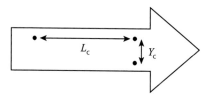

$$L_c = c\tau_c. \tag{5-74}$$

FIGURE 5-14. Relation between transverse and longitudinal coherence lengths for a beam traveling to the right.

The relationship between transverse and longitudinal coherence is shown in Figure 5-14. The coherence time τ_c is related to the line width Δv discussed in section 4.6. Just as for section 5.8, consider light from two sources falling together on a screen, this time with two slightly different frequencies. The first field is

$$E_1 = E_o \cos\left(\left(\omega_o + \frac{\Delta\omega}{2}\right)t\right)$$
$$= E_o \cos(\omega_o t)\cos\left(\frac{\Delta\omega}{2}t\right) - \sin(\omega t)\sin\left(\frac{\Delta\omega}{2}t\right), \tag{5-75}$$

where $\Delta\omega = 2\pi\,\Delta v$, and the spatial component has been suppressed. The second field is given by

$$E_2 = E_o \cos\left(\left(\omega_o - \frac{\Delta\omega}{2}\right)t\right). \tag{5-76}$$

Then

$$E_1 + E_2 = 2E_o \cos(\omega_o t)\cos\left(\frac{\Delta\omega}{2}t\right). \tag{5-77}$$

This sum is analogous to the "beating" of two sound waves of slightly different pitch, a rapidly varying oscillation at the central frequency multiplied by a slowly varying envelope with frequency Δv, as shown in Figure 5-15. If averaged over a time which is long compared with the period of the x-ray, $2\pi/\omega_o$, but short compared to the envelope variation,

$$I = 8I_o\left\langle \cos^2(\omega_o t)\cos^2\left(\frac{\Delta\omega}{2}t\right)\right\rangle \approx 8I_o\left\langle \cos^2(\omega_o t)\right\rangle\cos^2\left(\frac{\Delta\omega}{2}t\right)$$
$$= 4I_o\,\frac{1+\cos(\Delta\omega t)}{2}. \tag{5-78}$$

The coherence time is the period of this envelope, during which the intensity changes from $4I_o$ to 0 and back,

$$\tau_c = \frac{2\pi}{\Delta\omega} = \frac{1}{\Delta\nu}. \qquad (5\text{-}79)$$

For time scales which are short compared with the coherence time, $t \ll \tau_c$, the time average becomes

$$I = 4I_o \cos^2\left(\frac{\Delta\omega}{2}t\right) \approx 4I_1. \qquad (5\text{-}80)$$

The waves interact coherently, giving constructive interference. For long time scales, $t \gg \tau_c$, the time average becomes

$$I = 4I_o \left\langle \cos^2\left(\frac{\Delta\omega}{2}t\right)\right\rangle = 2I_1. \qquad (5\text{-}81)$$

The waves interact incoherently, and the intensity is the sum of the two individual intensities.

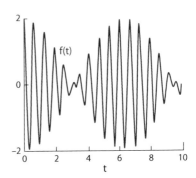

FIGURE 5-15. Beating function plotted for $\Delta\omega = \omega/10$.

EXAMPLE 5-14: COHERENCE TIME AND LONGITUDINAL COHERENCE LENGTH

If a 10 keV x-ray emission line has a width of 4.1 eV, what are the coherence time and coherence length and how do they compare with the period and wavelength of the x ray?

To compute the coherence time, we need the frequency width, which we can get from the relationship between energy and frequency,

$$U = h\nu \Rightarrow \Delta\nu = \frac{\Delta U}{h} = \frac{4.1\,\text{eV}}{4.1\times10^{-15}\,\text{eV}\cdot\text{s}} \approx 10^{15}\,\text{Hz}$$

$$\Rightarrow \tau \approx \frac{1}{\Delta\nu} \approx \frac{1}{10^{15}\,\text{Hz}} \approx 10^{-15}\,\text{s} = 1\,\text{fs}.$$

This is the same as the lifetime of the electron level from which the electronic transition created the photon, from equation 4-12. The longitudinal coherence length is

$$L_c = c\tau = \left(3\times10^8\,\frac{\text{m}}{\text{s}}\right)(10^{-15}\,\text{s}) = 0.3\,\mu\text{m}.$$

For comparison, the wavelength of the x ray is

$$\lambda = \frac{hc}{U} \approx \frac{12.4\,\text{keV}\times\text{Å}}{10\,\text{keV}} \approx 1.2\,\text{Å} \Rightarrow \frac{L_c}{\lambda} \approx \frac{3\times10^{-7}\,\text{m}}{1.2\times10^{-10}\,\text{m}} \approx 2.4\times10^3,$$

and the period is

$$\tau_{xray} = \frac{\lambda}{c} \approx \frac{1.2\times10^{-10}\,\text{m}}{3\times10^8\,\frac{\text{m}}{\text{s}}} \approx 4\times10^{-19}\,\text{s}$$

$$\Rightarrow \frac{\tau_c}{\tau_{xray}} \approx \frac{10^{-15}\,\text{s}}{4\times10^{-19}\,\text{s}} \approx 2.4\times10^3.$$

5.10 In-line phase imaging

Interference effects can be used to greatly increase edge visibility for low-contrast objects. As was discussed in chapter 1 and will be seen in chapter 9, x rays penetrate materials well. Conventional radiography depends on differences in transmission for contrast. As a result, it can be difficult to see much structure in the resulting shadow pictures such as in Figure 1-1. Interference effects can enhance the visibility of the edges, as shown in Figure 5-16.

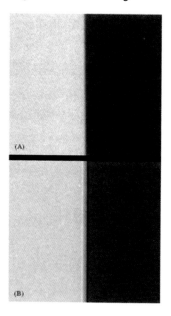

FIGURE 5-16. Example of edge enhancement. A 3 mm thick piece of plastic was placed in the left half of the screen. The image is a negative (bright is high density; dark is low density). The upper image was taken with the detector close to the sample, at $L=0$, the lower image, at $L=0.4$ m. Interference fringes can be seen as bright and dark vertical bands in the lower image. E. Donnelly et al., *Physics in Medicine and Biology* 51 (2006): 21–30. Copyright Institute of Physics and Engineering in Medicine. Reproduced by permission of IOP Publishing. All rights reserved.

If the beam is coherent over the entire object to be imaged, imaging the object at several distances allows the phase of the beam to be retrieved. The phase of the x-ray beam changes much more rapidly than its intensity. Coherence over large areas requires very small sources very far away. This requires intense directed sources such as lasers or synchrotrons. An example of phase-contrast tomography using a synchrotron beam is shown in Figure 5-17. More techniques for phase imaging are discussed in sections 14.10, 15.1, and 15.3.1.

A consequence of spatial coherence that can be a nuisance in x-ray measurements is speckle. *Speckle* is the coherent addition of randomly scattered light, as is seen when

FIGURE 5-17. A 100 million-year-old fossil insect virtually extracted from opaque amber from Charentes (France) using propagation phase-contrast X-ray synchrotron microtomography at the European Synchrotron Radiation Facility (ESRF). Image by Paul Tafforeau/ESRF.

trying to illuminate an object with a laser pointer. The resulting nonuniformity of the beam, due to interference between the incident and scattered rays, must be considered for some x-ray applications with very coherent sources. Phase issues and imaging are discussed further in section 14.9.

Problems

SECTIONS 5.1–5.3

1. Consider a Mo anode tube source with a tube voltage of 40 kV, current of 1 mA, and 2×10^{-4} x rays per electron when the tube voltage is twice the binding energy. The Kα energy is 17.5 keV, the line width is 7 eV, and the source has a diameter of 50 μm and is 100 mm away from a 5 mm diameter detector. Considering only the characteristic Kα emission, give the emission rate, power, intensity at 100 mm, angular intensity, photon intensity at 100 mm, angular photon intensity, brightness and brilliance, the power, and count rate into the detector. *Hint:* Recall equation 4-9.

SECTION 5.4

2. A synchrotron x-ray beam has a height of 2 mm at a distance of 20 m from the source, and is restricted by apertures to a width of 5 mm in the horizontal direction. Give the horizontal and vertical global divergence of the beam.

SECTION 5.5

3. What is the local divergence of the source of problem 1, as viewed from the detector?
4. What is the minimum image resolution for a 100 mm thick object placed 500 mm from the source of problem 1, but with the detector moved back to 600 mm?

SECTION 5.7

5. Show that the time average of the product of sine and cosine $\langle \sin(\omega t)\cos(\omega t) \rangle$, is zero.
6. Prove equation 5-54.

SECTION 5.8

7. What is the spacing of the interference fringes on a screen 10 m from slits separated by 0.5 μm and irradiated with 10 keV radiation?
8. How far do the fringes of problem 7 shift if the source, 10 cm from the slits, is moved 10 μm?
9. What is the maximum source size that can be employed to see the interference pattern of problem 7?
10. What is the transverse coherence length for the source of problem 1, at the detector distance of 100 mm?

SECTION 5.9

11. What is the longitudinal coherence length of the source in problem 1?

Further reading

Beam properties as in sections 5.1–5.3
See chapter 1 references.

Coherence
M. Born and E. Wolf, **Principles of Optics**, Cambridge University Press; 7th ed., 1999.

X-ray tubes
See chapter 1 references.

S. M. Lee and C. A. MacDonald, X-ray Tube Sources, chap. 54 in M. Bass, C. DeCusatis, J. Enoch, V. Lakshminarayanan, G. Li, C. MacDonald, V. N. Mahajan and E. Van Stryland, eds., **Handbook of Optics**, 3rd ed., vol. 5, McGraw-Hill, 2010.

Phase imaging
R. Fitzgerald, Phase sensitive x-ray imaging, *Physics Today*, July 2000, p. 23.

K. A. Nugent, T. E. Gureyev, D. F. Cookson, D. Paganin, and Z. Barnea, Quantitative phase imaging using hard x rays, *Physical Review Letters* 77 (1996): 2961–64.

Atsushi Momose, Recent advances in x-ray phase imaging, *Japanese Journal of Applied Physics* 44, no. 9A (2005): 6355–67.

6

BREMSSTRAHLUNG RADIATION AND X-RAY TUBES

6.1 Field from a moving charge

In chapter 4, we examined the emission from an x-ray tube when the incoming electrons knocked out core electrons in the target anode. However, other interactions are possible and in fact probable between the incoming electron and the anode atoms. As the incoming electron moves through the anode material it experiences the electric field of the electrons and positive atomic cores of the metal target, which tends to slow the electron, eventually bringing it to a stop. As the electron velocity is reduced some of the electron's kinetic energy is converted into electromagnetic energy, which is emitted as radiation called *bremsstrahlung*, from the German for "braking radiation."

In the simplest classical picture, a column of electrons with electron density ρ_e moving at a constant velocity $\vec{v}$ provide a current density $\vec{J}_A = \rho_e \vec{v}$, and, as shown in Figure 6-1, an induced magnetic field B, given by Maxwell's equations as

$$\vec{\nabla} \times \vec{B} = \mu_o \vec{J}_A, \qquad (6\text{-}1)$$

where μ_o is the vacuum permeability

$$\mu_o = \frac{1}{c^2 \varepsilon_o}, \qquad (6\text{-}2)$$

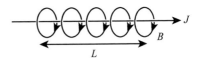

FIGURE 6-1. Magnetic field from a column of moving charge of length L and total current J.

c is the speed of light, and ε_o is the vacuum permittivity.

6.2 Radiation from an accelerating (or decelerating) charge

If the electron velocity changes, there is a time-dependent current and therefore a time-dependent magnetic field, which causes an electric field to radiate according to

$$\vec{\nabla} \times \vec{E} = -\frac{\partial \vec{B}}{\partial t}. \qquad (6\text{-}3)$$

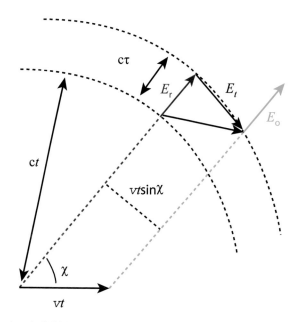

FIGURE 6-2. The electric field due to an electron moving at a constant velocity would have originated from the position vt and been directed radially outward, along the blue dashed line. This is the field E_o that exists for distances $r > c(t+\tau)$ away from the charge. After the charge stops, the field radiates from the origin, along the red line. This is the field that exists for $r < ct$. The field line "kinks" between these two distance intervals, following the solid black line.

A simple way to visualize the radiation field is to consider a charge moving at a constant (nonrelativistic) velocity v in the z direction.* Its field is that of a point charge, directed radially outward from the charge from its current position, at $z = vt$, as shown in blue in Figure 6-1. Consider the case that it is subjected to a deceleration for a time interval of length τ and comes to a stop at time $t = 0$. For later times, the field will be that of a stationary charge, radiating outward from the origin, along the red line in Figure 6-2. But the information about this change can propagate outward only at the speed of light. Thus for distances $r > c\,(t + \tau)$ the field should still look like that of the moving charge, centered at the position the charge would have had, $z = vt$. For distances $r < ct$, the field should be that of the stationary charge. In the interval of width $c\tau$, the field lines connect (bold black line in Figure 6-2. This kinked field line has both a radial component E_r and a tangential component E_t. From the geometry sketched in Figure 6-2, the ratio of the components is

$$\frac{E_t}{E_r} = \frac{vt \sin \chi}{c\tau}. \tag{6-4}$$

Making the substitution that the acceleration is

$$a = \frac{v}{\tau} \tag{6-5}$$

---

*This analysis is due to E. Purcell; see the further reading suggestions at the end of the chapter.

and that the position is

$$r = ct \Rightarrow t = \frac{r}{c},$$
(6-6)

and using the fact that the final field is simply that of a point charge,

$$E_r = \frac{q_e}{4\pi\varepsilon_0 r^2},$$
(6-7)

the tangential field becomes

$$E_{radiation} = E_t = \frac{q_e}{4\pi\varepsilon_0 c^2 r} a \sin\chi,$$
(6-8)

where χ is the angle between $\vec{a}$ and $\vec{r}$. The radiation field is a pulse which travels out from the decelerated electron at the speed of light. The radiation field of equation 6-3 and hence 6-8 arises from the change in the induced magnetic field, that is, a change in the current in equation 6-1. The current is proportional to the electron velocity, so the change in magnetic field is proportional to the acceleration. The radiation field is linear in the acceleration a and falls off as $1/r$.

Noting that $\dfrac{a}{c}\sin\chi = \left| \hat{r} \times \dfrac{\vec{a}}{c} \right|$ and that the field must be perpendicular to $\vec{r}$, that is, transverse to the direction of propagation and in the plane defined by $\vec{r}$ and $\vec{a}$, as shown in Figure 6-3, then the classical radiation field of equation 6-8 can be written

$$\vec{E}_{radiation} = \frac{q_e}{4\pi\varepsilon_0 c r} \hat{r} \times \left(\hat{r} \times \frac{\vec{a}}{c} \right), \qquad \textbf{Classical}$$
(6-9)

where $\hat{r} = \dfrac{\vec{r}}{r}$ is a unit vector in the observation direction.

The fully relativistic case can be computed in a fairly straightforward manner by writing the vector and scalar (Liénard-Wiechert) potentials, which depend on the location of the electron at the *retarded time* (the time when the acceleration happened as opposed to the time when the pulse reaches the observation location). The bremsstrahlung problem is not usually relativistic, but for completeness (and because it is used in the next chapter for accelerating electrons) the relativistic solution is presented here. The full calculation is described in many electrodynamics texts (e.g., see the further reading section at the end of this chapter). The radiation field due to a charge with a relativistic velocity $\vec{v}$ and acceleration $\vec{a}$ is

FIGURE 6-3. The electric field due to an electron with acceleration $\vec{a}$ is perpendicular to $\vec{r}$ and in the plane formed by $\vec{r}$ and $\vec{a}$.

$$\vec{E} = \frac{q_e}{4\pi\varepsilon_0 c} \frac{\hat{r} \times \left\{ \left(\hat{r} - \dfrac{\vec{v}}{c} \right) \times \dfrac{\vec{a}}{c} \right\}}{r \left(1 - \dfrac{\vec{v} \cdot \hat{r}}{c} \right)^3}.$$
(6-10)

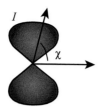

FIGURE 6-4. A "polar plot" of intensity versus angle of the dipole radiation from an accelerating charge. The map of directions of nonzero intensity forms a "doughnut" (coming out of the page) around the direction of acceleration.

For electron velocities much less than the speed of light, the expression reduces to equation 6-9. The instantaneous (not time averaged) intensity of the radiation is then

$$I = c\varepsilon_0 E_{rad}^2 = \frac{q_e^2}{16\pi^2\varepsilon_0 c^3 r^2} a^2 \sin^2\chi. \tag{6-11}$$

This intensity pattern, sketched in Figure 6-4, is the same as the field of an oscillating dipole. The intensity is zero in the direction of the acceleration. The total power in the field is found by integrating over a sphere,

$$P_1 = \int I r^2 \sin\chi \, d\chi \, d\varphi = \frac{q_e^2}{6\pi\varepsilon_0 c^3} a^2, \tag{6-12}$$

where the subscript 1 is a reminder that the result is for a single electron with a constant acceleration.

It can be difficult to use that equation for a tube source. In general, the incoming electron has a complicated trajectory, undergoing multiple scattering within the target anode material, so that the details of the acceleration are unknown. However, for thin targets or controlled trajectories, it is possible to compute the radiated electric field from equation 6-12. For example, the acceleration when the electron is at a distance r from a nuclear charge Zq_e, where Z is the atomic number, is

$$a = \frac{Zq_e^2}{4\pi\varepsilon_0 M_e r^2}. \tag{6-13}$$

Because the acceleration and hence radiated power depends on the mass, bremsstrahlung radiation is very much reduced for protons as compared with electrons. For that reason, protons are sometimes used to excite x-ray fluorescence, as they produce low background compared to excitation with an electron beam (the technique is called PIXE, particle-induced x-ray emission).

6.3 Emission from a very thin anode

The emitted radiation given by equations 6-12 and 6-13 depends on the details of the electron trajectory. A Fourier analysis by Kramers given a range of scattering angles for the electron yields the result that the frequency spectrum of the radiation pulse due to a decelerating electron is flat up to the maximum frequency,

$$P_\nu = \begin{cases} P_{\nu 0} & \nu \leq \nu_{max} \\ 0 & \nu > \nu_{max} \end{cases}, \tag{6-14}$$

as shown in Figure 6-5. The maximum photon frequency is produced if the electron gives up its entire kinetic energy to a single photon,

$$h\nu_{max} = K_e. \tag{6-15}$$

The kinetic energy of the electron is determined by the tube voltage Φ used to accelerate the electron from the cathode to the anode,

$$K_e = q_e \Phi. \tag{6-16}$$

The electron can give up some or all of its kinetic energy to a photon. For an average over many electrons, the amount of energy deposited in each photon frequency interval is a constant.

The total x-ray power is found from integrating over frequency,

$$P = \int P_v \, dv = \int_0^{v_{max}} P_{vo} \, dv = P_{vo} v_{max} = P_{vo} \frac{q_e \Phi}{h}. \tag{6-17}$$

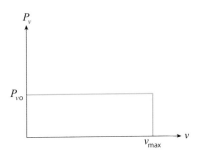

FIGURE 6-5. Theoretical bremsstrahlung spectrum for a thin anode in which each electron undergoes a single scattering event. The maximum photon energy is given by the kinetic energy of the incoming electron.

The total emitted x-ray power depends on the rate at which electrons enter the anode, Γ_e. This is related to the anode current,

$$J = q_e \Gamma_e. \tag{6-18}$$

The power also depends on how many atoms each electron encounters before passing through the anode material. If the anode thickness is x and the atomic density is ρ_{atom} atoms per volume, then the analysis gives

$$P = \Gamma_e \left(\frac{1}{4\pi\varepsilon_o} \right)^3 \frac{16\pi^2}{3\sqrt{3}} \frac{Z^2 q_e^6}{c^3 M_e h} \rho_{atom} x. \tag{6-19}$$

EXAMPLE 6-1

For an x-ray tube with a tube voltage of 100 kV, what are the maximum photon energy and frequency?

The maximum photon energy is $U_{max} = K_e = q_e \Phi = q_e \, (100 \text{ kV}) = 100 \text{ keV}$. Converting units gives

$$U_{max} \approx 100 \text{ keV} \left(\frac{10^3 \text{ eV}}{\text{keV}} \right) \left(\frac{1.6 \times 10^{-19} \text{ J}}{\text{eV}} \right) \approx 1.6 \times 10^{-14} \text{ J}.$$

The frequency is thus

$$v_{max} = \frac{U_{max}}{h} \approx \frac{1.6 \times 10^{-14} \text{ J}}{6.6 \times 10^{-34} \text{ J} \cdot \text{s}} \approx 2.4 \times 10^{19} \text{ Hz}.$$

For a current of 1 mA, what is the total bremsstrahlung x-ray power emitted from a 10 μm thick tungsten anode? Tungsten has atomic number 74, a density of 19.3 g/cm³, and a molar mass of 184 g/mole.

First, we need the atomic density,

$$\rho_{atom} = N_A \frac{\rho}{M_M} \approx \left(6.2 \times 10^{23} \frac{\text{atoms}}{\text{mole}} \right) \left(\frac{19.3 \times 10^3 \frac{\text{kg}}{\text{m}^3}}{184 \times 10^{-3} \frac{\text{kg}}{\text{mole}}} \right) \approx 6.3 \times 10^{28} \frac{\text{atoms}}{\text{m}^3}.$$

Then, combining $q_e\Gamma$ into J gives the power,

$$P = J \left(\frac{1}{4\pi\varepsilon_0} \right)^3 \frac{16\pi^2}{3\sqrt{3}} \frac{Z^2 q_e^5}{c^3 M_e h} \rho_a x \approx 10^{-3} \frac{\text{Coul}}{\text{s}} \left(\frac{1}{8.85 \times 10^{-12} \frac{\text{Coul}^2}{\text{N} \cdot \text{m}^2}} \right)^3$$

$$\times \left(\frac{1}{4\pi} \right) \frac{1}{3\sqrt{3}} \frac{(74)^2 (1.6 \times 10^{-19} \text{Coul})^5 \left(6.3 \times 10^{28} \frac{1}{\text{m}^3} \right) (10 \times 10^{-6} \text{m})}{\left(3 \times 10^8 \frac{\text{m}}{\text{s}} \right)^3 (9.1 \times 10^{-31} \text{kg}) (6.6 \times 10^{-34} \text{J} \cdot \text{s})}$$

$$\approx 0.5 \frac{\text{J}}{\text{s}} = 0.5 \text{ W}.$$

What is the efficiency, the ratio of x-ray power to electrical power, in this case?

$$P_{elec} = J\Phi = \left(10^{-3} \frac{\text{Coul}}{\text{s}} \right) (100 \times 10^3 \text{ V}) = 100 \text{ W}$$

$$\Rightarrow \eta = \frac{P}{P_{elec}} \approx \frac{0.5 \text{ W}}{100 \text{ W}} \approx 0.005 = 0.5\%.$$

An experiment is set up with a detector and electronics that selectively collect x-ray photons from only 48 keV to 52 keV in energy. What is the power emitted into this 4 keV wide energy bin?

$$P_{bin} = \int_{v_1}^{v_2} P_{vo} \, dv = P_{vo} (v_2 - v_1) = P_{vo} \left(\frac{U_2}{h} - \frac{U_1}{h} \right) = \frac{P_{vo}}{h} \Delta U.$$

$$P = P_{vo} \frac{q_e\Phi}{h}$$

$$\Rightarrow \frac{P_{vo}}{h} = \frac{P}{q_e\Phi} \approx \frac{0.5 \text{ W}}{q_e (100 \times 10^3 \text{ V})} = \frac{5 \times 10^{-3}}{q_e} \frac{\text{W}}{\text{kV}} = 5 \times 10^{-3} \frac{\text{W}}{\text{keV}}.$$

$$P_{bin} = \frac{P_{vo}}{h} \Delta U \approx 5 \times 10^{-3} \frac{\text{W}}{\text{keV}} (4 \text{ keV}) \approx 0.02 \text{ W}.$$

What is the emission rate for photons into this energy bin?

$$\Gamma_{bin} \approx \frac{P_{bin}}{\overline{U}} \approx \frac{0.02 \frac{\text{W}}{\text{bin}}}{50 \frac{\text{keV}}{\text{photon}}} \left(\frac{1 \text{ keV}}{10^3 \text{eV}} \right) \left(\frac{1 \text{ eV}}{1.6 \times 10^{-19} \text{J}} \right) \approx 2.5 \times 10^{12} \frac{\text{photons}}{\frac{\text{s}}{\text{bin}}}.$$

6.4 Emission from a thick anode

Unless the anode is very thin, an electron which has given up some of its energy to a photon is still traveling within the material. The electron will be decelerated again, this time starting with a lower kinetic energy. The total range of an electron with a few kiloelectronvolts of energy is hundreds to thousands of nanometers. The resulting spectrum is a superposition of flat spectra with decreasing maximum photon energy, as shown in Figure 6-6, so that the frequency distribution can be written

$$P_v = [C_K Z \, \Gamma_e] \, (v_{max} - v), \tag{6-20}$$

where the prefactor in square brackets is a constant, written as a product to explicitly show the dependence on the atomic number Z of the anode material and the rate at which electrons enter the anode, Γ_e. Kramers' constant, C_K, is

$$C_K = \frac{1}{4\pi\varepsilon_0} \frac{8\pi q_e^2 h}{3\sqrt{3}\ell c^3 M_e}, \tag{6-21}$$

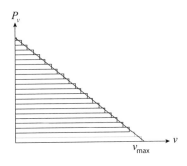

FIGURE 6-6. Triangular shape of bremsstrahlung intensity from a thick target, showing the superposition of flat spectra from thin-target bremsstrahlung with decreasing kinetic energy as the electrons penetrate into the material.

where ℓ is a unitless geometric constant approximately equal to 5. The total x-ray power is found from integrating over frequency,

$$P = \int P_v \, dv = \int_0^{v_{max}} C_K \Gamma_e \, Z \, (v_{max} - v) \, dv = \frac{C_K Z \Gamma_e}{2} v_{max} = \frac{C_K Z \Gamma_e}{2} \left(\frac{q_e \Phi}{h} \right)^2. \tag{6-22}$$

This result says the radiated bremsstrahlung power is proportional to the number of electrons being decelerated, the charge on the nucleus decelerating them, and the square of the tube voltage (which is proportional to the square of the initial kinetic energy of the electrons).

6.5 Efficiency

The efficiency η of conversion of electrical energy into the bremsstrahlung continuum radiation is the x-ray power divided by the electrical power,

$$\eta = \frac{P}{P_{elec}} = \frac{\dfrac{C_K \Gamma_e \, Z}{2} \left(\dfrac{q_e \Phi}{h} \right)^2}{\Phi J} = \left(C_k \frac{q_e}{2h^2} \right) Z\Phi = C_\eta Z\Phi, \tag{6-23}$$

where the constant C_η is defined as the quantity in parentheses,

$$C_\eta \approx \frac{1.1 \times 10^{-9}}{\text{Volt}}. \tag{6-24}$$

This allows us to rewrite equation 6-22 as

$$P = \eta J \Phi = C_\eta J Z \Phi^2.$$

(6-25)

EXAMPLE 6-2

What is the power conversion efficiency for a thick tungsten anode at 100 kV?

$$\eta = C_\eta Z \Phi \approx \frac{1.1 \times 10^{-9}}{V} (74)(100 \times 10^3 \text{ V}) \approx 0.008 = 0.8\%.$$

As expected, this result for a thick anode is slightly higher than found in example 6-1 for a thin anode, but the efficiency of bremsstrahlung production for even a bulk high-Z material at high voltage is less than 1%. Most of the remaining power in the incident electron beam is converted to heat.

6.6 Thick-target photon emission rate modeling

The spectra shown in Figures 6-5 and 6-6 are given in units of power per frequency range. Since the energy carried by a photon is $h\nu$, the number of photons per second emitted in each frequency range is given by

$$\Gamma_\nu \, d\nu = \frac{P_\nu}{h\nu} \, d\nu.$$

(6-26)

The emission rate diverges for very low energy photons. However, zero-energy "virtual" photons are not detectable. The spectrum of Figure 6-6 cannot be measured unless all the bremsstrahlung photons emitted in the target reach the detector. In

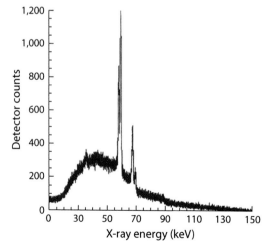

FIGURE 6-7. Measured spectrum from W anode at 150 kV. The peaks at around 57 and 69 keV are the characteristic emission discussed in chapter 4. The intensity falls off rapidly below 30 keV as a result of absorption. Courtesy of GE CR&D.

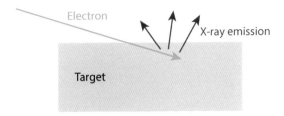

FIGURE 6-8. Electron beam striking a thick anode at an oblique angle to minimize the depth at which x rays are emitted.

fact, many of the photons are absorbed in the target, in the window between the x-ray tube and the measurement system, in the air, and in the detector window. As we saw in chapter 1, and will investigate more fully in chapter 9, "soft," low-energy photons are more easily absorbed than higher-energy photons. For that reason, the low-energy part of the spectrum tends to be less intense, as shown in the measured spectrum of Figure 6-7. To minimize x-ray losses in the target, the electron beam is usually brought in at an angle to the target, as shown in Figure 6-8.

The resultant measured bremsstrahlung photon emission rate, shown in Figure 6-7, can often be approximated as a polynomial. Using the simplest approximation, the triangle shown in Figure 6-9,

$$\Gamma_U = \begin{cases} \dfrac{\Gamma_P}{U_P} U & U \leq U_P \\ \dfrac{\Gamma_P}{U_{max} - U_P} (U_{max} - U) & U_P < U \leq U_{max} \\ 0 & U > U_{max} \end{cases}.$$

(6-27)

The total photon emission rate is

$$\Gamma = \int_0^{U_{max}} \Gamma_U dU = \Gamma_P \frac{U_{max}}{2}.$$

(6-28)

The total power after absorption is

$$P_d = \int_0^{U_{max}} \Gamma_U U dU = \Gamma_P \frac{U_{max}}{6} (U_{max} + U_P),$$

(6-29)

which gives

$$\Gamma_P = \frac{6P_d}{U_{max}(U_{max} + U_P)},$$

(6-30)

so that the emission rate becomes

$$\Gamma = \frac{6P_d}{U_{max}(U_{max} + U_P)} \frac{U_{max}}{2} = \frac{3P_d}{(U_{max} + U_P)}.$$

(6-31)

The average detected photon energy $\bar{U}$ is then given by

$$\bar{U} = \frac{P_d}{\Gamma} = \frac{P_d}{\dfrac{3P_d}{(U_{max}+U_P)}} = \frac{U_{max}+U_P}{3}.$$

(6-32)

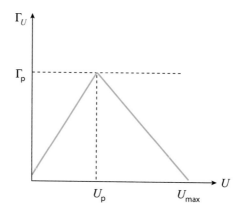

FIGURE 6-9. Approximate measured bremsstrahlung photon emission rate as a function of photon energy.

EXAMPLE 6-3:

Assuming the total bremsstrahlung power emitted by an x-ray tube with a thick tungsten anode at 100 kV is given by the ideal equation (assume $P_d = P$) and that the photon emission spectrum can be approximated by a triangle like that of Figure 6-9, with the peak at $U_p = 0$, what is the total photon emission rate per milliampere? per electron? What is the emission rate into a 100 eV wide energy at the average photon energy if the current is 1 mA?

The power per milliampere is

$$\frac{P}{J} = \eta\Phi = C_\eta Z\Phi^2 \approx (0.008)(100\,\text{kV}) = 0.8\,\frac{\text{W}}{\text{mA}}.$$

The maximum photon energy is $U_{max} = q_e\Phi = 100\,\text{keV}.$

The emission rate per milliampere is

$$\frac{\Gamma}{J} = \frac{3\dfrac{P}{J}}{(U_{max}+U_P)} \approx \frac{3\left(0.8\dfrac{\text{J}}{\text{s}\cdot\text{mA}}\right)}{(100\,\text{keV}+0)}\left(\frac{1\,\text{keV}}{10^3\,\text{eV}}\right)\left(\frac{1\,\text{eV}}{1.6\times10^{-19}\,\text{J}}\right)$$

$$\approx 1.5\times10^{14}\,\frac{\text{photons}}{\text{s}\cdot\text{mA}}.$$

The current with 1 electron per second is

$$J \approx 1.6 \times 10^{-19} \frac{\text{Coul}}{\text{s}} = 1.6 \times 10^{-19} \text{ A} = 1.6 \times 10^{-16} \text{ mA}.$$

Thus the emission rate per electron is

$$\Gamma_{per\ elec} \approx \left(1.6 \times 10^{-16} \frac{\text{mA}}{\text{electron}}\right)\left(1.5 \times 10^{14} \frac{\text{photons}}{\text{s} \cdot \text{mA}}\right)$$

$$\approx 0.024 \frac{\text{photons}}{\text{electron}} = \frac{24\ \text{photons}}{1000\ \text{electrons}}.$$

The average photon energy is $U_{max}/3$, or 33 keV.
The emission rate into a 100 eV bin at this energy is

$$\Gamma_{bin} = \Gamma_U\ dU = \frac{\Gamma_P}{U_{max} - U_P}(U_{max} - U)\Delta U$$

$$= \frac{6P}{U_{max}(U_{max} + U_P)(U_{max} - U_P)}(U_{max} - U)\Delta U$$

$$\approx \frac{6(0.8\ \text{W})}{(100\ \text{keV})^3}(67\ \text{keV})(0.1\ \text{keV})\left(\frac{1\ \text{keV}}{10^3\ \text{eV}}\right)\left(\frac{1\ \text{eV}}{1.6 \times 10^{-19}\ \text{J}}\right)$$

$$\approx 2 \times 10^{11} \frac{\text{photons}}{\text{s}}.$$

6.7 Spectral shaping

The tube voltage can be used to shape the spectrum. Both the characteristic and the continuum radiation power increase monotonically with tube voltage. The resultant ratio of the *power* in the characteristic line to the total bremsstrahlung power is

$$Ratio = \frac{P_{char}}{P_{brem}} = \frac{\Gamma_{char}\ U_{char}}{C_\eta Z\Phi^2 J} = \frac{U_{char}}{C_\eta Z\Phi^2 J} C_{Char} \frac{J}{q_e}\left(\frac{q_e\Phi - q_e\Phi_b}{q_e\Phi_b}\right)^{O_{Char}}. \qquad (6\text{-}33)$$

where U_{char} is the characteristic photon energy and Γ_{char} is taken from equation 4-9. The maximum can be found by differentiation,

$$\frac{\partial Ratio}{\partial \Phi} = 0 = -2\frac{U_{char}}{C_\eta Z\Phi^3 J} C_{char} \frac{J}{q_e}\left(\frac{q_e\Phi - q_e\Phi_b}{q_e\Phi_b}\right)^{O_{char}}$$

$$+ \frac{U_{char}}{C_\eta Z\Phi^2 J} C_{char} \frac{J}{q_e} O_{char}\left(\frac{q_e\Phi - q_e\Phi_b}{q_e\Phi_b}\right)^{O_{char}-1} \frac{q_e}{q_e\Phi_b} \qquad (6\text{-}34)$$

$$\Rightarrow 0 = \frac{-2}{\Phi} + O_{char} \frac{q_e}{q_e\Phi - q_e\Phi_b} \Rightarrow 2q_e\Phi - O_{char}q_e\Phi = 2q_e\Phi_b$$

$$\Rightarrow \Phi = \frac{2}{2 - O_{char}} \Phi_b.$$

Because the exponent O_{char} is typically about 1.6, the highest ratio of characteristic to bremsstrahlung *power* occurs at a tube voltage corresponding to roughly five times the ionization energy, for example, at about 45 kV for Cu K radiation. However, the ratio of the characteristic *emission rate* to the bremsstrahlung emission rate increases monotonically with tube voltage. The relative importance of the bremsstrahlung for a particular experiment depends on the measurement, the optics, and whether the detector counts photons or integrates power. For some applications, operating at a tube voltage that does not excite the characteristic energies simplifies analysis.

The tube voltage may not be stable, but will oscillate slightly, and many tubes use a pulsed voltage supply to produce a brief burst of x rays. For that reason, the tube voltage Φ is usually specified in terms of its peak value, for example, 100 kV$_p$.

Problems

SECTION 6.2

1. Show that equation 6-10 is the same as equation 6-9 when $v \ll c$, and find the direction of polarization of the electric field if the velocity and acceleration are in the z direction, and the observer is in the y direction.

SECTION 6.4–6.6

2. X rays are produced by an x-ray tube with a voltage of 30 kV, a current of 1 mA, a source size of 0.5 mm $\times$ 0.5 mm, and a thick molybdenum anode.
   a) What is the minimum x-ray wavelength produced?
   b) What is the bremsstrahlung power?
   c) Assuming the bremsstrahlung has a typical triangular frequency spectrum, with its peak at half the maximum energy, what is the emission rate within an energy window of width 0.5 keV centered on the Kα line?
   d) What is the brilliance of the bremsstrahlung emission in that energy window?
   e) Sketch the x-ray energy spectrum, including both the bremsstrahlung and the characteristic lines (assume doublets are not resolved) measured with a detector with 0.5 keV energy bins. Assume the Kβ line has one-fifth the emission rate of the Kα line, and use $C_{char} = 3 \times 10^{-4}$. The energies and the heights of the bremsstrahlung and the characteristic lines should be correct.
   f) Repeat (e) with the voltage increased to 60 kV.

3. An x-ray tube has a thick tungsten, W, anode, a tube voltage of 100 kV, a current of 20 mA, and an electron spot size of 0.2 mm $\times$ 2 mm, viewed at an angle of 10°. Assume the emission rate spectrum peaks at 0 keV, as in Figure 6-10.
   a) What is the bremsstrahlung emission rate within an energy window of width 1 keV centered on 50 keV?

b) What is the brilliance of the bremsstrahlung emission at 50 keV?

4. A plot of the spectrum from a silver, Ag, anode tube with a tube voltage of 40 kV and an energy bin of 0.5 keV is shown in Figure 6-10. (The K, L, and M energies of silver are approximately 25.5, 3.5, and 0.7 keV, respectively)

   a) Give the numerical values of U_A, U_B, and U_C (located at the labeled features on the plot).

   b) Fill in table 6-1 with the approximate factor. (For example, if the tube voltage doubles, U_C goes up by a factor of 2, so a 2 has been entered for that item.)

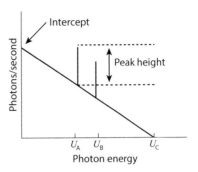

FIGURE 6-10. Flux spectrum for problem 4.

TABLE 6-1.

By what factor does the item below change if:	The tube voltage doubles?	The tube current doubles?	The bin size doubles?	The atomic number doubles?
U_A				
U_C	2			
y axis intercept				
The peak height for the peak at U_A				

Further reading

H. A. Kramers, On the theory of x-ray absorption and of the continuous x-ray spectrum, *Philosophical Magazine* 46 (1923): 836.

A. G. Michette and C. J. Buckley, **X-ray Science and Technology**, Institute of Physics Publishing, 1993.

J. D. Jackson, **Classical Electrodynamics**, 2nd ed., Wiley & Sons, 1975.

E. Purcell, **Electricity and Magnetism**, 2nd ed., McGraw-Hill Science/Engineering/Math, 1984.

X-ray data booklet, http://xdb.lbl.gov/.

S. M. Lee and C. A. MacDonald, X-ray Tube Sources, chap. 54 in M. Bass, C. DeCusatis, J. Enoch, V. Lakshminarayanan, G. Li, C. MacDonald, V. N. Mahajan, and E. Van Stryland, eds., **Handbook of Optics**, 3rd ed., vol. 5, McGraw-Hill, 2010.

7

SYNCHROTRON RADIATION

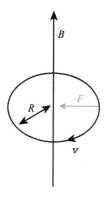

Figure 7-1. Electrons in a circular orbit of radius R at a speed v in a magnetic field B.

Chapter 6 discussed the bremsstrahlung radiation field produced by the electrons decelerating in an x-ray tube. The same type of radiation is produced when electrons are deliberately moved in a circular track by a magnetic field, as shown in Figure 7-1. The first circular particle accelerators, or cyclotrons, were developed 80 years ago for nuclear physics experiments. Synchrotron sources were originally parasitic to particle experiments, a consequence of unavoidable energy loss from an accelerated particle. One example was the Cornell High Energy Synchrotron Source (CHESS), which was originally a byproduct of the Cornell high energy particle experiment. Synchrotrons generate highly collimated x-ray beams of high intensity and coherence.

7.1 Classical (nonrelativistic) orbits

An electron with a low velocity v under the influence of a constant magnetic field B, treated as a classical particle, experiences a force

$$\vec{F} = q_e \vec{v} \times \vec{B}. \tag{7-1}$$

Because the force is perpendicular to the velocity, it results in a circular orbit. For nonrelativistic velocities the acceleration due to the circular motion is

$$a = \frac{v^2}{R}. \tag{7-2}$$

The radius R of the orbit can be found classically by balancing the magnetic force with the required centripetal force,

$$F = q_e v B = \frac{M_e v^2}{R} \Rightarrow R = \frac{M_e v}{q_e B} = \frac{p_e}{q_e B}, \tag{7-3}$$

where p_e is the electron momentum. If the electron speed is increased, the radius will increase in a spiral, as shown in Figure 7-2.

Even if the speed and radius are kept constant, when the electron goes around a ring of radius R it must be accelerated toward the center, and hence it emits radiation. As before, the field is described by equation 6-10,

$$\vec{E} = \frac{q_e}{4\pi\varepsilon_0 c} \frac{\hat{r} \times \left\{\left(\hat{r} - \dfrac{\vec{v}}{c}\right) \times \dfrac{\vec{a}}{c}\right\}}{r\left(1 - \dfrac{\vec{v}\cdot\hat{r}}{c}\right)^3}. \quad (7\text{-}4)$$

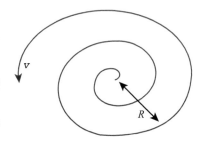

FIGURE 7-2. Electron with increasing v spiraling with increasing radius in a magnetic field.

Recall that if v is small, this implies

$$E = \frac{q_e}{4\pi\varepsilon_0 c^2 r} a \sin \chi, \quad (7\text{-}5)$$

where χ is the angle between $\vec{a}$ and $\vec{r}$, as shown in Figure 7-3. The observer is at a location $\vec{r}$, at an angle χ from the acceleration. This again results in dipole-like radiation, as shown in Figures 7-4 and 6-4.

From equation 6-12, for classical electron velocities the total emitted power from a single electron is

$$P_1 = \frac{q_e^2}{6\pi\varepsilon_0 c^3} a^2 = \frac{q_e^2 v^4}{6\pi\varepsilon_0 c^3 R^2} = \frac{q_e^2 v^4}{6\pi\varepsilon_0 c^3 \left(\dfrac{M_e v}{q_e B}\right)^2} = \frac{B^2 q_e^4 v^2}{6\pi\varepsilon_0 M_e^2 c^3}. \quad (7\text{-}6)$$

Since the classical kinetic energy of the electron is

$$K_e = \frac{1}{2} M_e v^2, \quad (7\text{-}7)$$

the emitted power from a single electron can be written

$$P_1 = \frac{B^2 q_e^{\,4} K_e}{3\pi\varepsilon_0 M_e^{\,3} c^3}. \quad (7\text{-}8)$$

As the electron travels around the ring it loses energy, in the radiation it is emitting, as well as to collisions with stray gas molecules. To maintain the electron velocity, the cyclotron must impart some energy to the electron on each orbit. The electron travels around the circumference in a time

$$t_e = \frac{2\pi R}{v}. \quad (7\text{-}9)$$

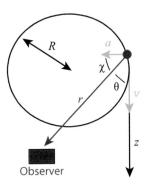

Observer

FIGURE 7-3. An observer at a location r of the radiation from an electron moving at a classical velocity.

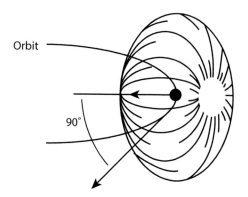

F<small>IGURE</small> 7-4. Radiation from a low-velocity electron in a circular track. From S. L. Hulbert and G. P. Williams, Synchrotron Sources, chap. 55, *Handbook of Optics*, 3rd ed., vol. 5, McGraw-Hill, 2010.

Thus the frequency with which the cyclotron electric field must cycle to provide energy to maintain the electron speed is

$$f_e = \frac{v}{2\pi R} = \frac{v}{2\pi \left(\dfrac{M_e v}{q_e B} \right)} = \frac{1}{2\pi} \frac{q_e B}{M_e}, \tag{7-10}$$

which is called the *cyclotron frequency*. For classical (nonrelativistic) velocities, this frequency is independent of the radius R and also the electron velocity, so a single frequency works as the electron is spiraled up in speed, as shown in Figure 7-2. The cyclical force, supplied, for example, by microwave radiation applied at one point on the orbit, results in electron bunching. The force is plotted in Figure 7-5. For instance, assume that the amplitude of the force (amplitude of the microwave source) is adjusted so that the acceleration is just enough to counteract losses for an electron arriving at time t_2. The net increase in speed is zero, and the electron arrives at the same point 2 one period later, on the next orbit. Next, consider an electron which arrives early, at time t_1. It is traveling with too much speed around the ring but experiences less force to counteract losses, loses some speed on the next orbit, and arrives back at the force point at a time closer to electron 2. Conversely, the electron which arrived at time t_3 was going too slow but received an extra speed boost. These electrons become bunched together and tend to travel together around the ring. An electron which arrives on the downslope, for example at t_4, is unstable, receives too much or too little acceleration, and will be lost to collision with the walls. After some time the only electrons in the ring will be in the one bunch. The cyclotron frequency is the frequency with which this one bunch passes a point on the orbit.

The period of the orbit also defines the current for a single electron,

$$J_1 = \frac{q_e}{t_e} = \frac{q_e v}{2\pi R}. \tag{7-11}$$

Substituting into the expression in equation 7-6, the total emitted radiation power for an electron current J, is

$$P = \frac{q_e^2 v^4}{6\pi \varepsilon_0 c^3 R^2} = \frac{q_e v^3}{3\varepsilon_0 c^3 R} J. \tag{7-12}$$

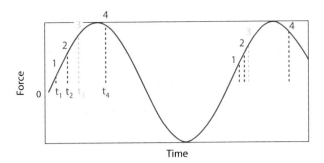

FIGURE 7-5. Supplied electrical force as a function of time. Fast electrons, which arrive sooner at the point at which the force is applied, see less acceleration than slower electrons, causing electron bunching.

EXAMPLE 7-1

Electrons with a kinetic energy of 10 keV are traveling in a magnetic field of 0.2 T. What is the velocity of the electrons, the radius of the orbit, the period of the orbital motion, and the emitted power for a 1 mA beam?

The velocity is $v = \sqrt{\dfrac{2K_e}{M_e}} \approx \sqrt{\dfrac{2(10 \times 10^3 \text{ eV})}{(9.1 \times 10^{-31} \text{ kg})} \left(\dfrac{1.6 \times 10^{-19}}{\text{eV}} \right)} \approx 5.9 \times 10^7 \dfrac{\text{m}}{\text{s}}$. Because this value is enough less than the speed of light, we can use classical equations.

The radius is $R = \dfrac{M_e v}{q_e B} \approx \dfrac{(9.1 \times 10^{-31} \text{ kg}) \left(5.9 \times 10^7 \dfrac{\text{m}}{\text{s}} \right)}{(1.6 \times 10^{-19})(0.2 \text{ T})} \approx 1.7 \text{ mm}.$

The period is $t_e = \dfrac{2\pi R}{v} \approx \dfrac{2\pi(1.7 \times 10^{-3} \text{ m})}{5.9 \times 10^7 \dfrac{\text{m}}{\text{s}}} \approx 0.2 \text{ ns}.$

The radiated power is

$$P = \dfrac{q_e v^3}{3\varepsilon_0 c^3 R} J \approx \dfrac{(1.6 \times 10^{-19} \text{ Coul}) \left(5.9 \times 10^7 \dfrac{\text{m}}{\text{s}} \right)^3}{3 \left(8.85 \times 10^{-12} \dfrac{\text{Coul}^2}{\text{N} \cdot \text{m}^2} \right) \left(3 \times 10^8 \dfrac{\text{m}}{\text{s}} \right)^3 (1.7 \times 10^{-3} \text{ m})}$$

$$\times \left(10^{-3} \dfrac{\text{Coul}}{\text{s}} \right) \approx 3 \times 10^{-11} \text{ W}.$$

The power is much too small to be useful.

7.2 Semiclassical analysis

To obtain significant radiation, it is necessary to have much more energetic electrons. For relativistic velocities, it is still the case that

$$R = \frac{p_e}{q_e B},$$

(7-13)

as before, but now

$$p_e = \gamma M_e v \approx \gamma M_e c,$$

(7-14)

where γ is the Lorentz factor

$$\gamma = \frac{1}{\sqrt{1 - \frac{v^2}{c^2}}}.$$

(7-15)

The relationship between the total electron energy and momentum is

$$U_e^2 = U_{eo}^2 + p_e^2 c^2,$$

(7-16)

where U_{eo} is the rest mass of the electron,

$$U_{eo} = M_e c^2 \approx 0.51\,\text{MeV},$$

(7-17)

so that

$$U_e = \sqrt{U_{eo}^2 + \gamma^2 M_e^2 v^2 c^2} = U_{eo} \sqrt{\left(1 + \gamma^2 \frac{v^2}{c^2}\right)}$$

$$= U_{eo} \sqrt{\left(1 + \frac{\frac{v^2}{c^2}}{\left(1 - \frac{v^2}{c^2}\right)}\right)}.$$

$$= \gamma U_{eo}$$

(7-18)

The kinetic energy is

$$K = U_e - U_{eo}.$$

(7-19)

Now the radius, which is

$$R = \frac{p_e}{q_e B} \approx \gamma \frac{M_e c}{q_e B},$$

(7-20)

can be expressed in terms of a ratio of electron energy to magnetic field,

$$R \approx \gamma \frac{M_e c}{q_e B} = \frac{U_e}{U_{eo}} \frac{M_e c^2}{q_e B c} = \frac{1}{q_e c} \frac{U_e}{B} \approx \left(3.3 \frac{\text{T·m}}{\text{GeV}}\right) \frac{U_e}{B}.$$

(7-21)

Because of the high magnetic fields, large radii, and large radiation losses, synchrotrons for gigaelectronvolt energy beams are not circular paths around a single magnet but consist of long straight pieces with bending magnets spaced at the ends, so that the beam path is more polygonal than circular. The radiation is emitted from small regions near the bending magnets, as shown in Figure 7-6. Accelerators designed specifically to produce synchrotron radiation (as opposed to using radiation from particle beam experiments) are known as second generation light sources, such as the original National Light Source (NLS) at Brookhaven National Labs in New York.

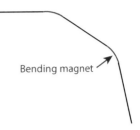

Bending magnet

FIGURE 7-6. A beam path in a synchrotron consists of curved arcs at the bending magnets, which bend the beam a few degrees, separated by long straight pieces. There is no synchrotron radiation from the straight paths.

EXAMPLE 7-2

Electrons with a kinetic energy of 1 MeV are traveling in a magnetic field of 0.2 T. What are the total energy of the electrons, the velocity of the electrons, the relativistic factor γ, and the radius of the orbit?

The total energy is $U_e = K + U_{eo} \approx 1\text{ MeV} + 0.51\text{ MeV} \approx 1.5\text{ MeV}$.

The relativistic factor is then $\gamma = \dfrac{U_e}{U_{eo}} \approx \dfrac{1.51\text{ MeV}}{0.51\text{ MeV}} \approx 3.$

The velocity is $v = c\sqrt{1 - \dfrac{1}{\gamma^2}} \approx \left(3 \times 10^8\,\dfrac{\text{m}}{\text{s}}\right)\sqrt{1 - \dfrac{1}{3^2}} \approx 2.8 \times 10^8\,\dfrac{\text{m}}{\text{s}}.$

The radius is

$$R = \gamma\frac{M_e v}{q_e B} \approx (3)\frac{(9.1 \times 10^{-31}\text{ kg})\left(2.8 \times 10^8\,\dfrac{\text{m}}{\text{s}}\right)}{(1.6 \times 10^{-19})(0.2\text{ T})} \approx 2.4\text{ cm}.$$

This value was estimated as

$$R \approx \left(3.3\,\frac{\text{T} \cdot \text{m}}{\text{GeV}}\right)\frac{U}{B} \approx \left(3.3\,\frac{\text{T m}}{\text{GeV}}\right)\frac{1.5 \times 10^{-3}\text{ GeV}}{0.2\text{ T}} \approx 2.5\text{ cm}.$$

The approximation is better for higher γ.

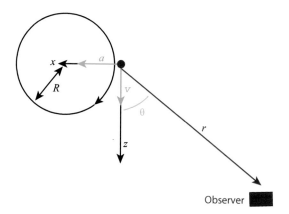

FIGURE 7-7. An observer of the synchrotron radiation at a location r.

7.3 Relativistic bremsstrahlung

For these relativistic electrons we must now consider the full radiative term in equation 6-10,

$$\vec{E}=\frac{q_e}{4\pi\varepsilon_o c}\frac{\hat{r}\times\left\{\left(\hat{r}-\dfrac{\vec{v}}{c}\right)\times\dfrac{\vec{a}}{c}\right\}}{r\left(1-\dfrac{\vec{v}\cdot\hat{r}}{c}\right)^3}. \tag{7-22}$$

For convenience, let the electron be instantaneously moving in the z direction, with its acceleration in the x direction, as shown in Figure 7-7. The observer is at a distance r, at an angle θ from the z axis, as shown in Figure 7-7. The term in parentheses in the denominator of the field equation is

$$1-\frac{\vec{v}\cdot\hat{r}}{c}=1-\frac{v}{c}\cos\theta. \tag{7-23}$$

For large γ, v/c approaching 1, the denominator is a minimum for small θ. The field is therefore a maximum near $\theta=0$. In that case,

$$1-\frac{\vec{v}\cdot\hat{r}}{c}=1-\cos\theta\sqrt{1-\frac{1}{\gamma^2}}\approx1-\cos\theta\left(1-\frac{1}{2\gamma^2}\right)$$

$$\approx1-\left(1-\frac{\theta^2}{2}\right)\left(1-\frac{1}{2\gamma^2}\right)\approx\frac{1}{2}\left(\theta^2+\frac{1}{\gamma^2}\right). \tag{7-24}$$

The term doubles, and the field falls off by a factor of 8, if θ is as large as about $1/\gamma$. Thus the intensity dipole becomes narrowly focused in the forward direction, as shown in Figure 7-8. In other words, there is relativistic length contraction between the lab frame and the electron frame, which causes the radiation to be forward directed within a cone angle of $1/\gamma$, since the vector which is at 45° in the electron frame is at 45°/γ in the lab frame. Instead of radiating in all directions, the radiation acts like a narrow flashlight beam, shining a burst of light toward the observer for a brief time as the cone swings past the direction of observation, as shown in Figure 7-9.

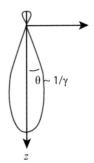

F<small>IGURE</small> 7-8. A polar plot of the dipole radiation Lorentz contracted when the electron has large γ. The "doughnut" of Figure 7-4 has become very elongated in the direction of the electron velocity.

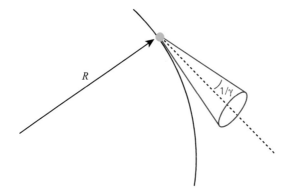

F<small>IGURE</small> 7-9. A relativistic electron traveling on a circular path and the resulting narrow cone of radiation, of half width $1/\gamma$.

At $\theta=0$, the direction of observation is parallel to the velocity (in the z direction), and the numerator in the field becomes

$$\hat{r}\times\left\{\left(\hat{r}-\frac{\vec{v}}{c}\right)\times\frac{\vec{a}}{c}\right\}=\hat{z}\times\left\{\left(1-\frac{v}{c}\right)\hat{z}\times\frac{a}{c}\hat{x}\right\}=-\left(1-\frac{v}{c}\right)\frac{a}{c}\hat{x}. \tag{7-25}$$

The field is linearly polarized in the plane of the electron orbit. If the observation direction is slightly out of the plane, the beam is elliptically polarized.

Thus in the electron beam direction ($\theta=0$),

$$\begin{aligned}\vec{E}&=\frac{-q_e}{4\pi\varepsilon_o c}\frac{\left(1-\frac{v}{c}\right)\frac{a}{c}\hat{x}}{r\left(1-\frac{v}{c}\right)^3}=\frac{-q_e a\hat{x}}{4\pi\varepsilon_o c^2 r}\frac{1}{\left(1-\frac{v}{c}\right)^2}\\[2mm]&=\frac{-q_e a\hat{x}}{4\pi\varepsilon_o c^2 r}\frac{1}{\left(1-\sqrt{1-\frac{1}{\gamma^2}}\right)^2}\approx\frac{-q_e\gamma^4 a}{\pi\varepsilon_o c^2 r}\hat{x}.\end{aligned} \tag{7-26}$$

The full computation of the radiated power in the relativistic case is somewhat complex, and is detailed in the further reading the end of the chapter. In computing the radiated intensity from the square of the field, it is necessary to consider the contraction of the acceleration and the time interval elapsed at the observer, rather than that of the electron, and to integrate over observer locations. Further, the calculation for the field is for a single electron, so an expression for the current must be computed, as was done for the classical case in equation 7-11.

In the classical limit, the power per radian of arc can be found from dividing the expression in equation 7-12 by 2π,

$$P_{\chi, class} = \frac{q_e}{6\pi\varepsilon_o R}\left(\frac{v}{c}\right)^3 J. \tag{7-27}$$

For the relativistic case, v/c is approximately unity, but factors of γ arise from the transformation of the field and time. The radiated electromagnetic power per linear angle of circular arc is

$$P_{\chi} = \frac{q_e}{6\pi\varepsilon_o R}\gamma^4 J, \tag{7-28}$$

which is proportional to γ^4. The power can be written

$$P_{\chi} = \frac{hc\alpha}{3\pi q_e}\gamma^4\frac{J}{R} = \left(14.1\frac{W\cdot m}{A\,(mrad)\,GeV^4}\right)\frac{JU^4}{R}, \tag{7-29}$$

where α is the fine-structure constant,

$$\alpha = \frac{q_e^2}{2\varepsilon_o hc}. \tag{7-30}$$

EXAMPLE 7-3

Electrons with a kinetic energy of 3 GeV are traveling in a magnetic field of 1.2 T with a current of 200 mA. Find R, γ, the angular divergence of the beam out of the plane of the orbit, the x-ray power into a detector which subtends 1 mrad in plane of the circular arc, and the angular intensity.

The relativistic factor is

$$\gamma = \frac{U_e}{U_{eo}} = \frac{U_{eo} + K_e}{U_{eo}} \approx \frac{3.0005\,GeV}{0.51\,MeV} \approx 5.9\times10^3.$$

The radius is

$$R \approx \gamma\frac{M_e c}{q_e B} \approx (5.9\times10^3)\frac{(9.1\times10^{-31}\,kg)\left(3\times10^8\,\frac{m}{s}\right)}{(1.6\times10^{-19})(1.2\,T)} \approx 8.3\,m.$$

The power per linear angle is

$$P_{\chi} \approx 14.1\frac{W\cdot m}{A\,(mrad)\,(GeV)^4}\frac{(200\times10^{-3}\,A)(3\,GeV)^4}{8.3\,m} \approx 27\frac{W}{mrad}.$$

The power into the detector is

$$P_{detector} = P_\chi \Delta\chi = \left(27\,\frac{W}{\text{mrad}}\right)(1\,\text{mrad}) \approx 27\,W.$$

The emission cone has a half width of $1/\gamma$, so we consider the full width of $2/\gamma$ in computing angular divergence of the beam out of the plane of the orbit, $\Delta\theta = 2/\gamma \approx 0.34$ mrad.

The angular intensity of the beam is

$$I_\chi \approx \frac{P_\chi}{\Delta\theta} \approx \frac{27\,\dfrac{W}{\text{mrad}}}{0.34\,\text{mrad}} \approx 80\,\frac{W}{\text{mrad}^2}.$$

7.4 Synchrotrons

The power increases as the fourth power of the Lorentz factor γ, so high electron energies are essential for a cyclotron to become a radiation source. A problem that develops as the electron is accelerated by the cyclotron up to relativistic speeds is that the orbital period is time dilated, so the cyclotron frequency is compressed. The required frequency then becomes

$$f_e = \frac{v}{2\pi R} = \frac{v}{2\pi\left(\dfrac{\gamma M_e v}{q_e B}\right)} = \frac{1}{2\pi}\frac{q_e B}{\gamma M_e}, \tag{7-31}$$

which changes with γ as the speed is increased. To achieve high electron energies, it is necessary to change either the frequency of the electric accelerating field or the magnetic field B in the cyclotron to keep the acceleration synchronous with the electron orbit. Cyclotrons with this feature are called *synchrotrons*. Synchrotron radiation was first observed at the General Electric Research Lab in Schenectady, New York, in 1947.

7.5 Pulse time and spectrum

In the same way that a beam from a lighthouse appears to flash on and off when observed from far away, the synchrotron beam will appear pulsed when seen by a fixed observer. The distance in Figure 7-10 from A to B traveled by the electron in the lab frame while the beam is pointed at the observer is

$$dz = R\frac{2}{\gamma}, \tag{7-32}$$

which takes the electron a time

$$t_e = \frac{2R}{v\gamma}. \tag{7-33}$$

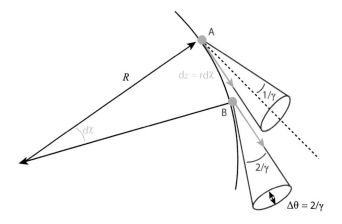

FIGURE 7-10. An electron sweeping out an arc $d\chi$. An observer in the direction marked by the arrows will see the radiation for the time that the electron travels from point A (for which the direction is at one edge of the cone) to B (for which the direction is at the other edge, an angular distance of $2/\gamma$).

However, that is not the apparent time of the flash, because a photon leaving from B has less distance to travel toward the observer than did the photon emitted from A, saving the light a time

$$t_{ph} \approx \frac{2R}{c\gamma}. \tag{7-34}$$

The pulse duration for the observer is then the time difference

$$\Delta t = t_e - t_{ph} \approx \frac{2R}{v\gamma} - \frac{2R}{c\gamma} = \frac{2R}{v\gamma}\left(1 - \frac{v}{c}\right) = \frac{2R}{v\gamma}\left(1 - \sqrt{1 - \frac{1}{\gamma^2}}\right)$$

$$\approx \frac{R}{v\gamma^3} \approx \frac{R}{c\gamma^3}. \tag{7-35}$$

As was seen in sections 4.6 and 5.9, a short time duration produces a frequency broadening. The very short pulse duration of the synchrotron radiation causes a wide frequency broadening, out to approximately a frequency of

$$v_{max} = \frac{1}{\Delta t} \approx \frac{c\gamma^3}{R}. \tag{7-36}$$

The result is approximately white radiation. The maximum frequency is many times higher than the cyclotron frequency of the electron, because the very short pulse generates extremely high harmonics of that fundamental frequency. The characteristic frequency is defined as

$$V_{char} = \frac{3}{4\pi} V_{max}.$$ (7-37)

Half the total power is radiated at frequencies below V_{char}, and half above. The beam brightness is fairly flat for long wavelengths, and falls off rapidly for shorter wavelengths, as shown in Figure 7-11. The power spectrum, plotted in Figure 7-12, is similar to the thick-anode bremsstrahlung spectrum of section 6.6. The total power per angular arc emitted below the characteristic frequency can be very roughly approximated as if that half of the spectrum were a rectangle, so that

$$\frac{1}{2} P_\chi = \int_0^{V_{char}} P_{\chi v} \, dv \approx V_{char} (P_{\chi v})_{max}.$$ (7-38)

Putting in the formulas for P_χ and V_{char},

$$(P_{\chi v})_{max} \approx \frac{\frac{1}{2} P_\chi}{V_{char}} = \frac{\frac{1}{2}\left(\frac{q_e}{6\pi\varepsilon_0 R} \gamma^4 J \right)}{\left(\frac{3}{4\pi} \frac{c\gamma^3}{R} \right)} = \frac{q_e}{9\varepsilon_0 c} \gamma J.$$ (7-39)

Alternatively, in terms of photon energy,

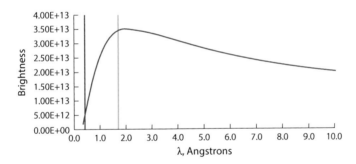

FIGURE 7-11. Brightness plotted versus wavelength for a 3 GeV beam in a 1.2 T magnet, computed from figure 7-13. The light blue line is at the characteristic wavelength, and the red line corresponds to the approximate "maximum" frequency.

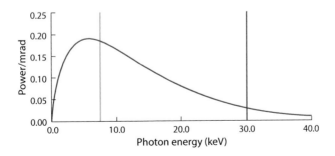

FIGURE 7-12. Power per milliradian into a 0.1 keV energy bin, computed from Figure 7-11. The light blue line is at the characteristic energy, and the red line is at the approximate maximum energy.

$$\frac{1}{2}P_\chi \sim \int_0^{U_{char}} P_{\chi U}\,dU \approx (P_{\chi U})_{max}\,U_{char}$$

$$\Rightarrow (P_{\chi U})_{max} = \frac{\frac{1}{2}P_\chi}{U_{char}} = \frac{q_e}{9\varepsilon_o hc}\gamma \text{ J.} \tag{7-40}$$

More detailed calculations of the spectra and angular dependence are available online, as shown in Figure 7-13. Figures 7-11 and 7-12 were derived from this online calculator. It uses exact solutions of Maxwell's equations that involve modified Bessel functions.

EXAMPLE 7-4

For the beam of example 7-3, calculate the characteristic x-ray photon energy, the power per mrad into the detector for a 0.1 keV energy bin centered on the characteristic energy, and the brightness of the beam at that energy.

The characteristic energy is

$$U_{char} = h\nu_{char} \approx \frac{3}{4\pi}\frac{hc\gamma^3}{R} \approx \frac{3}{4\pi}\frac{12.4\text{ keV}\cdot\text{Å}}{\left(8.3\text{ m}\frac{10^{10}\text{ Å}}{\text{m}}\right)}(5.9\times10^3)^3 \approx 7.2\text{ keV.}$$

The spectrum in Figure 7-13 peaks around this energy. The power per milliradian per keV is

$$(P_{\chi U})_{max} = \frac{\frac{1}{2}P_\chi}{U_{char}} \approx \frac{\left(27\frac{\text{W}}{\text{mrad}}\right)}{2\,(7.2\text{ keV})} \approx \left(2\frac{\text{W}}{\text{mrad}\cdot\text{keV}}\right).$$

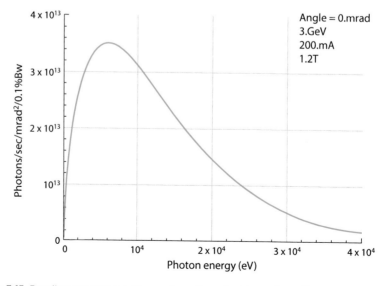

Angle = 0.mrad
3.GeV
200.mA
1.2T

FIGURE 7-13. Bending-magnet brightness, from the calculator at http://henke.lbl.gov/optical _constants/bend2.html.

The power per milliradian into the detector in the desired photon energy bin is then

$$P_{bin} \approx (P_{\chi U})_{max} \, \Delta U \approx \left(2 \, \frac{W}{mrad \cdot keV} \right) (0.1 \, keV) \approx 0.2 \, \frac{W}{mrad}.$$

This result is consistent with the plot in Figure 7-12.

The brightness is computed for an energy bin for a bandwidth of 0.1%. The power per milliradian into that energy bin is $(P_{\chi U})_{max} \, \Delta U_{bw}$. The photon flux requires dividing the power by the photon energy, and finally the brightness requires considering the angular divergence out of the plane, so that the brightness is

$$\Psi_{\chi Bndwth} = \frac{\left((P_{\chi U})_{max} / \Delta\theta \right) \Delta U_{bw}}{U} = \frac{(P_{\chi U})_{max}}{\Delta\theta} \frac{\Delta U_{bw}}{U}$$

$$\approx \left(2 \, \frac{W}{mrad \cdot keV} \, \frac{10^{-3} \, keV}{eV} \, \frac{eV}{1.6 \times 10^{-19} \, J} \right) \frac{1}{0.34 \, mrad} (0.001)$$

$$\approx 3.5 \times 10^{13} \, \frac{photons}{s \cdot mrad^2 \, (0.1\% \, bw)},$$

which is consistent with Figure 7-13. This brightness is about 8 orders of magnitude higher than for the characteristic emission seen in section 5.3.

7.6 Insertion devices

In addition to the bending magnets at the corners of the straight beam pipes, required to bring the beam around in a rough circle, magnets can be inserted into the straight parts of the path to produce additional radiation. A *wiggler* is a set of magnets designed to move the beam in successive arcs, as shown in Figure 7-14. The device consists of N_{mag} bending magnets spaced closely together. Such insertion devices are characterized in terms of a deflection parameter which can be estimated from the force supplied by the magnet, as shown in Figure 7-15. The magnetic field is assumed to be sinusoidal,

$$\vec{B} = B_o \hat{y} \sin\left(\frac{2\pi}{d} z \right), \tag{7-41}$$

where d is the period of the wiggler magnet array in the lab frame. The transverse deflection of the electron beam is also sinusoidal,

$$x = x_o \sin\left(\frac{2\pi}{d} z \right). \tag{7-42}$$

In the lab frame,

$$z \approx ct, \tag{7-43}$$

so that

$$x \approx x_o \sin\left(\frac{2\pi}{d} ct \right) \tag{7-44}$$

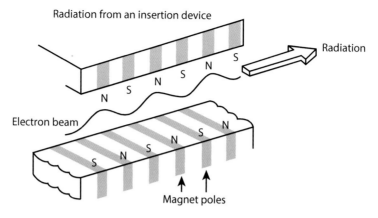

FIGURE 7-14. Wiggler device. From S. L. Hulbert and G. P. Williams, Synchrotron Sources, chap. 55, *Handbook of Optics*, 3rd ed., vol. 5, McGraw-Hill, 2010.

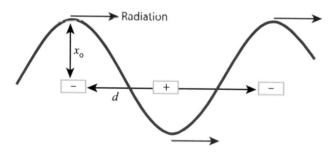

FIGURE 7-15. Sinusoidal path of the electron past a sinusoidal magnetic field represented by magnets with their fields into and out of the page. Radiation is emitted from the regions of highest curvature.

and

$$v_\perp \approx \frac{2\pi c}{d} x_o \cos\left(\frac{2\pi}{d} ct\right). \tag{7-45}$$

The transverse momentum is then

$$p_\perp \approx \gamma M_e \frac{2\pi c}{d} x_o \cos\left(\frac{2\pi}{d} ct\right). \tag{7-46}$$

Setting the change in momentum equal to the force gives

$$|F| \approx q_e B_o c \sin\left(\frac{2\pi}{d} ct\right) = \left|\frac{dp_\perp}{dt}\right| \approx \gamma M_e \left(\frac{2\pi c}{d}\right)^2 x_o \sin\left(\frac{2\pi}{d} ct\right). \tag{7-47}$$

Solving for the maximum deflection of the electron gives

$$x_o = \frac{q_e B_o c}{\gamma M_e}\left(\frac{d}{2\pi c}\right)^2 = \frac{q_e B_o d^2}{4\pi^2 \gamma M_e c} = \frac{d}{2\pi} \frac{K_{insert}}{\gamma}, \tag{7-48}$$

where the deflection parameter for the insertion device, K_{insert}, is defined as

$$K_{insert} = \frac{q_e B_o d}{2\pi M_e c}.$$ (7-49)

The maximum deflection angle of the electron path from the z axis is

$$\theta \approx \frac{v_{\perp max}}{v} = \frac{K_{insert}}{\gamma}.$$ (7-50)

This value is also the angular width in the deflection plane for radiation from the wiggler.

In general, the emission from successive periods of the wiggler is incoherent, since the fields do not overlap. For that reason the power from the wiggler is proportional to the number of magnets N_{mag}. To evaluate the power from a single wiggler magnet from equation 7-29 it is necessary to estimate the bending radius. Using the acceleration from equation 7-47, and comparing that value with $a = v^2/R$ from equation 7-2, yields

$$a = x_o \left(\frac{2\pi}{d} c\right)^2 = \frac{c^2}{R_u} \Rightarrow R_u \approx \frac{d^2}{4\pi^2 x_o} = \frac{d^2}{4\pi^2}\left(\frac{2\pi}{d}\frac{\gamma}{K_{insert}}\right) = \frac{d}{2\pi}\frac{\gamma}{K_{insert}}.$$ (7-51)

Applying that result to the power per arc from a single magnet with an electron current J yields

$$P_{\chi,1} \approx \frac{q_e}{6\pi\varepsilon_o R}\gamma^4 J = \frac{q_e}{6\pi\varepsilon_o}\left(\frac{2\pi}{d}\frac{K_{insert}}{\gamma}\right)\gamma^4 J = \frac{q_e K_{insert}}{3\varepsilon_o d}\gamma^3 J.$$ (7-52)

The total power for each turn around a magnet then depends on integrating over the angle which contributes to the beam. The angle is proportional to the deflection angle, $\Delta\theta \approx K_{insert}/\gamma$. The angular integral, and a more exact calculation of the radius of deflection, which varies as the electron oscillates, yields an additional factor of $\pi/2$. There are then $2N$ arcs (for a wiggler with N periods), so that the power in the beam for the wiggler

$$P \approx \frac{\pi q_e}{3\varepsilon_o}\frac{JN}{d}\gamma^2 K_{insert}^2. \qquad \textbf{For wigglers}$$ (7-53)

The spectrum is similar to that for an individual magnet.

Wigglers with small deflection parameters, $K \ll 1$, are called *undulators*. In that case the cone angle is small, and the fields from successive magnets overlap, and so can interfere. The total power is then compressed into a narrow beam, so that the peak angular intensity can be N^2 as large as for a single magnet.

The electron has a root mean square (rms) average transverse velocity proportional to the angular deflection,

$$v_{\perp rms} = v\frac{K_{insert}}{\sqrt{2}\gamma},$$ (7-54)

where v is the total electron velocity. The average velocity in the z direction is thus, from the Pythagorean theorem,

$$v_z = \sqrt{v^2 - \left(v\frac{K_{insert}}{\sqrt{2}\gamma}\right)^2} = v\sqrt{1 - \frac{K_{insert}^2}{2\gamma^2}} \approx v\left(1 - \frac{K_{insert}^2}{4\gamma^2}\right). \tag{7-55}$$

The velocity in the z direction is not constant, but oscillates, which gives rise to higher harmonics. The time for the electron to travel to the next magnet is

$$t_e \approx \frac{d}{v\left(1 - \frac{1}{4}\frac{K_{insert}^2}{\gamma^2}\right)} \approx \frac{d}{c}\left(1 + \frac{1}{2}\frac{1}{\gamma^2}\right)\left(1 + \frac{1}{4}\frac{K_{insert}^2}{\gamma^2}\right), \tag{7-56}$$

where v has been approximated in terms of c and γ. The time required for a photon emitted at an angle θ from the z axis to travel from one magnet to the plane of the emission from the next magnet is somewhat shorter,

$$t_{ph} = \frac{d\cos\theta}{c} \approx \frac{d}{c}\left(1 - \frac{\theta^2}{2}\right). \tag{7-57}$$

The distance that the photon travels in the extra time must be an integral number of wavelengths of the light for the light to constructively interfere,

$$m\lambda_m = c(t_e - t_{ph}) \approx d\left[\left(1 + \frac{1}{2}\frac{1}{\gamma^2}\right)\left(1 + \frac{1}{4}\frac{K_{insert}^2}{\gamma^2}\right) - \left(1 - \frac{\theta^2}{2}\right)\right]$$

$$\lambda_m \approx \frac{d}{2\gamma^2 m}\left[1 + \frac{K_{insert}^2}{2} + \gamma^2\theta^2\right]. \tag{7-58}$$

The factor $1/\gamma^2$ can be considered to result from two Lorentz contractions. First, the period of the undulator is d/γ in the electron frame, then the radiation wavelength is Doppler shifted into the lab frame. For radiation observed on axis ($\theta = 0$), with $K_{insert} \ll 1$, the frequency of the radiation is

$$v_m = \frac{c}{\lambda_m} = 2\gamma^2\frac{c}{d}m. \tag{7-59}$$

The spectrum consists of a fundamental frequency and very many higher harmonics. The fundamental frequency can be tuned by changing the spacing between magnets ($d/2$).

The power in the central cone of radiation, which is often more important than the total power, in the case of undulators with small K_{insert} is somewhat complicated. An undulator acts as a temporal grating, so that the frequency width of the mth harmonic, in analogy with the spot size produced by a spatial grating, is approximately

$$\frac{\Delta v_m}{v_m} = \frac{1}{Nm}, \tag{7-60}$$

where N is the number of undulator periods. Similarly, the cone angle of the emission for the central cone is

$$\Delta\theta \approx \frac{1}{\gamma}\sqrt{\frac{1}{N}}. \tag{7-61}$$

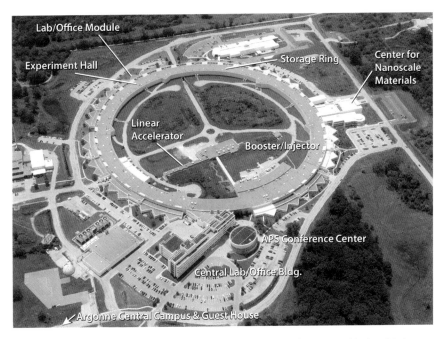

FIGURE 7-16. The Advanced Photon Source, photo courtesy of Argonne National Lab, managed by the University of Chicago, for the DOE.

In practice, the angle and frequency width are broadened by the size and energy distribution of the electron bunch traveling through the device.

To compute the power in the central cone of an undulator it is necessary to take into account the fact that the arc over which radiation is emitted is quite small, so that the solid angle subtended by that cone falls with $1/N$. However, the total number electrons passing through the undulator in the time of travel is given by

$$N_e = \frac{J}{q_e} t = \frac{J}{q_e} \frac{Nd}{c}. \tag{7-62}$$

The two dependences on the number of periods N cancel, so that the power in the central cone is

$$P_{central} \approx \frac{\pi q_e \gamma^2}{\varepsilon_o d} J \frac{K_{insert}^2}{\left(1 + \dfrac{K_{insert}^2}{2}\right)^2}. \qquad \textbf{For undulators} \tag{7-63}$$

Because the beam is very narrowly collimated and very monochromatic, it is very brilliant.

Third-generation sources are designed to have optimized insertion devices. Examples include APS, shown in Figure 7-16, the Advanced Light Source (ALS) at Lawrence Berkeley Labs, the European Synchrotron Research Facility (ESRF), and Spring-8 in Japan.

7.7 Collimation and coherence

We saw in example 7-4 that the beam from a synchrotron has high brightness. Because the source size—the size of the electron bunch moving around the ring—is typically a few tens of microns, it also has high brilliance. While the spot size is no smaller than that of many conventional tubes, an advantage of the high collimation of the synchrotron beam is that the experiment can be performed tens of meters downstream of the magnet or insertion device. This results in very small local divergence and hence high spatial coherence lengths. Fourth-generation sources such as x-ray lasers are designed to have even better coherence.

Problems

SECTION 7.2
1. What is the electron velocity if $\gamma = 3$?
2. Consider a 2000 MeV beam and a magnet with a field of 0.5 T. Find the orbit radius and the relativistic factor γ.

SECTION 7.3–7.5
3. Show that equation 7-28 implies both sides of equation 7-29.
4. Repeat examples 7-3 and 7-4 for a 2000 MeV beam, a magnet with a field of 0.5 T, and a current of 200 mA (with an observation angle of $\theta = 0$).
5. Use the Web resource of Figure 7-13 to check your answers for problem 4.
6. For the beam of problem 5, compute the brilliance, source divergence, and transverse coherence length if the electron bunch size is 50 μm and the experiment is performed at the characteristic energy at a distance of 20 m.

SECTION 7.6
7. An insertion device used with a beam of 2000 MeV electrons at 200 mA is 10 m long. Each magnet has a magnetic field of 0.5 T. Calculate the deflection parameter K_{insert}, the cone angle, and the effective radius when the device has a 1 m period and then when it has a 5 mm period. Note that one is a wiggler, and one is an undulator.

Further reading

David Attwood and A. Sakdinawat, **X-rays and Extreme Ultraviolet Radiation: Principles and Applications**, Cambridge University Press, 2016.

D. Halliday, R. Resnick, and J. Walker, **Fundamentals of Physics**, 10th ed., John Wiley & Sons, 2013, chap 37.

A. G. Michette and C. J. Buckley, **X-ray Science and Technology**, Institute of Physics Publishing, 1993.

Jens Als-Nielsen and Des McMorrow, **Elements of Modern X-ray Physics**, John Wiley & Sons, 2001.

S. L. Hulbert and G. P. Williams, Synchrotron Sources, chap. 55 in M. Bass, C. DeCusatis, J Enoch, V. Lakshminarayanan, G. Li, C. MacDonald, V. N. Mahajan, and E. Van Stryland, eds., **Handbook of Optics**, 3rd ed., vol.5, McGraw-Hill, 2010.

A. Thompson et al., **X-ray Data Booklet**, Lawrence Berkeley National Laboratory, 2009, http://xdb.lbl.gov/.

L. D. Landau and E. M. Lifshitz, **Mechanics and Electrodynamics: A Shorter Course of Theoretical Physics,** vol.1, Pergamon Press, 1972.

8

X-RAY LASERS

X-ray lasers fall into three classes: conventional lasers which radiate at visible wavelengths but whose output wavelength has been drastically shortened by high harmonic generation; conventional lasers using highly ionized plasmas as their gain medium to radiate at soft x-ray wavelengths; and accelerator-based free-electron lasers. To begin the discussion of lasers, the first two sections of this chapter provide a brief introduction to stimulated emission and laser cavities.

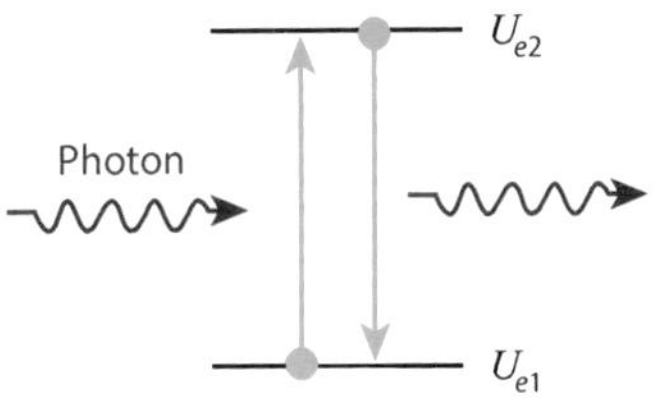

FIGURE 8-1. A two-level "atom" with energy levels U_{e1} and U_{e2}. The vertical arrow on the left corresponds to absorption of the incoming photon. On the right is spontaneous emission of the outgoing photon.

8.1 Stimulated and spontaneous emission

A simple discussion of stimulated emission can begin by considering an "atom" with two electronic levels, as shown in Figure 8-1. A photon with energy

$$U = U_{e2} - U_{e1} \tag{8-1}$$

can be absorbed by the atom, causing an electron to move to the upper level. The electron in the upper level can fall to the lower level, emitting a photon of the same energy U. The rate at which electrons in the upper level will spontaneously fall and cause the emission of photons clearly depends on the number of electrons in the upper level, N_2

$$\Gamma_{\downarrow, spon} = AN_2. \tag{8-2}$$

The proportionality constant is known as the Einstein A coefficient. The rate of absorption depends on the number of electrons in the lower level, N_1, and the rate of arrival of photons of the appropriate energy Γ_U,

$$\Gamma_{\uparrow} \propto \Gamma_U N_1. \tag{8-3}$$

Since for a monochromatic beam the photon arrival rate is proportional to the intensity I_U, this rate can be written

$$\Gamma_\uparrow = BI_U N_1. \tag{8-4}$$

The proportionality constant is known as the Einstein B coefficient. In thermal equilibrium the ratio of the population of electrons in the two energy levels is given by the Boltzmann relationship (we used this relationship before, for equation 3-14),

$$N_2 = N_1 e^{-\frac{U}{k_B T}}. \tag{8-5}$$

In thermal equilibrium the rate of emission should be equal to the rate of absorption, which makes it possible to solve for the intensity as a function of temperature,

$$\Gamma_\uparrow = BI_U N_1 \overset{?}{=} \Gamma_{\downarrow, spon} = AN_2 \Rightarrow I_U \overset{?}{=} \frac{AN_2}{BN_1} = \frac{A}{B} e^{-\frac{U}{k_B T}}, \tag{8-6}$$

where the question mark indicates that the assignment is tentative. The problem is that the computed intensity distribution is incorrect. The intensity should be the blackbody distribution of equation 3-1, and it is not. The assumption that the only emission is the spontaneous emission is flawed. There is an additional emission term which is proportional to the number of photons already present to "stimulate" further emission,

$$\Gamma_{\downarrow, stim} = B_2 I_U N_2, \tag{8-7}$$

as shown in Figure 8-2. Then, the condition for equilibrium becomes

$$\Gamma_\uparrow = BI_U N_1 = \Gamma_{\downarrow, spon} + \Gamma_{\downarrow, stim} = AN_2 + B_2 I_U N_2$$

$$\Rightarrow I_U = \frac{AN_2}{BN_1 - B_2 N_2} = \left(\frac{A}{B}\right) \frac{1}{\dfrac{N_1}{N_2} - \dfrac{B_2}{B}} = \frac{A/B}{\left(e^{\frac{U}{k_B T}}\right) - \left(\dfrac{B_2}{B}\right)}. \tag{8-8}$$

By comparison with equation 3-1, which is

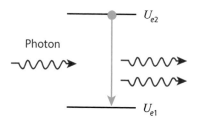

U_{e2}

Photon

U_{e1}

FIGURE 8-2. In stimulated emission, the incoming photon causes the electron to fall out of the upper state, emitting another photon in phase with the first.

$$I_U = \frac{8\pi}{h^3 c^2} \frac{U^3}{\left(e^{\left(\frac{U}{k_B T}\right)} - 1\right)},$$ (8-9)

consistency with the blackbody equation implies that

$$B_2 = B$$ (8-10)

and

$$\frac{A}{B} = \frac{8\pi U^3}{h^3 c^2}. *$$ (8-11)

The equality between the coefficient B for absorption and B_2 for stimulated emission implies a sort of microscopic reversibility for these two processes, both of which depend on the number of photons present. Stimulated emission has another unique and important property. When a photon is emitted by stimulated emission, it is in phase with and in the same direction as the original "stimulating" photon. Thus the stimulated emission will be coherent and well collimated, typical of a laser beam.

Starting from zero intensity, the relative rate of emission compared to absorption is

$$\frac{\Gamma_{\downarrow, spon} + \Gamma_{\downarrow, stim}}{\Gamma_{\uparrow}} = \frac{(A + B(0))N_2}{AN_1} = \frac{N_2}{N_1}.$$ (8-12)

Normally, the emission rate is less than the absorption rate because in thermal equilibrium, $N_2 \ll N_1$. The emission rate will become larger than the absorption rate only if the population is inverted. Population inversion is achieved in a laser gain medium by exciting electrons up to the higher level, using electric fields, light, or collisions with hot atoms.

The relationship between A and B gives the relative rates of spontaneous and stimulated emission,

$$\frac{\Gamma_{\downarrow, spon}}{\Gamma_{\downarrow, stim}} = \frac{AN_2}{BI_U N_2} = e^{\left(\frac{U}{k_B T}\right)} - 1.$$ (8-13)

This implies that even if the population is inverted, most of the emission will be random spontaneous events, not the coherent stimulated emission required for lasers (light amplification by the stimulated emission of radiation). For conventional lasers, this problem is solved by using a three-level system, as shown in Figure 8-3. Electrons are excited up to an unstable highest level at energy U_{e3} by some pumping mechanism. They fall to the middle level. However, not all transitions from upper energy levels

*Some care with units is necessary here. For blackbody radiation, the quantity is the energy emitted per frequency bin, per second, per surface area of the hot object. Here for absorption, the intensity is taken as the arrival rate of photons within the frequency line width onto the single two-level atom. The actual coefficients A and B are thus proportional to the cross section for absorption, discussed in section 9.1, which falls rapidly with photon energy—another problem with creating an x-ray laser.

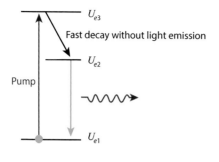

FIGURE 8-3. A three-level "atom" with emission photon energy $U = U_{e2} - U_{e1}$. The rapid decay of level 3 before spontaneous emission can occur results in a high, "inverted" population of level 2. In an alternative scheme, the fast nonradiative decay may happen between the lower two levels, similarly resulting in a population inversion between levels 3 and 2 (since 2 is rapidly emptied).

FIGURE 8-4. A visible-light laser cavity. The increasing thickness of the arrows from top to bottom is meant to illustrate increasing intensity after successive passes through the gain medium.

result in photon emission. In particular, atoms excited to the upper state must have some way of rearranging energy internally by some fast nonradiative mechanism, for example by emission of a phonon, a vibration of the lattice. In that case the atom rapidly falls to the metastable second level with energy U_{e2}. If the decay from 3 to 2 is fast enough, it can happen before the electrons which were pumped to level 3 can spontaneously fall back to level 1. If level 2 has a long lifetime (i.e., a slow rate of spontaneous decay to level 1), population inversion can be obtained, and any spontaneous decay will cause an intense pulse of stimulated emission. Finding atoms that have a fairly long-lived metastable level reachable by a nonradiative process can be challenging.

8.2 Laser cavities

Another problem with x-ray lasers is providing a laser cavity. For most lasers, the amount of population inversion afforded by the gain medium is not high, so that a single passage of the beam through the medium does not provide much amplification of the original light. This problem is generally solved by placing mirrors on either side of the gain medium, so that most of the light is reflected back through the medium, being repeatedly amplified until a high intensity is reached, as shown in Figure 8-4. To keep the repeated amplification in phase, the length of the cavity must be tuned to be an integral multiple of the laser wavelength. Maintaining this alignment is, of course, more difficult for x-ray wavelengths. Even more problematic is that mirrors which operate at normal (90°) incidence, as in Figure 8-4, do not exist for hard x rays (reflectivity is discussed in chapter 11, and reflective optics, in chapter 12). This obstacle can be circumvented by creating a ring of multiple optics to return the radiation to the cavity, generally including some diffractive optics which can operate at large angles to the surface (diffractive optics are described in chapter 15), but this adds a degree of complexity. Instead, some free-electron x-ray lasers are designed to have high gain in order to operate in a single pass, with no cavity. The mechanism is referred to as *self-amplified spontaneous emission*, or SASE.

8.3 Highly ionized plasmas

A number of suitable gain media can be found for visible light lasers, for example, the Ne gas in HeNe lasers, which is "pumped" by collisions with the hot He gas; the Cr impurity levels in a ruby laser, which are pumped by a flash lamp; or the electronic states in diode lasers, which are pumped by electric currents. Finding gain media is more challenging for x-ray lasers, as both the unstable and metastable levels depicted in Figure 8-3 must be at very high energies. As was seen in chapter 3, fairly high energies can be accessible in highly ionized plasmas. Conventional lasers using hot plasmas as gain media are successful up to EUV energies, with photon energies of a few hundred electronvolts.*

8.4 High-harmonic generation

Because of the issues involved in attempting to find suitable gain media and optics for x-ray lasers, an alternative scheme for producing laser beams at x-ray energies is to start with powerful conventional lasers and use nonlinear optical techniques to create harmonics of the original laser frequency. When an electric field E enters a material, it creates a displacement field, usually written as

$$\vec{D} = \varepsilon \vec{E}. \tag{8-14}$$

However, this linear relationship is actually an approximation, and at high enough fields nonlinear terms become important,

$$D = \varepsilon_1 E + \varepsilon_2 E^2 + \cdots + \varepsilon_m E^m + \cdots, \tag{8-15}$$

which for simplicity has been written as a scalar equation. Because the second-order term depends on the square of the field,

$$D_2 = \varepsilon_2 E^2 = \varepsilon_2 (E_o e^{i(\kappa r_1 - \omega t)})^2 = \varepsilon_2 E_o^2\, e^{i((2\kappa)r_1 - (2\omega)t)}, \tag{8-16}$$

that term describes a field which oscillates at twice the original frequency. This effect is known as *frequency doubling*. The amount of harmonic generation depends strongly on the field strength. The field at the second harmonic frequency depends on the square of the incoming field, so the intensity of the second harmonic depends on the square of the incident intensity. From a quantum perspective, frequency doubling can be viewed as the nearly simultaneous absorption of two photons, as shown in Figure 8-5. Absorption of a photon causes the electron energy to increase. As we will see again in chapter 9, this energy fluctuation can be sketched as a "virtual" level. As was seen in section 4.6, a large energy fluctuation must have a short lifetime, which means the electron will quickly fall back to the original level. For second harmonic generation, the second photon must be absorbed before the electron can fall out of the short-lived

---

*Care should be taken in terminology: chapter 3 discussed *laser-generated* plasmas, while this chapter is discussing lasing *of* hot plasmas (which are often, in fact, generated by visible light lasers, although they can be created by electrical discharge).

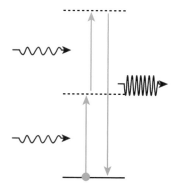

FIGURE 8-5. Second-harmonic generation by the absorption of two photons. The dotted lines represent "virtual" electron levels—temporary fluctuations of the electron energy level.

virtual level, which requires a high photon density in the incoming beam. The mth harmonic requires the nearly simultaneous absorption of m photons. To have the necessary extremely high incident intensities (with a reasonable total energy input), very short pulses are required. High-harmonic generation is typically performed with picosecond (10^{-12} s) or femtosecond (10^{-15} s) laser pulses.

The nonlinear coefficients ε_m are zero for materials with high symmetry, but nonzero for certain classes of asymmetric crystals. Even in those crystals the coefficients are small and initially decrease rapidly with m. However, for odd harmonics a plateau is reached in which the strength of the nonlinear coefficient becomes independent of the order m. This allows for generation of very high harmonics, up to a few hundred electronvolts, in the EUV or soft x-ray regime.

An alternative view is to note that the electrons are accelerated by the laser beam, and, as was seen in section 6.2, accelerated charges radiate. The maximum photon energy that can be achieved is thus limited to the order of magnitude of the kinetic energy of the oscillating electron. Per the discussion of equations 3-7 and 3-8 (and as will be seen in section 11.1), the amplitude of the electron oscillating in a field at frequency ω, for $x = x_o \cos(\omega t)$, is

$$x_o = \frac{q_e E_o}{M_e \omega^2}, \qquad (8\text{-}17)$$

where ω is the frequency of the incident laser beam. The kinetic energy of the electron is then

$$K_e = \frac{1}{2} M_e \langle v^2 \rangle = \frac{1}{2} M_e \frac{1}{2} (x_o \omega)^2 = \frac{1}{4} M_e \left(\frac{q_e E_o}{M_e \omega^2} \right)^2 \omega^2 = \frac{q_e^2}{4 M_e} \frac{E_o^2}{\omega^2}. \qquad (8\text{-}18)$$

EXAMPLE 8-1

Higher harmonic generation is performed with visible light of photon energy 3 eV, with a laser pulse energy of 10 mJ of 100 fs duration, focused to a diameter of 20 μm. What is the approximate maximum possible generated photon energy?

We need to find the field, which we can get from the intensity. The intensity is

$$I = \frac{U_{pulse}}{\pi \left(\frac{W}{2} \right)^2 t} = \frac{10 \times 10^{-3}\,\text{J}}{\pi \left(\frac{20 \times 10^{-6}\,\text{m}}{2} \right)^2 (100 \times 10^{-15}\,\text{s})} = 3.2 \times 10^{20} \, \frac{\text{W}}{\text{m}^2}.$$

The field is thus

$$I = \frac{1}{2} c \varepsilon_o E^2$$

$$\Rightarrow E = \sqrt{\frac{2I}{c\varepsilon_o}} \approx \sqrt{\frac{2\left(3.2 \times 10^{20} \frac{W}{m^2}\right)}{\left(3 \times 10^8 \frac{m}{s}\right)\left(8.85 \times 10^{-12} \frac{Coul^2}{N \cdot m^2}\right)}} \approx 4.9 \times 10^{11} \frac{N}{Coul}.$$

We also need the laser frequency:

$$\omega = \frac{U}{\hbar} \approx \frac{(3\,eV)\left(\frac{1.6 \times 10^{-19}\,J}{eV}\right)}{1.05 \times 10^{-34}\,keV} \approx 4.6 \times 10^{15} \frac{rad}{s}.$$

Finally, the kinetic energy of the electron oscillating in the laser beam, and hence the approximate maximum generated photon energy, is

$$K_e = \frac{q_e^2}{4M_e} \frac{E^2}{\omega^2} \approx \frac{(1.6 \times 10^{-19}\,Coul)^2}{4(9.1 \times 10^{-31}\,kg)} \frac{\left(4.9 \times 10^{11} \frac{N}{Coul}\right)^2}{\left(4.6 \times 10^{15} \frac{rad}{s}\right)^2} \left(\frac{eV}{1.6 \times 10^{-19}\,J}\right)$$

$$\approx 0.5\,keV.$$

Thus this laser beam can in principle produce soft x-ray emission.

8.5 Free-electron lasers

Free-electron lasers (FELs) use an entirely different mechanism. Electrons traveling down undulators, described in section 7.6, are the gain medium. As noted in that section, the emission from each turn of the undulator path overlaps and adds coherently. However, in a normal undulator, the radiation from different individual electrons is not in phase and adds incoherently, so that the x-ray intensity is simply proportional to the number of electrons N_e circulating in the ring (and hence is proportional to the electron current). If the electron bunches are sufficiently small, *microbunched* to a diameter less than the wavelength of the x-ray radiation, the emission from separate electrons can add in phase. In that case the x-ray intensity can rise as N_e^2. The microbunching is the result of a feedback interaction between the emitted radiation and the electron beam, and occurs if the undulator is long enough (the required length is tens or even hundreds of meters). The force which creates the electron bunching arises from the emitted radiation itself, not an external field, but the process is similar to that shown in Figure 7-5 in that the slower electrons receive more acceleration than the faster ones. In the free-electron laser, the electrons with the largest amplitude of oscillation away from the beam axis have the slowest longitudinal velocity but absorb the most energy from the radiation beam. Conversely, the fastest electrons radiate more energy to the beam and reduce their velocity. Once the microbunch begins to form, the radiation becomes more intense as a result of the constructive interference, and the bunching becomes more pronounced. Eventually, if the undulator is long enough, all the electrons become part of microbunches, and there is no further gain. Because the electrons must stay within the highly collimated x-ray beam for tens or even hundreds of meters, the electron beam must be small and have very low divergence. This narrow electron beam is created by a linear accelerator. There is no "recycling" of the beam, as in the circular path of the synchrotron of chapter 7, as the exiting beam would have too large a divergence to be reused. Another issue which needs to be circumvented to create intensity gain is that the electrons are traveling slightly slower than the speed of light and in an oscillatory, longer path than the x-ray beam, so the phase of the electron microbunch tends to slip with respect to the x-ray beam. This phase mismatch is resolved by designing the undulator so that the electron phase slips by one x-ray wavelength for each undulator magnet period,

$$v_z t = d - \lambda, \tag{8-19}$$

where λ is the wavelength of the x rays, d is the undulator period, and t is the time required for the x-ray beam to traverse one period,

$$t = \frac{d}{c}. \tag{8-20}$$

Substituting the average longitudinal velocity of the electron in the undulator from equation 7-55 yields

$$v_z = c \left(1 - \frac{\lambda}{d} \right) \approx v \left(1 - \frac{K_{insert}^2}{4\gamma^2} \right). \tag{8-21}$$

Using

$$v = c \sqrt{1 - \frac{1}{\gamma^2}} \approx c \left(1 - \frac{1}{2\gamma^2} \right), \tag{8-22}$$

and dropping terms of order $1/\gamma^4$, gives the required Lorentz factor (electron kinetic energy) for resonance in the FEL,

$$\gamma = \sqrt{\frac{d}{2\lambda} \left(1 + \frac{K_{insert}^2}{2} \right)}. \tag{8-23}$$

This equation also gives the result that the x-ray emission wavelength can be tuned by changing the electron beam energy. The resonant wavelength is the same as the fundamental wavelength of the undulator. At resonance the absorption balances the emission, and the net gain is zero. If the electron kinetic energy is just slightly higher than resonance, there is more emission from the electrons than absorption and net gain. The energy must be close to resonance, however, to maintain the phase matching.

Low-gain FELs require multiple passes, and hence a laser cavity which generally comprises several optics. High-gain FELs are single-pass devices and so rely on

<small>Figure</small> 8-6. The SLAC FEL, courtesy of the SLAC National Accelerator Laboratory.

self-amplified spontaneous emission. Because SASE is a stochastic process, it results in a noisy start-up to the laser beam. Instead, the FEL can be "seeded" by introducing coherent radiation from a conventional x-ray laser (as in sections 8.3 or 8.4), or even some filtered radiation from the FEL itself.

The free-electron laser at the Stanford Linear Accelerator is shown in Figure 8-6. The short pulse and high brilliance of the x-ray laser beam was first used for protein structure measurements. Currently, free-electron lasers are limited to soft x rays, with wavelengths of tens of nanometers.

8.6 Novel sources

Other sources of x-ray radiation include channeling radiation, parametric radiation, and transition radiation. Channeling radiation occurs when a high-velocity electron beam moves through a crystal. The field of the regular array of atoms deflects the beam much like a miniature undulator. Parametric radiation occurs when a relativistic beam encounters Bragg diffraction (diffraction is described in chapter 13). Transition radiation results when a relativistic beam encounters an abrupt change in the index of refraction of the material. Inverse Compton sources are described in section 10.3.

Problems

SECTION 8.1

1. What is the relative rate of stimulated to spontaneous emission at 100 eV for a plasma in equilibrium at 500,000 K?

SECTION 8.5

2. Consider an undulator with a small insertion parameter, to be used to build a free-electron laser. What is the resonant wavelength of an undulator with $d = 15$ mm and $\gamma = 3000$?

Further reading

David Attwood and A. Sakdinawat, **X-rays and Extreme Ultraviolet Radiation: Principles and Applications**, Cambridge University Press, 2016.

http://photon-science.desy.de/facilities/flash/index_eng.html.

Z. Huang and K. Kim, Review of x-ray free-electron laser theory, *Physical Review Accelerators and Beams* (2007): doi:10.034801.

PART III
X-RAY INTERACTIONS WITH MATTER

9

PHOTOELECTRIC ABSORPTION, ABSORPTION SPECTROSCOPY, IMAGING, AND DETECTION

9.1 Absorption coefficients

When a beam of light or x rays is incident on a material, photons can be absorbed, causing electrons to be emitted. From the point of view of the electron, this process is called *photoelectric emission*. From the point of view of the photon, it is termed *photoelectric absorption*. This knock-out of an electron is the same process discussed in section 4.7 on x-ray fluorescence, although in that case the emphasis was on what happened next if the emitted electron came from a core level. Absorption can be regarded as causing a fluctuation of the electron energy up to a virtual level, as shown in Figure 9-1. For an electron to be emitted, the incoming photon must have an energy greater than the binding energy of the elec-

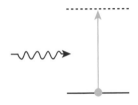

FIGURE 9-1. Absorption of a photon can be regarded as the fluctuation of the electron energy up to a "virtual" electron level.

tron, but it is not necessary that the *incoming* photon energy match a difference in electron energy levels, as is the case for the *emitted* photon in x-ray fluorescence, illustrated in Figure 4-3.

The rate at which photons will be absorbed by a single atom is given in terms of a cross section σ_{ab} for absorption,

$$\Gamma_{ab,1} = \sigma_{ab}\Psi, \qquad (9\text{-}1)$$

where $\Gamma_{ab,1}$ is the absorption rate for a single atom, and Ψ is the photon intensity, the number of photons per area per second in the beam. For comparison, the rate Γ at which photons would hit a target of area A placed in the beam is

$$\Gamma = A\Psi, \qquad (9\text{-}2)$$

so that the cross section can be seen to have units of area. Cross sections are described in units of *barns*, where 1 barn $= 10^{-24}$ cm². (The unit is a reference to an old saying referring to someone with poor aim as not being able to hit "the broad side of a barn." A barn is a large cross section in nuclear physics.)

A sense of the dependence of the cross section on photon energy and atomic number can be found from a rough estimate based on a quantum mechanical calculation similar to that of equation 4-6 from section 4.3 on characteristic spectra. According to Fermi's golden rule, the transition probability is proportional to the Hamiltonian matrix element H_{if}, where i and f refer to the initial and final state, respectively,*

$$\sigma_{ab} \propto \left| H_{if} \right|^2. \tag{9-3}$$

The initial state has the bound electron and an incident photon, and the final state has the free electron and no photon. The matrix element H_{if} depends on the integral over the initial bound state of the electron (for x rays generally a K or L shell electron), and so depends on the coulomb energy, which, according to equation 4-2 is proportional to Z^2, so that $H_{if} \propto Z^2$. The Hamiltonian for photon absorption includes the photon annihilation operator, which has a normalization factor of $\dfrac{1}{\sqrt{\omega}} \propto \dfrac{1}{\sqrt{U}}$. The condition that the kinetic energy of the ejected electron be equal to the difference between the photon energy and the electron binding energy introduces a factor of $1/U$ in the matrix element after the integration, so that it is expected that

$$H_{if} \propto Z^2 \frac{1}{\sqrt{U}} \frac{1}{U} \Rightarrow \sigma_{ab} \propto Z^4 \frac{1}{U^3}. \tag{9-4}$$

Empirically, the atomic cross section for photoelectric absorption depends on the photon energy U and the atomic number Z of the material as

$$\sigma_{ab} \approx \sigma_o \left(\frac{U_o}{U} \right)^3 Z^4, \tag{9-5}$$

where $U_o = 1$ keV (so that U/U_o is a unitless energy), and σ_o is a constant. For the ejection of K shell electrons, $\sigma_o \approx 38.8$ barns.

The energy and atomic number dependences in equation 9-5 mean that high-energy photons have a lower cross section and hence are more penetrating, and that high-Z materials stop photons more effectively, as was asserted in chapter 1.

EXAMPLE 9-1: CROSS SECTION

Nickel has atomic number 28. Silicon has atomic number 14. For photon energies of 10 and 20 keV, estimate the atomic cross section for photoelectric absorption for nickel and silicon.

---

*As for section 4.3, the details of the quantum analysis here are not necessary for the rest of the development in this chapter, but the results are again useful. More discussion of quantum calculations can be found in the texts listed in the further reading for chapter 4.

First, we need to check whether K shell electrons can be knocked out by the incoming photons, to determine whether the constant given below equation 9-5 is applicable to this problem. For silicon the "K-edge" energy, the minimum energy needed to knock out a K shell electron, is 1.8 keV. For nickel, it is 8.3 keV. Both photon energies are high enough for both elements.

Ni at 10 keV:

$$\sigma_{ab} \approx \sigma_o \left(\frac{U_o}{U} \right)^3 Z^4 \approx (38.8 \, \text{barns}) \left(\frac{1 \, \text{keV}}{10 \, \text{keV}} \right)^3 (28)^4$$

$$\approx 2.4 \times 10^4 \, \text{barns} \approx 24 \, \text{kbarns.}$$

Ni at 20 keV:

$$\sigma_{ab} \approx \sigma_o \left(\frac{U_o}{U} \right)^3 Z^4 \approx (38.8 \, \text{barns}) \left(\frac{1 \, \text{keV}}{20 \, \text{keV}} \right)^3 (28)^4 \approx 3 \, \text{kbarns.}$$

As expected, the cross section at 20 keV falls by a factor of 8 compared with that for the same element at 10 keV.

Si at 10 keV:

$$\sigma_{ab} \approx \sigma_o \left(\frac{U_o}{U} \right)^3 Z^4 \approx (38.8 \, \text{barns}) \left(\frac{1 \, \text{keV}}{10 \, \text{keV}} \right)^3 (14)^4$$

$$\approx 1.5 \times 10^3 \, \text{barns} \approx 1.5 \, \text{kbarns.}$$

The cross section has fallen by a factor of 16 relative to that for nickel at the same energy.

Si at 20 keV:

$$\sigma_{ab} \approx \sigma_o \left(\frac{U_o}{U} \right)^3 Z^4 \approx (38.8 \, \text{barns}) \left(\frac{1 \, \text{keV}}{20 \, \text{keV}} \right)^3 (14)^4 \approx 186 \, \text{barns.}$$

As expected, the high-energy photon with the low-Z material has the smallest cross section.

EXAMPLE 9-2: ABSORPTION RATE FOR A SINGLE ATOM

Find the rate of photoelectric absorption and the power loss for a single nickel atom in a beam of 10 keV photons with an intensity of 1.6 mW/cm^2.

$$I = \Psi U \Rightarrow \Psi = \frac{I}{U} \approx \frac{1.6 \times 10^{-3} \, \text{W/cm}^2}{10 \times 10^3 \, \dfrac{\text{eV}}{\text{ph}} \dfrac{1.6 \times 10^{-19} \, \text{J}}{\text{eV}}} \approx 10^{12} \, \frac{\text{ph}}{\text{cm}^2 \cdot \text{s}}.$$

Hence the rate at which photons are absorbed is

$$\Gamma_{1,ab} = \Psi \sigma_{ab} \approx \left(10^{12}\, \frac{ph}{cm^2 \cdot s}\right)(2.4 \times 10^4\, barns)\frac{10^{-24}\, cm^2}{barn} \approx 2.4 \times 10^{-8}\, \frac{ph}{s},$$

and the rate at which power is lost from the beam is

$$P_{loss} = U\Gamma_1 \approx \left(10 \times 10^3\, \frac{eV}{ph}\, \frac{1.6 \times 10^{-19}\, J}{eV}\right)\left(2.4 \times 10^{-8}\, \frac{ph}{s}\right) \approx 4 \times 10^{-23}\, W.$$

This is not an easy way to detect a single nickel atom.

Generally, one is interested in the absorption from a collection of atoms in a material. From the calculations of example 9-2, the change in power if a beam passes through a material with N_{atom} atoms is

$$\Delta P = -N_{atom}P_{loss,1} = -N_{atom}U\Gamma_1 = -N_{atom}U\Psi \sigma_{ab} = -N_{atom}I\sigma_{ab}. \tag{9-6}$$

The loss of intensity if a beam passes through a material of thickness Δz is then

$$\Delta I = \frac{\Delta P}{A} = -\frac{N_{atom}}{A}I\sigma_{ab} = -I\sigma_{ab}\left\{\frac{N_{atom}}{A\Delta z}\right\}\Delta z$$
$$= -(I)\sigma_{ab}\left\{\rho_{atom}\right\}\Delta z. \tag{9-7}$$

The atomic density, atoms per unit volume, ρ_{atom} can be computed from

$$\rho_{atom} = \frac{\rho N_A}{M_m}, \tag{9-8}$$

where ρ is the mass density, M_m is the molar mass, and N_A is Avogadro's number. Considering an infinitesimal slice of material,

$$dI = -I\sigma_{ab}\rho_{atom}\, dz. \tag{9-9}$$

The resulting depth dependence is exponential,

$$I = I_o e^{-\mu_{ab}z}, \tag{9-10}$$

where I_o is the incident intensity, and the constants are combined into an absorption coefficient μ_{ab},

$$\mu_{ab} = \rho_{atom}\sigma_{ab} = \frac{\rho N_A}{M_m}\sigma_{ab}. \tag{9-11}$$

Absorption coefficients are usually tabulated without the density factor, as the density can depend on the detailed preparation of the material, but rather as the mass absorption coefficient μ_{ab}/ρ,

$$\frac{\mu_{ab}}{\rho} = \frac{N_A}{M_m}\sigma_{ab}, \tag{9-12}$$

so that none of the factors are preparation dependent. The intensity can also written in terms of an absorption length μ_{ab}^{-1},

$$I = I_o e^{-\frac{z}{\mu_{ab}^{-1}}}. \tag{9-13}$$

Thus the absorption length is the thickness of material through which the x-ray beam can pass before falling to about one-third of its original intensity. The transmission T through a thickness L of material is the ratio of intensity out to the intensity in,

$$T = \frac{I}{I_o} = e^{-\mu_{ab} L}. \tag{9-14}$$

EXAMPLE 9-3: ABSORPTION

Nickel has density of 8.9 g/cm³. For nickel and a photon energy of 20 keV, estimate the absorption coefficient, the mass absorption coefficient, the absorption length, and the transmission through a 30 μm thick foil.

The atomic density is

$$\rho_{atom} = \frac{\rho N_A}{M_m} \approx \frac{\left(8.9\frac{g}{cm^3}\right)\left(6.02\times10^{23}\frac{atoms}{mole}\right)}{58.7\frac{g}{mole}} \approx 9.1\times10^{22}\frac{atoms}{cm^3}.$$

The absorption coefficient is

$$\mu_{ab} = \rho_{atom}\sigma_{ab} \approx \left(9.1\times10^{22}\frac{atoms}{cm^3}\right)(3\times10^3\ barns)\left(\frac{10^{-24}\ cm^2}{barn}\right) \approx 272\ cm^{-1}.$$

The absorption length is

$$\mu_{ab}^{-1} = \frac{1}{\mu_{ab}} \approx \frac{1}{272\ cm^{-1}} \approx 3.7\times10^{-3}\ cm = 37\ \mu m.$$

The mass absorption coefficient is

$$\frac{\mu_{ab}}{\rho} \approx \frac{273\ cm^{-1}}{8.9\frac{g}{cm^3}} \approx 30.6\frac{cm^2}{g}.$$

The transmission is

$$T = e^{-\mu_{ab}L} \approx e^{-\left(273\frac{1}{cm}10^{-4}\frac{cm}{\mu m}\right)(30\,\mu m)} \approx 44\%.$$

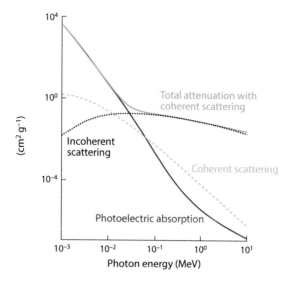

FIGURE 9-2. Mass attenuation coefficients for water. From D. Pfeiffer, Requirements for Medical Imaging and X-Ray Inspection, chap. 31, *Handbook of Optics*, 3rd ed., vol. 5, McGraw-Hill, 2010.

9.2 Attenuation versus absorption

The energy dependence of the photoelectric absorption coefficient is shown in Figure 9-2 for water. In addition to photoelectric absorption, photons can be removed from the x-ray beam by incoherent and coherent scattering, which will be discussed in

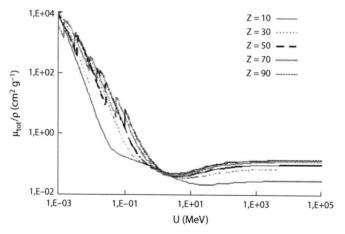

FIGURE 9-3. Comparison of mass attenuation coefficients for a variety of atomic numbers as a function of photon energy in megaelectronvolts. From I. Akkurt et al., *Journal of Quantitative Spectroscopy and Radiative Transfer* 94, nos. 3–4 (1 September 2005): 379–85. Copyright Elsevier.

chapters 10 and 11. The total attenuation coefficient is given by the sum of the coefficients for absorption, scatter, and pair production,

$$\mu_{tot} = \mu_{ab} + \mu_{s,\,coh} + \mu_{s,\,incoh} + \mu_P. \tag{9-15}$$

Pair production (generation of an electron and a positron) is not possible at a photon energy below $U = 2U_{eo}$, twice the rest mass energy of the electron, about 1 MeV, an energy higher than in most of the applications discussed in this text. Pair production is included in Figure 9-3.

9.3 Index of refraction

As was seen in equation 5-44, the electric field amplitude of a plane wave can be written as

$$E = \mathrm{Re}\left\{ E_o e^{i(\kappa z - \omega t)} \right\}. \tag{9-16}$$

For convenience we will leave the time dependence and the real part as understood, and write

$$E = E_o e^{i\kappa z}. \tag{9-17}$$

If, instead of being in vacuum, the wave is traveling in a material of index of refraction n, the wavevector κ becomes

$$\kappa = \frac{2\pi}{\lambda} n, \tag{9-18}$$

where λ is the vacuum wavelength. The index of refraction is complex and is generally written

$$n = 1 - \delta + i\beta. \tag{9-19}$$

As will be seen shortly for β (and in chapter 11 for δ), β and δ are both small, and $|n| \sim 1$. Because the refractive index is close to 1, x rays mostly do not refract when passing through materials. The result is expected, as it implies that shadow images, such as that in Figure 1-1, are sharp. The consequences for refractive optics are discussed in section 12.1. The real part of the index of refraction is less than 1, as will be computed in section 11.1. The consequences for phase velocity and for reflectivity and reflective optics are discussed in sections 11.3, 11.9, and 12.2.

Concentrating now on the absorption term, and using equation 9-19 in the equation for the field gives

$$E = E_o e^{i\kappa z} = E_o e^{i\frac{2\pi}{\lambda}(1-\delta+i\beta)z} = E_o e^{i\frac{2\pi}{\lambda}(1-\delta)z} e^{-\frac{2\pi}{\lambda}\beta z}. \tag{9-20}$$

The intensity is then

$$I = \frac{c\varepsilon}{2}|E|^2 = \frac{c\varepsilon}{2} EE^* = \left[\frac{c\varepsilon}{2} E_o^2 \right]\left(e^{-\frac{2\pi}{\lambda}\beta z} \right)^2 = \left[I_o \right]\left(e^{-\frac{4\pi}{\lambda}\beta z} \right), \tag{9-21}$$

where I_o is the incident intensity.* Therefore, by comparison with equation 9-10, the absorption coefficient μ_{ab} is related to the complex part of the index of refraction β by

$$\mu_{ab} = \frac{4\pi}{\lambda}\beta. \tag{9-22}$$

The index of refraction can also be written in terms of a dimensionless quantity called the atomic scattering factor†

$$f = f_1 - if_2 \tag{9-23}$$

as

$$n = 1 - \rho_{atom}\frac{r_e\lambda^2}{2\pi}(f_1 - if_2), \tag{9-24}$$

where r_e is the classical electron radius, which is a somewhat arbitrary constant obtained from setting the rest mass of the electron equal to the potential energy of a charge at a distance r_e,

$$M_ec^2 = \frac{q_e^2}{4\pi\varepsilon_o r_e} \Rightarrow r_e = \frac{q_e^2}{4\pi\varepsilon_o M_e c^2}. \tag{9-25}$$

Thus, by comparison with equation 9-18,

$$\beta = \rho_{atom}\frac{r_e\lambda^2}{2\pi}f_2. \tag{9-26}$$

Combining equations 9-11, 9-22, and 9-26, gives

$$\sigma_{ab} = 2r_e\lambda f_2. \tag{9-27}$$

EXAMPLE 9-4: INDICES

For nickel and a photon energy of 20 keV, estimate the imaginary parts of the index of refraction and the atomic scattering factor.

First, we need the wavelength $\lambda = \frac{hc}{U} \approx \frac{12.4\,\text{keV}\cdot\text{Å}}{20\ \text{keV}} \approx 0.6\,\text{Å}$ to compute the complex

part of the index of refraction, $\beta = \frac{\lambda}{4\pi}\mu_{ab} \approx \left(\frac{0.6\times10^{-8}\,\text{cm}}{4\pi}\right)(272\,\text{cm}^{-1}) \approx 1.3\times10^{-7}$. As

---

*There is a sign convention to be careful of here. In equation 9-16 the exponential is written with a positive i, in which case the imaginary part of the index of refraction must be taken as positive as well. The other convention, in which both are negative, is also seen. Either is fine, but they should not be mixed. Taking one positive and one negative will result in an unphysical expression in which the intensity grows with distance instead of decaying.

†The positive choice of sign convention for the imaginary part of the index of refraction forces the negative choice for the imaginary part of the atomic scattering factor.

expected, β is quite small. Next, we need the electron radius,

$$r_e = \frac{q_e^2}{4\pi\varepsilon_o M_e c^2} \approx 2.8\times 10^{-5}\,\text{Å} = 2.8\times 10^{-13}\,\text{cm},$$

to compute the imaginary part of the atomic scattering factor,

$$f_2 = \frac{\sigma_{ab}}{2r_e\lambda} \approx \frac{3\times 10^3\,\text{barns}\left(\dfrac{10^{-24}\,\text{cm}^2}{\text{barn}}\right)}{2(2.8\times 10^{-13}\,\text{cm})(0.6\times 10^{-8}\,\text{cm})} \approx 0.85.$$

The imaginary part of the atomic scattering factor is a unitless number of order unity.

9.4 Absorption coefficient of compounds and broadband radiation

In the previous sections, the cross section, absorption coefficient, and index were calculated for pure elements and monochromatic radiation. To compute the absorption coefficient for a compound (for example, the water of Figure 9-2), it is necessary to take into account the number of each kind of atom. If a compound is given as $A_{m1}B_{m2}$, with $m1$ and $m2$ integers, then the mass of a "molecule" of the compound is

$$M_{molecule} = m_1 M_A + m_2 M_B, \tag{9-28}$$

where M_A and M_B are the elemental molar masses, so that the number of molecules per unit volume is

$$\rho_{molecule} = \frac{\rho N_A}{M_{molecule}}, \tag{9-29}$$

where ρ is the mass density of the compound. Since there are m_1 A atoms per molecule and m_2 B atoms, the absorption coefficient for the compound is

$$\mu = \rho_{molecule}(m_1\sigma_A + m_2\sigma_B) = \frac{\rho N_A}{m_1 M_A + m_2 M_B}(m_1\sigma_A + m_2\sigma_B). \tag{9-30}$$

EXAMPLE 9-5: COMPOUND ABSORPTION

Find the absorption coefficient for nickel silicide, Ni_2Si, with a density of 7 g/cm^3 at 20 keV.

The molecular molar mass is

$$M_{molecule} = 2M_{Ni} + M_{Si} \approx 2\left(58.7\,\frac{g}{\text{mole}}\right) + \left(28\,\frac{g}{\text{mole}}\right) \approx 145\,\frac{g}{\text{mole}}.$$

The molecular density is

$$\rho_{molecule} = \frac{\rho N_A}{M_m} \approx \frac{\left(7\,\dfrac{g}{cm^3}\right)\left(6.02\times 10^{23}\,\dfrac{\text{atoms}}{\text{mole}}\right)}{145\,\dfrac{g}{\text{mole}}} \approx 3\times 10^{22}\,\frac{\text{molecules}}{cm^3}.$$

The absorption coefficient is then

$$\mu_{ab} = \rho_{molecule}(2\sigma_{Ni} + \sigma_{Si})$$

$$\approx \left(3 \times 10^{22} \frac{atoms}{cm^3}\right)\left[2\,(3 \times 10^3\ barns) + 186\ barns\right]\left(\frac{10^{-24}\ cm^2}{barn}\right)$$

$$\approx 178\ cm^{-1}.$$

The absorption coefficient is reduced relative to pure nickel because the silicon in the Ni_2Si takes up some of the volume but is less absorptive.

If the radiation is not monochromatic, the absorption coefficient must be a weighted average of the coefficients at different photon energies,

$$\bar{\mu} = \frac{\displaystyle\int_0^{U_{max}} \mu(U)P_U\,dU}{\displaystyle\int_0^{U_{max}} P_U\,dU} = \frac{\displaystyle\int_0^{U_{max}} \mu(U)P_U\,dU}{P}, \qquad (9\text{-}31)$$

where $P_U\,dU$ is the power in the photon energy interval from U to $U+dU$, $\mu(U)$ is the absorption coefficient at the photon energy U, and P is the total power. For broadband radiation, the average photon energy will change after passing through a block of material. The average photon energy after a block of thickness L is

$$\bar{U} = \frac{\displaystyle\int_0^{U_{max}} U\,P_U\,dU}{P} = \frac{1}{P}\int_0^{U_{max}} U\,P_{oU}\,e^{-\mu(U)L}\,dU, \qquad (9\text{-}32)$$

where $P_{oU}\,dU$ is the initial power in the energy interval. Because the absorption coefficient drops off rapidly with photon energy, the power in the intervals at high photon energy will be only weakly attenuated compared with those at low photon energies. The average photon energy will increase. This phenomenon is known as *beam hardening*. It is a complication for image analysis, especially in computed tomography, as the back face of an object (the part farthest from the source) is subject to a different average photon energy and different average attenuation coefficients than is the front face of the object.

9.5 Absorption edges

The smooth $1/U^3$ dependence of the absorption coefficient on x-ray photon energy is true only if the x-ray photon energy is well above the binding energy for the electrons. As noted in chapter 4 for characteristic photon emission, a core electron will not be knocked out if the photon energy does not exceed the binding energy. If the photon energy is, for example, between the binding energies for K and L shell electrons, only L electrons can be knocked out, and the cross section is about a factor of 10 lower than if K electrons can also participate (the constant σ_o below equation 9-5 is lower for L shell electrons). Plotted as a function of photon energy, the cross section makes a series of sharp, almost step-function upward transitions at each energy equal to an electron shell binding energy, as shown in Figure 9-4.

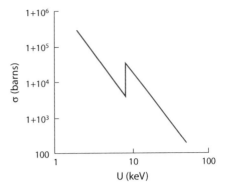

FIGURE 9-4. Atomic photoelectron cross section, in barns, versus photon energy in kiloelectronvolts, for nickel. Data from http://henke.lbl.gov/optical_constants/.

9.6 Absorption spectroscopy

Because of the jump in absorption cross sections at photon energies corresponding to core binding levels, as shown in Figure 9-5, it is possible to deduce the elements present in a sample by plotting the absorption, or the transmission, as a function of incident photon energy. The principle is similar to that of x-ray fluorescence, as discussed in section 4.7. Because the absorption coefficient depends on the number of atoms of that element in the path of the beam, quantitative analysis of composition is possible for uniform samples. With sufficient energy resolution, it is possible to measure small shifts in the binding energy due to the chemical state of the atom. In addition, it is even possible to see subtle interference effects due to scattering of the photoelectrons off neighboring atoms in the sample.

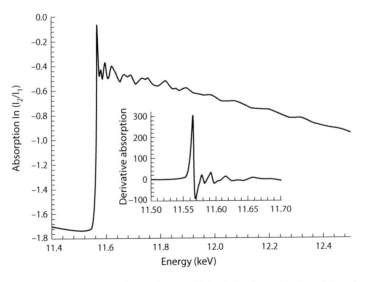

FIGURE 9-5. Oscillations in the Pt absorption coefficient due to scattering of the photoelectrons from neighboring atoms. From D. Lützenkirchen-Hecht and R. Frahm, Requirements for X-Ray Spectroscopy, chap. 30, *Handbook of Optics*, 3rd ed., vol. 5, McGraw-Hill, 2010.

When these subtle effects are addressed, the measured absorption coefficient is usually expressed in terms of the relative difference with respect to the coefficient μ_o for an isolated atom,

$$\mu' \equiv \frac{\mu - \mu_o}{\mu_o}. \tag{9-33}$$

Measuring $\mu'(U)$ allows information to be obtained about the local environment of the ionized atom. A detailed discussion of absorption spectroscopy can be found in the references, but, roughly, changing the input photon energy U changes the photoelectron kinetic energy K_e,

$$K_e = U - U_o. \tag{9-34}$$

where U_o is the binding energy, for example, for K shell electrons. The wavevector κ_e of the photoelectron is related to its kinetic energy by

$$K_e = \frac{p_e^2}{2M_e} = \frac{(\hbar \kappa_e)^2}{2M_e}, \tag{9-35}$$

so that the wavefunction for the electron propagating through the material is

$$\psi_i = \psi_o e^{i\vec{\kappa}_e \cdot \vec{r}}. \tag{9-36}$$

(This is similar to equation 5-44, as was employed in section 9.3.) The wavefunction for an electron which has been coherently scattered from a location R is thus

$$\psi = \psi_s e^{i\vec{\kappa}_e \cdot (\vec{r} - \vec{R})}. \tag{9-37}$$

The sum of the wavefunctions for backscattering from all atomic locations is then

$$\psi_B(\vec{r}) = \psi_s \int g(\vec{R}) e^{i\vec{\kappa}_e \cdot (\vec{r} - \vec{R})} d^3\vec{R}, \tag{9-38}$$

where, $g(\vec{R})$, known as the radial distribution function, is the probability that there is an atom at $\vec{R}$. The total electron wavefunction, including both the original and backscattered terms, is

$$\psi_{tot}(\kappa) = \psi_o e^{i\vec{\kappa}_e \cdot \vec{r}} + \psi_s \int g(\vec{R}) e^{i\vec{\kappa}_e \cdot (\vec{r} - \vec{R})} d^3\vec{R}. \tag{9-39}$$

The probability for an electron to be present with a wavevector κ is

$$P(\kappa) = \psi_{tot}^* \psi_{tot}. \tag{9-40}$$

The probability includes an interference term, similar to the cross term from adding two light waves in equation 5-57,

$$\psi_i^* \psi_B \propto e^{-i\vec{\kappa} \cdot \vec{r}} \int g(\vec{R}) e^{i\vec{\kappa} \cdot (\vec{r} - \vec{R})} d^3\vec{R} = \int g(\vec{R}) e^{-i\vec{\kappa} \cdot \vec{R}} d^3\vec{R}. \tag{9-41}$$

The integral on the right is the Fourier transform of $g(R)$. The probability of a photo-electron being created is equal to the probability of the photon being absorbed, which is proportional to the absorption coefficient μ. The normalized deviation μ' from the absorption coefficient for an isolated atom is therefore related to the Fourier transform of $g(R)$,

$$\mu' \propto \int g(\vec{R}) e^{-i\vec{\kappa}\cdot\vec{R}} d^3\vec{R}. \tag{9-42}$$

The radial distribution function of atoms in the neighborhood of the ionized atom can then be found from the inverse Fourier transform of the measured quantity,

$$g(\vec{R}) \propto \int \mu'(\vec{\kappa}) e^{i\vec{\kappa}\cdot\vec{R}} d^3\vec{\kappa}. \tag{9-43}$$

Measuring the oscillations in the absorption coefficient gives detailed information about the distribution of atoms in the material. This is similar to the information contained in a diffraction pattern, as will be discussed in chapters 13 and 14. Studies of oscillations extending up to 0.5 keV above the ionization energy are known as extended x-ray absorption fine structure, or EXAFS. Investigation in a smaller energy range around the ionization "edge" is referred to as x-ray absorption near-edge structure, or XANES. To perform absorption spectroscopy, it is necessary to have a beam with a very narrow, tunable bandwidth. This requires the use of a high-resolution crystal diffraction monochromator, of the type discussed in section 15.3.1.

9.7 Filtering

An x-ray tube emits nearly white radiation. A high-resolution monochromator has a very narrow energy bandwidth, at the expense of loss of intensity. Often, a more moderate energy bandwidth is desirable. For low-resolution diffraction studies, it is often adequate to remove the Kβ line with an absorption filter. For example a copper x-ray

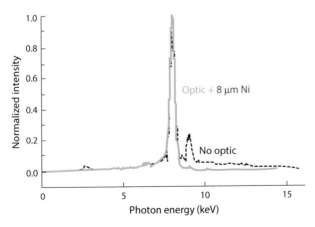

FIGURE 9-6. Suppression of the Cu Kβ peak with an 8 μm Ni filter. There is additional suppression of the high-energy bremsstrahlung from the use of x-ray optics, described in section 12.2. From S. Owens et al., *Proceedings SPIE* 2859 (1996): 205.

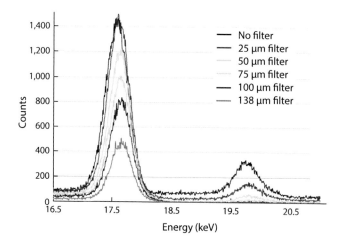

FIGURE 9-7. Mo anode emission with and without a thin Zr filter. The Mo Kβ peak at 20 keV is nearly totally absorbed. Courtesy of Katie Kern.

tube produces a Kα line at 8 keV and a Kβ line at 9 keV. If diffraction is performed with the unfiltered radiation, there will be confusion owing to peaks from each of these bright lines. A nickel filter, which has an absorption edge at 8.3 keV (as shown in Figure 9-4), can be used to preferentially remove the Kβ radiation, since the absorption is about a factor of 10 larger just above the absorption edge than it is below the edge. The effect is shown in Figure 9-6 for a copper source. A similar Kβ suppression is shown in Figure 9-7 for a molybdenum x-ray tube.

9.8 Imaging

9.8.1 CONTRAST

The change in transmission of an x-ray beam with the atomic number, density, and thickness of a sample is responsible for the contrast seen in radiographic imaging. If a measurement is made through the simple object sketched in Figure 9-8, the two intensities after the object will be

$$I_1 = I_o e^{-\mu_1 L}, \tag{9-44}$$

for the path above the embedded object and

$$I_2 = I_o e^{-\mu_1(L-z_o)} e^{-\mu_2 z_o} = I_o e^{-\mu_1 L} e^{-(\mu_2-\mu_1)z_o} \tag{9-45}$$
$$= I_1 e^{-\Delta\mu z_o},$$

for the path through the object. The subject contrast C is defined as the relative difference in intensity,

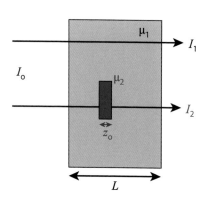

FIGURE 9-8. Subject contrast due to an object of thickness z_o and absorption coefficient μ_2.

$$C = \left| \frac{I_1 - I_2}{I_1} \right| = 1 - e^{-|\Delta\mu|z_o}, \tag{9-46}$$

where the choice of which path is I_1 and which is I_2 is made so that C is positive and less than 1. C is called the "subject" contrast because it depends only on the subject being imaged, not on the characteristics of the detector (or the amount of Compton scattering that is also detected, discussed in chapter 10). For low-contrast objects, that is, those with $z_o\Delta\mu \ll 1$,

$$C = 1 - e^{-|\Delta\mu|z_o} \approx z_o|\Delta\mu|. \tag{9-47}$$

Because of the $1/U^3$ dependence in equation 9-5, the difference between absorption coefficients, as shown in Figure 9-3, is higher for lower-energy photons. The contrast is thus also higher for lower-energy photons, as discussed in chapter 1. However, the absorption is also higher, which results in a lower fraction of the photons being available to create the image. Too few photons can result in unacceptable quantum noise, as was discussed in section 2.5, and is explored further in section 9.8.3. Higher absorption also results in higher subject dose.

A typical image is shown in Figure 9-9. The bones are denser and have higher atomic number elements than the surrounding flesh, so they have a lower transmission. They appear bright in the image because radiographs are typically presented

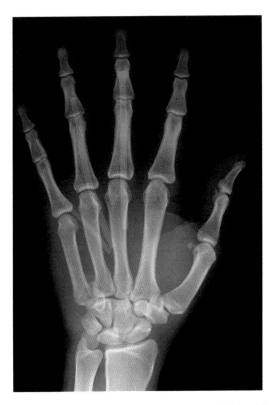

FIGURE 9-9. Modern radiograph of a human hand. Courtesy of Michael Flynn, University of Michigan.

as negatives (originally, to save the time required for the printing step from film, and to increase the visibility by allowing viewing in transmission).

9.8.2 DOSE

Dose is the amount of radiation absorbed by the object, divided by the mass of the object within the beam. For a uniform object like the box of thickness L imaged in Figure 9-8, the absorbed dose is

$$Dose = \frac{U_{absorbed}}{M} = \frac{I_o A t (1 - e^{-\mu_l L})}{\rho L A} = \frac{I_o t (1 - e^{-\mu_l L})}{\rho L}, \tag{9-48}$$

where I_o is the incident intensity, t is the exposure time, A is the area of the beam on the object, and ρ is its density. For very high energy photons, for example those used in megaelectronvolt radiation therapy, a correction must be made to account for the photoelectrons that have enough energy to escape the region of interest where the absorption occurs. This correction is unimportant for radiographic energies. The unit of absorbed dose is the *gray*, where 1 Gy = 1 J/kg. For transparent objects for which $\mu_l L \ll 1$,

$$Dose \approx \frac{I_o t \mu_l L}{\rho L} = \frac{I_o t \mu_l}{\rho}. \tag{9-49}$$

Dose is highest for low photon energies, because the absorption coefficient μ varies as $1/U^3$. Because different particles (e.g., photons, α particles, and neutrons) have different biological effects, dose is sometimes expressed in terms of *equivalent dose* QD, where Q (also called W_R) is a "quality factor." The Q for photons of all energies is 1. The unit of equivalent dose is the *sievert*, where 1 Sv = 1 J/kg. A typical chest x-ray image produces a patient dose of about 0.1 mSv. Background, natural radiation gives a dose of about 0.01 mSv per day. The older units of dose and effective dose were the *rad* and the *roentgen equivalent man*, or *rem*. 1 Gy = 100 rad, and 1 Sv = 100 rem. The *roentgen* (R) is a unit of exposure, power times time, defined as the amount of radiation required to produce 2.58×10^{-4} Coul of ionization in 1 kg of air. The exposure of tissue to 1 R of x-ray radiation gives a dose of about 1 rad. The term *effective dose* takes into account tissue weighting factors since, for example, bone marrow is more sensitive than skin to radiation damage.

9.8.3 NOISE

Because minimizing dose is usually a consideration, the total number of photons incident on an object must be limited, and noise becomes a significant issue in imaging. If the area of the detector (or a pixel of an imaging detector) shown in Figure 9-8 is A_D, the number of photons reaching the detector in the initial path, given a beam with initial photon intensity Ψ_o, is

$$N_1 = \Psi_o t A_D e^{-\mu_l L}. \tag{9-50}$$

If the beam is monochromatic with photon energy U, the initial photon intensity is

$$\Psi_o = \frac{I_o}{U}. \tag{9-51}$$

The number of photons reaching a pixel behind the embedded object is

$$N_2 = N_1 e^{-\Delta \mu z_o}. \tag{9-52}$$

The signal is the difference between these two values,

$$S = N_2 - N_1 = N_1 \left(1 - e^{-\Delta \mu z_o}\right) = N_1 \boldsymbol{C} = \Psi_o t A_D e^{-\mu_1 L} \boldsymbol{C}. \tag{9-53}$$

As discussed in section 2.5, the noise is $\sqrt{N_1}$. The signal-to-noise ratio, SNR, is then

$$SNR = \frac{S}{\sqrt{N_1}} = \sqrt{N_1} \boldsymbol{C} = \sqrt{\Psi_o t A_D e^{-\mu_1 L}} \boldsymbol{C}. \tag{9-54}$$

A commonly used quality factor is the ratio of SNR to dose, or if it is desirable to cancel out the impact of the initial intensity–time product, SNR^2 to dose.

EXAMPLE 9-6: CONTRAST

The absorption length of Lucite (regularly used as a "tissue-equivalent material" for modeling), also known as poly(methyl methacrylate) or PMMA, at 20 keV is 17.3 mm.

a) What is the subject contrast of a 1 mm thick void in a 5 cm thick slab of the material at 20 keV?

b) What is the signal-to-noise ratio for detecting the void if the detected count rate per pixel through the material is 1000 counts/s, and the measurement is done for one second?

a) The subject contrast is

$$\boldsymbol{C} = \frac{N_2 - N_1}{N_2} = \frac{\Psi_o A_d t e^{-\mu(L-z_o)} - \Psi_o A_d t e^{-\mu L}}{\Psi_o A_d t e^{-\mu(L-z_o)}}$$

$$= 1 - e^{-\mu z_o} \approx \mu z_o = \frac{z_o}{\mu^{-1}} \approx \frac{1 \, mm}{17.3 \, mm} \approx 0.058.$$

b) The noise is the standard deviation from Poisson statistics,

$$\boldsymbol{N} = \sqrt{N} = \sqrt{\left(1000 \, \frac{counts}{s}\right)(1 \, s)} \approx 32.$$

The signal is the difference between N_2 and N_1, $S = \boldsymbol{C} N_1 \approx (0.058) 1000 \approx 58$.

Thus, $SNR = \dfrac{S}{\sqrt{N}} = \boldsymbol{C} \sqrt{N} = \dfrac{58}{32} = 1.8.$

For all imaging, the intensity must be captured and then displayed. For a digital display, the signal, which is usually proportional to the number of detected photons, must be converted to a scale of color intensities or brightness, or grayscale in the displayed image.

9.9 Detectors

Detectors can roughly be divided into imaging, spectroscopic, and current (or count-rate) categories. Imaging detectors, such as the one shown in Figure 9-10, have two-dimensional spatial resolution; spectroscopic detectors provide information on the photon energy distribution; and current detectors, the simplest kind, give information on the intensity or photon intensity in the beam. Detectors can also be classed as direct detectors, which accept x rays, or indirect detectors, which detect the light emitted from a phosphor when an x ray strikes it. Finally, detectors can be classified according to their mechanism of operation, for example, as ionization detectors, electron multipliers, or microcalorimeters. An indirect detector using a scintillating phosphor and a multichannel plate was discussed in section 2.4. An x ray striking the phosphor activates hundreds or thousands of electrons, which fall from a metastable state to produce visible light.

FIGURE 9-10. GE DXR250 imaging detector. Courtesy of GE Healthcare.

One of the easiest detectors to construct is some form of ion chamber. These are simple containers, filled with a gas, which can be just air, between flat plates across which a high voltage is applied, as illustrated in Figure 9-11. X rays incident on the gas

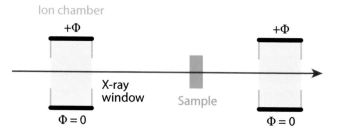

FIGURE 9-11. Two ion chambers monitoring synchrotron beam intensity before and after a sample. Ions created by the radiation are attracted to the charged plates at top and bottom.

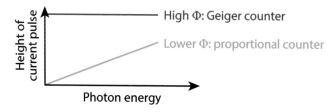

FIGURE 9-12. The height of the current pulse is linear in photon energy when generated by a single x-ray photon absorbed in a proportional counter similar to the ion chamber in Figure 9-11. If the voltage on the plates is raised, the current pulse becomes larger but is no longer proportional to photon energy.

ionize atoms, and the free electrons then travel to the positively charged plate, creating a measurable current. Thin ion
chambers can be employed before and after a sample in a synchrotron beam to measure x-ray absorption. Lower-intensity
beams, such as from low-power tube sources, do not produce
measurable currents. Instead, the gas pressure and voltage are
increased sufficiently so that cascades of electrons are formed
from the collisions of the original free electron. In this case
the anode is usually a thin wire at the center of a cylindrical detector, and the pulse produced is proportional to the photon energy, as shown in Figure 9-12. A multiwire proportional counter
can have spatial as well as energy sensitivity. If the electric field is

FIGURE 9-13. Hamamatsu direct imaging detector. Courtesy of Hamamatsu Photonics K.K.

increased by increasing the voltage or decreasing the wire diameter, the electron cascade
becomes large and then is no longer proportional to photon energy. This is called *Geiger
mode operation*. The large electron pulse allows for simplified electronics in Geiger tube
monitoring devices, which are used to measure the photon intensity.

In semiconductor detectors, an x ray produces thousands of electron-hole pairs in
the depletion region of a reverse-biased pn junction diode. Electrons created in the
p-doped side of the diode are then rapidly swept across the region to create a charge
pulse on a capacitor. The basic physics is similar for the CCD imagers used in visible
light digital cameras or the CMOS cameras in most cell phones. Of course, not all the
electrons created by photoelectric absorption are collected onto the capacitor and read
out. The ratio of the number of electrons collected to the
number which could have been created as electron-hole
pairs is called the *detector quantum efficiency* (sometimes
written as DQE, although that acronym is sometimes reserved for the *detective* quantum efficiency described
later). Semiconductor detectors can be pixelated for imaging, with or without a phosphor coating for indirect
detection. A direct detector is shown in Figure 9-13.

For silicon detectors, a phosphor is often necessary for
imaging with high-energy x rays, because the lithography process for making the pixels results in a thin
structure, which is often less than the absorption length
for the x ray in the semiconductor. High-Z semiconductors such as CdTe or CdZnTe have been employed for direct detection because of their shorter absorption lengths,
which results in higher efficiency for photon capture. The
ratio of the number of detected to incident photons is
called the *quantum detector efficiency*, QDE.

Semiconductor detectors with a single large pixel
can be used to provide energy sensitivity. In that case,
as for the proportional counter of Figure 9-12, the
number of electrons collected from the capture of a
single x-ray photon is proportional to the photon energy. High-resistivity material is used to allow for large
depletion regions, and cooling, as shown in Figures 9-14

FIGURE 9-14. Cryogenic x-ray detector sits on a liquid nitrogen
dewar. The detector is a silicon
crystal through which have been
diffused lithium ions to passivate
impurities and create a high-
resistivity device. Picture courtesy of Canberra Inc.

FIGURE 9-15. Small silicon pin diode with Pelletier cooling. Courtesy of Amptek Inc.

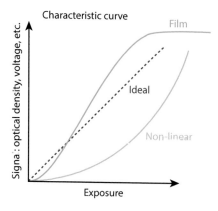

FIGURE 9-16. The signal as expressed by the detector as a function of x-ray exposure.

and 9-15, is generally required to reduce thermal noise. The energy sensitivity is typically on the order of 100 eV. Silicon drift detectors collect the charge onto an area of small capacitance, and so allow for higher count rates. The detectors with the finest energy resolution, in the range of a few electronvolts, are superconducting devices such as microcalorimeters and superconducting tunnel junction devices.

Some imaging detectors, such as the relatively old multiwire proportional counters, the gamma cameras described in section 2.4, or charge injection device (CID) detectors, combine spatial resolution with energy resolution. Newer imaging detectors which threshold the energy spectrum into a small number of bins are termed *photon counting* imaging systems.

Film is the oldest x-ray detector, used originally for most medical applications. Film employs an emulsion containing silver halides. The x ray photon causes the halide to dissociate; for example, silver bromide dissociates into silver and bromide ions. The silver ions are trapped in the emulsion, creating a latent image, and are reduced to silver atoms during the development process, turning the film opaque in that region. Because the emulsion is thin, it is not very efficient at detecting x rays. Dose reductions of two orders of magnitude were achieved when it became standard to use film as an indirect detector with thin phosphor plates. A thicker phosphor has a higher efficiency in stopping high energy x rays, but the visible light has a longer path and hence spreads out farther before reaching the detector. There is a trade-off between resolution and efficiency in designing phosphor thickness.

Film was first replaced in medical applications by the use of restimulable phosphor plates (also called computed radiography plates). In the restimulable phosphor, the metastable trap state has a lifetime of several minutes to hours. After exposure to x rays the excited phosphor retains a memory in the form of electrons locally trapped in the metastable state. The plate is read by scanning with a red laser beam, which excites the electrons back up to the conduction band, from which they fall, emitting blue light. The record of the intensity of the blue light as a function of the laser position is used to create the digital image.

Finally, as large-area digital detectors have become more inexpensive, these have become standard on many imaging systems. Direct detectors are made with high-Z semiconducting materials, generally bonded to an amorphous silicon thin-film transistor array for reading out the individual pixels. Indirect detectors have phosphor materials and thicknesses optimized for the intended energy range. They can be used with small-area light detectors, such as CCD cameras, in which case large-area indirect detectors, such as those used for radiography, must have optics designed to carry the visible light emitted from the large-area phosphor to the smaller-area camera. Alternatively, the visible-light detection can take place in a large-area amorphous silicon photodiode array.

Generally, digital imaging detectors have a much higher dynamic range than that of film and are not sensitive to the errors that could be created during chemical development. They also tend to have a more linear characteristic curve, a description of the detector response to the exposure, as shown in Figure 9-16. *Dynamic range* is the range of input exposures over which the output signal is linear. A nonlinear response reduces contrast.

In addition to the linearity of the response, the detector pixel size is an important consideration. The Nyquist frequency theorem requires a pixel size of $z_o/2$ to image a sinusoidal object of spatial frequency $1/z_o$. Larger pixels blur the image, as illustrated in Figure 9-17. A plot of the measured image contrast of a high-contrast sinusoidal object as a function of the input spatial frequency κ is called the *modulation transfer function* (MTF) of the imaging system. The MTF is never better than the Fourier transform of the pixel modulation, which for a pixel of width W is

$$Pix(y) = rect\left(\frac{y}{W}\right) = \begin{cases} 1 & |y| < \dfrac{W}{2} \\ 0 & else \end{cases}$$

$$\Rightarrow Pix(\kappa) = \left| \int_{-\infty}^{\infty} Pix(y) e^{-i\kappa y} \, dy \right| = \left| \frac{\sin\left(\kappa \dfrac{W}{2}\right)}{\kappa \dfrac{W}{2}} \right|. \tag{9-55}$$

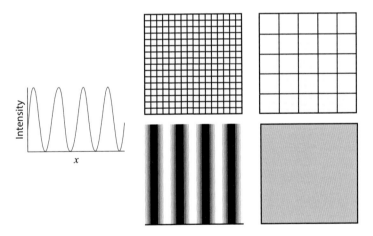

FIGURE 9-17. The sinusoidal pattern shown at left will be faithfully reproduced by the small pixels (top middle) to form the sinusoidal image (lower middle), but if the larger pixels at top right are used, the pattern will blur into the flat gray image at lower right.

The Fourier transform of the pixel modulation has a width in spatial frequency proportional to $1/W$.

While smaller pixels give better resolution, the small pixel area results in fewer photons per pixel for the same image dose, and hence can result in unacceptable quantum noise. This can be measured in terms of the signal-to-noise ratio (SNR) of the output image. The square of the ratio of the output to input SNR values is termed the *detective quantum efficiency*, which is also sometimes called DQE. It depends on the MTF, the noise, and the detector efficiency.

9.10 Tomosynthesis and tomography

Standard radiography produces a planar shadow image, reducing a three-dimensional object to a two-dimensional projection. A shadow image gives an intensity in each pixel which is the product of the attenuation values of all the object voxels in the path of the beam. Similar to equation 9-45, the expression for the intensity on the detector in a pixel at location x, y is

$$I_{x,y} = I_o \prod_z e^{-\mu_{x,y,z}z_o} = I_o e^{-\sum_z \mu_{x,y,z}z_o} \Rightarrow \ln\left(\frac{I}{I_o}\right) = -z_o \sum_z \mu_{x,y,z}, \qquad (9\text{-}56)$$

where $\mu_{x,y,z}$ is the attenuation coefficient of the object voxel at location x,y,z. The voxel is a cube of edge z_o, and z is the beam direction. The image contains no information about the distribution of voxels in the z direction, perpendicular to the detector. There are a variety of techniques for extracting that information from multiple images. Computed tomography is similar to the SPECT back projection discussed in section 2.6. In the case of tomography, the goal is to associate each voxel in the simulated object with a value corresponding to its attenuation coefficient μ. As in SPECT, the information from a single image is back projected to every voxel in the line of the beam, and the resulting matrices from multiple images are summed. Filtering is usually performed on a Fourier transform of the image to reduce the effects of noise and streaking artifacts.

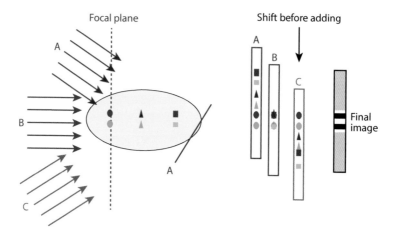

FIGURE 9-18. Three sequential images are taken of the subject (the ellipse containing the six embedded objects). The detector is shown behind the subject only for image A but is rotated with the beam for images B and C. The rectangular boxes represent the images recorded on each detector. The images are shifted before adding, as shown. The images of the triangles and squares are in different locations on each image and so blur to gray when summed in the final image at right. The images of the blue dots are, after shifting, at the same location in each image and so stay sharp. The blue dashed line indicates the plane that is imaged for that amount of image shifting.

Tomosynthesis and computed tomography are both techniques that use multiple images to produce depth information. Tomosynthesis requires fewer images and less computation but does not produce a full three-dimensional reconstruction. The principle of tomosynthesis is shown for a two-dimensional object in Figure 9-18. Three images are taken (one at a time) from three different angles. The images are then added digitally. If the pixels are not shifted, then the triangles in all three images are at the two center pixels, and only the triangle images stand out. The effect is stronger with a larger number of images. The summed image shows only the central plane of the subject. If the images are shifted before summing, as shown at right in Figure 9-18, the blue ovals are in the same position for each image, only the blue ovals stand out, and the "focal plane" of the image is the blue dotted line.

Problems

SECTIONS 9.1–9.4

1. Aluminum has atomic number 13 and density 2.7 g/cm³

   a) For photon energies of 10 and 20 keV, estimate the atomic cross section for photoelectric absorption.
   b) For a photon energy of 20 keV, estimate the absorption coefficient, the mass absorption coefficient, and the absorption length.
   c) Estimate the transmission through a 60 µm foil at 20 keV.
   d) Estimate the imaginary parts of the index of refraction and the atomic scattering factor at 20 keV.

2. Lucite, $C_5H_8O_2$, has a density of 0.9 g/cm³. For a photon energy of 30 keV, estimate the atomic cross sections for photoelectric absorption for elemental C, H, and O, and use that information to estimate the absorption coefficient and hence the transmission through a 5 cm thick block.

SECTION 9.7

3. A system designer wants to build an x-ray fluorescence system to use filtered Kα x rays to excite the fluorescence. Table 9-1 lists (make-believe) elements with their energies. Identify the best choice among the elements on the list (and give a reason) for each of the following:

TABLE 9-1.

Element	Kα, keV	Kβ, keV	Ionization Energy, keV
A	5	7	10
B	20	23	24
C	22	25	26
D	32	35	40
E	60	70	80

a) The anode

b) The filter

c) The sample emitting the fluorescence

d) Identify a reasonable tube voltage.

SECTION 9.8

4. Use the absorption coefficient of Al computed in problem 1 to answer the following:

   a) What is the subject contrast of a 1 mm thick void in a 1 cm thick slab of Al at 20 keV?

   b) What is the signal-to-noise ratio for detecting the void if the detected count rate through the material is 10,000 counts/s, and the measurement is done for 1 s?

   c) What is the dose, in gray, to the $1 \, \text{cm} \times 1 \, \text{cm} \times 1 \, \text{cm}$ Al block?

Further reading

Jens Als-Nielsen and Des McMorrow, **Elements of Modern X-ray Physics,** John Wiley & Sons, 2001.

David Attwood and A. Sakdinawat, **X-Rays and Extreme Ultraviolet Radiation: Principles and Applications,** Cambridge University Press, 2016

Jerrold T. Bushberg, J. Anthony Seibert, Edwin M. Leidholdt Jr, and John M. Boone, **The Essential Physics of Medical Imaging,** 3rd ed., Lippincott Williams & Wilkins, 2012.

W. Gibson and P. Siddons, Introduction to X-ray Detectors, chap. 60, and A. Couture, Advances in Imaging Detectors, chap. 61, both in M. Bass, C. DeCusatis, J. Enoch, V. Lakshminarayanan, G. Li, C. MacDonald, and V. N. Mahajan, and E. Van Stryland, eds., **Handbook of Optics,** 3rd ed., vol. 5, McGraw-Hill, 2010.

10

COMPTON SCATTERING

In chapter 9, we saw that a photon incident on an atom can be absorbed, causing the atom to eject an electron with a kinetic energy equal to the excess energy; that is, the kinetic energy is the difference between the photon energy and the binding energy of the electron. Alternatively, some or all of the excess energy can be given to a second photon, as shown in Figure 10-1. An example of a photon giving up energy and momentum to an electron is shown in Figure 10-2 from a more classical point of view. In both diagrams, the residual energy and momentum is carried away by a second photon.

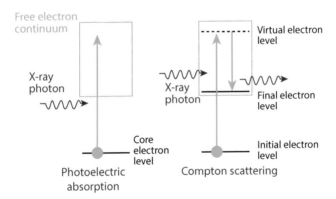

FIGURE 10-1. Absorption (left) or Compton scatter (right) of a photon. In absorption, the electron makes a transition from its initial level to another level or to the free continuum. Scatter can be considered to involve the fluctuation of an electron to a virtual level followed by an immediate downward transition, with the emission of a second photon. For Compton scattering, the final electron level is usually in the continuum, so that the electron is ejected from the atom.

FIGURE 10-2. Compton scattering of an x ray by an electron, initially at rest in this frame.

10.1 Conservation laws

If the wavelength of the incident photon is λ, then the wavelength of the scattered photon, λ', is related to the scattering angle of the x-ray, θ. There are two additional unknown variables, the final speed and angle of the electron. These four variables are related by three equations expressing conservation of energy and momentum. The energy of the photon is $U = hc/\lambda$, and if the frame for observing the event is chosen as the rest frame of the electron, then its initial energy is $U_{eo} = M_e c^2$. Equating the energy before and after the collision gives

$$\frac{hc}{\lambda} + M_e c^2 = \frac{hc}{\lambda'} + \sqrt{(M_e c^2)^2 + p_e^2 c^2}\,, \tag{10-1}$$

where p_e is the final momentum of the electron, and for completeness we have used the relativistic equation for the energy of the electron,

$$U_e^2 = (M_e c^2)^2 + p_e^2 c^2, \tag{10-2}$$

although for most of the applications of interest the electron velocity is well within the classical limits.

The momentum along an axis parallel to the original direction of travel is

$$\hbar\kappa + 0 = \frac{h}{\lambda} + 0 = \frac{h}{\lambda'}\cos\theta + p_e \cos\theta_e. \tag{10-3}$$

The momentum perpendicular to that axis, in the plane which includes the final directions of the electron and photon, is

$$0 = \frac{h}{\lambda'}\sin\theta - p_e \sin\theta_e, \tag{10-4}$$

from which the first of the electron variables can be eliminated,

$$\sin\theta_e = \frac{h}{\lambda' p_e}\sin\theta. \tag{10-5}$$

Applying this result and $\cos^2(\theta_e) = 1 - \sin^2(\theta_e)$ to equation 10-3 yields

$$\frac{h}{\lambda} = \frac{h}{\lambda'}\cos\theta + p_e\sqrt{1 - \frac{h^2\sin^2\theta}{\lambda'^2 p_e^2}} = \frac{h}{\lambda'}\cos\theta + \sqrt{p_e^2 - \frac{h^2\sin^2\theta}{\lambda'^2}}$$

$$\Rightarrow p_e^2 = \frac{h^2}{\lambda'^2} + \frac{h^2}{\lambda^2} - 2\frac{h^2}{\lambda\lambda'}\cos\theta. \tag{10-6}$$

Squaring the energy equation 10-1 and substituting into it the preceding expression for the electron final momentum p_e gives

$$\left(\frac{hc}{\lambda} - \frac{hc}{\lambda'} + M_e c^2\right)^2 = (M_e c^2)^2 + c^2\left[\frac{h^2}{\lambda'^2} + \frac{h^2}{\lambda^2} - 2\frac{h^2}{\lambda\lambda'}\cos\theta\right]$$

$$\Rightarrow -2\frac{h^2 c^2}{\lambda\lambda'} + 2M_e c^2 hc\left(\frac{1}{\lambda} - \frac{1}{\lambda'}\right) = -2\frac{h^2 c^2}{\lambda\lambda'}\cos\theta. \tag{10-7}$$

Multiplying by $\lambda\lambda' / 2M_ec^2hc$ gives

$$-\frac{h}{M_ec}+(\lambda'-\lambda)=-\frac{h}{M_ec}\cos\theta \Rightarrow \lambda'=\lambda+\frac{h}{M_ec}(1-\cos\theta), \qquad (10\text{-}8)$$

Because the electron gains kinetic energy, the scattered photon has less energy, and the final wavelength is always longer than or equal to the initial wavelength. The $\theta=0$ case corresponds to no scattering, and the wavelength is unchanged. The maximum change in wavelength occurs for backscattering, $\theta=180°$.

EXAMPLE 10-1: ENERGY LOSS

a) What is the maximum wavelength change due to Compton scattering?
b) What is the energy change for a backscattered photon originally at 20 keV?
c) What is the energy change for a backscattered photon originally at 200 keV?

a) The maximum wavelength change, also known as the *Compton wavelength*, is

$$\Delta\lambda = \frac{2h}{M_ec} = \frac{2hc}{M_ec^2} \approx 2\frac{12.4\,\text{keV}\,\text{Å}}{511\,\text{keV}} \approx 0.0485\,\text{Å}.$$

b) For $U=20$ keV, the initial wavelength is

$$\lambda = \frac{hc}{U} \approx 0.62\,\text{Å} \Rightarrow \lambda' \approx 0.67\,\text{Å} \Rightarrow U' = \frac{hc}{\lambda'} \approx 18.5\,\text{keV} \Rightarrow \Delta U \approx 1.5\,\text{keV}.$$

c) For $U=200$ keV,

$$\lambda = \frac{hc}{U} \approx 0.062\,\text{Å} \Rightarrow \lambda' \approx 0.11\,\text{Å} \Rightarrow U' = \frac{hc}{\lambda'} \approx 112\,\text{keV} \Rightarrow \Delta U \approx 88\,\text{keV}.$$

The energy loss is larger, and also a larger fraction of the original energy, for high-energy photons. This has important consequences for nuclear medicine, since, as we saw in section 2.4, the energy loss is used to distinguish against scattered photons. In nuclear medicine, high-energy photons are usually used, to produce sufficient energy change to be detectable. As a result, the collimators and detectors are generally thick and bulky.

10.2 Compton cross section

The cross section for scattering, and its angular dependence, is discussed in section 11.6. Scattering can be coherent, in which no electron is ejected from the atom and the output photon energy is unchanged from the input, or incoherent, generally known as Compton scattering. Compton scattering dominates at high energies and low Z. At lower energies or for higher-Z materials, binding effects dominate, and electrons are less apt to be ejected from the atom. In that case, coherent scatter dominates. Coherent and incoherent cross sections for carbon and lead are shown in Figures 10-3 and 10-4. Incoherent, Compton, scattering becomes higher at higher energies for both

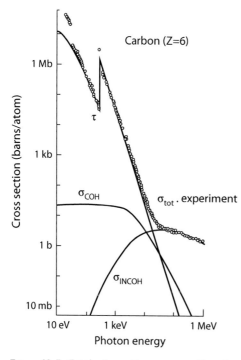

FIGURE 10-3. Total attenuation cross section, including coherent and incoherent scatter, for carbon. Photoelectric absorption is denoted by τ. From J. H. Hubbell et al., *Journal of Physical and Chemical Reference Data* 9, 1023. Reproduced by permission of AIP Publishing.

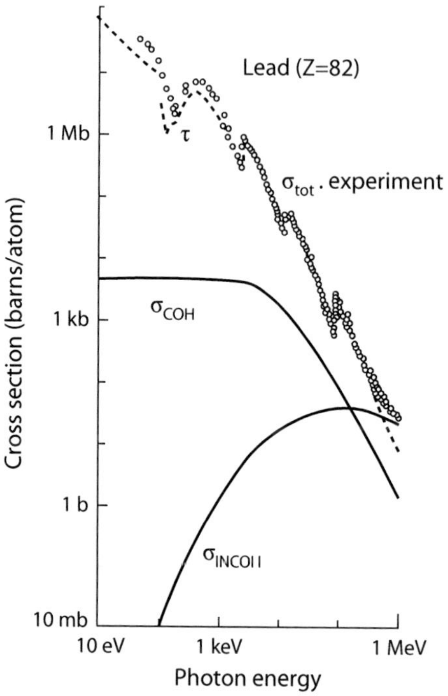

FIGURE 10-4. Total attenuation cross section, including coherent and incoherent scatter, for lead. From J. H. Hubbell et al. *Journal of Physical and Chemical Reference Data* 9, 1023. Reproduced by permission of AIP Publishing.

elements. The energy at which the two are equal is about 10 times higher for the high-Z material, for which electron binding to the atoms is more significant.

10.3 Inverse Compton sources

In the previous section we considered the electron to be at rest, which is a reasonable approximation for electrons in a solid in a laboratory frame. In that case, the incoming photon imparts energy to the electron. However, if the electron is initially at high velocity, it can lose energy, and the output photon energy can be higher than the input. This phenomenon is known as *inverse Compton scattering*, and it can be used to convert visible light photons to x-ray photons, using the scheme diagramed in Figure 10-5. An electron moving at a speed v with respect to a green laser light sees the frequency Doppler shifted from v_o to

$$v = v_o \sqrt{\frac{1 + \dfrac{v}{c}}{1 - \dfrac{v}{c}}}. \tag{10-9}$$

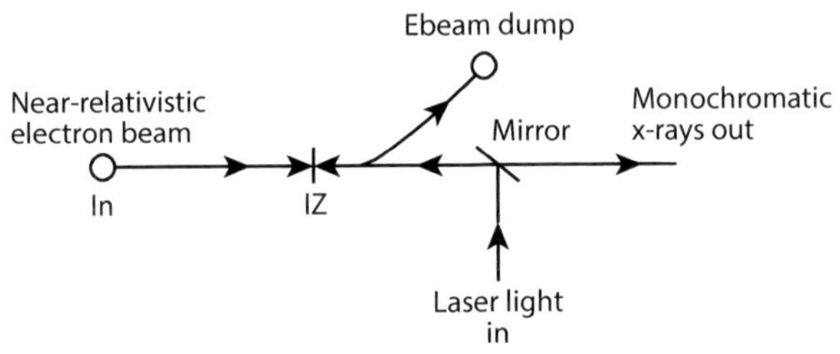

FIGURE 10-5. In an inverse Compton source, a relativistic electron beam is collided with a high-intensity visible-light laser pulse at the interaction zone (IZ). From F. Carroll, Inverse Compton X-Ray Sources, chap. 59, *Handbook of Optics*, 3rd ed., vol. 5, McGraw-Hill, 2010.

If the electron is moving a relativistic speeds, then

$$\frac{v}{c} = \sqrt{1 - \frac{1}{\gamma^2}} \approx 1 - \frac{1}{2\gamma^2} \Rightarrow v \approx 2\gamma v_o. \tag{10-10}$$

EXAMPLE 10-2: DOPPLER SHIFT

A typical electron energy for these sources is 50 MeV.
a) What is γ?
b) What is the Doppler-shifted wavelength seen by the electron if the original photon is green, $\lambda_o = 0.5\,\mu m$, in the lab frame?
c) What are the wavelength and energy change for backscattering for this photon, in the frame in which the electron is at rest?

a) $\gamma = \dfrac{K_e + U_{eo}}{U_{eo}} \approx \dfrac{50.5\,\text{MeV}}{0.5\,\text{MeV}} \approx 99.$

b) $\lambda = \dfrac{c}{v} = \dfrac{c}{2\gamma v_o} = \dfrac{\lambda_o}{2\gamma} \approx \dfrac{0.5\,\mu m}{198} \approx 25\,\text{Å}.$

c) Computed in the frame in which the electron is at rest, the energy loss and wavelength change due to Compton scattering is negligible for this wavelength,

$$U = \frac{hc}{\lambda} \approx \frac{12.4\,\text{keV Å}}{25.29\,\text{Å}} = 0.490\,\text{keV}$$

$$U' = \frac{hc}{\lambda'} = \frac{12.4\,\text{keV Å}}{25.34\,\text{Å}} = 0.489\,\text{keV} \Rightarrow \Delta U = 1\,\text{eV}.$$

In the electron's frame, the backscattered light is approximately the same frequency as the incident light, $v' \sim v$. However, the light is again Doppler shifted when observed in the lab frame,

$$v'' = v' \sqrt{\frac{1 + \dfrac{v}{c}}{1 - \dfrac{v}{c}}} \approx 2\gamma v' \approx 4\gamma^2 v_o. \tag{10-11}$$

EXAMPLE 10-3: DOPPLER SHIFT

For a 50 MeV electron energy, what is the backscattered photon energy observed in the lab frame if the initial visible light photon wavelength was $\lambda_o = 0.5\,\mu m$?

$$\lambda'' = \frac{\lambda_o}{4\gamma^2} \approx \frac{0.5\,\mu m}{4(99)^2} \approx 0.128\,\text{Å} \Rightarrow U'' = \frac{hc}{\lambda''} \approx \frac{12.4\,\text{keV Å}}{0.128\,\text{Å}} \approx 97\,\text{keV}.$$

As discussed in chapter 7 for synchrotron sources, relativistic contraction results in the Compton-scattered beam being highly forward directed, within an angle of $1/\gamma$. For a particular observation direction, the Compton-scattered radiation is monochromatic if the light input is monochromatic and the electrons all have the same energy. To obtain adequate intensity of Compton-scattered photons, the electron beam current must be fairly high, and the laser light must be an intense pulsed source, coordinated to arrive at the interaction region at the same time as the electron bunch. As a result, inverse Compton sources are pulsed sources of x rays, with short-duration pulses. In addition, the x-ray wavelength of the source can be tuned by changing the electron energy. A few research sources of this design have been built, more are contemplated, and commercial sources are now available.

10.4 Scatter in radiography

Conventional Compton scatter is an important effect in radiography. After a diagnostic x-ray beam passes through a human chest, approximately 90% of the photons will have been Compton scattered rather than passed directly through the object. The Compton-scattered photons no longer contribute information to the shadow image but instead create a uniform background fog. The amount of scattered radiation is measured by observing the intensity scattered into a detector placed behind a lead shield, as shown in Figure 10-6. The measured *scatter ratio*, the ratio of the intensity measured behind the lead shield to that measured in the open region, will depend on the size of the lead shield, as shown in Figure 10-7.

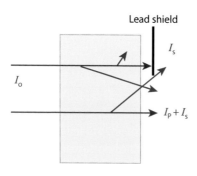

FIGURE 10-6. Compton-scattered photons are detected in the region behind a lead shield with intensity I_s. The sum of the primary and scattered beams is collected where there is no lead shield.

The limiting ratio as the shield diameter is extrapolated to zero is the scatter fraction (SF),

$$SF = \frac{I_s}{I_s + I_p}, \tag{10-12}$$

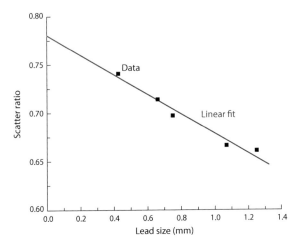

FIGURE 10-7. Measured scatter ratio as a function of lead shield diameter at 45 keV with a 5 cm thick Lucite object. The scatter fraction, found by a linear fit, was 0.77 for this case. Courtesy of Mihir Roy.

where I_s is the intensity of the scattered radiation, and I_p is the intensity of the primary, unscattered beam. The scatter fraction is always less than 1. The ratio of scatter to primary is thus

$$\frac{I_s}{I_p} = \frac{SF}{1-SF}.$$ (10-13)

This ratio can be greater than 1.

10.5 Contrast with scatter

The background of scatter fog will reduce the contrast from the original value of **C** seen in Figure 9-8 to that of Figure 10-8,

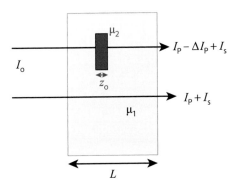

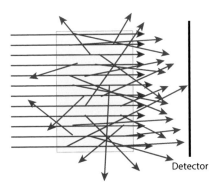

FIGURE 10-8. Subject contrast due to an object of thickness z_o and absorption coefficient μ_2 is reduced in the presence of scatter.

FIGURE 10-9. Compton-scattered photons showing scatter from a large-area beam. The sketch shows a case in which the Compton scatter is somewhat forward directed.

$$C_s = \frac{(I_p + I_s) - (I_p - \Delta I_p + I_s)}{(I_p + I_s)} = \frac{\Delta I_p}{(I_p + I_s)} = \frac{\Delta I_p}{I_p} \frac{I_p}{(I_p + I_s)} = C(1 - SF), \quad (10\text{-}14)$$

where C_s is the contrast in the presence of scatter. Because the contrast is decreased, the dose must be increased to give the same signal-to-noise ratio.

EXAMPLE 10-4: CONTRAST

If the total count rate measured into a detector of area A is 1000 counts/s, and the scatter count rate is 900 counts/s,
a) What is the scatter fraction?
b) What is the scatter-to-primary ratio?
c) If the original subject contrast is $C = 0.058$, what is the contrast in the presence of scatter?

a) $SF = \dfrac{900}{1000} = 0.9 = 90\%.$

b) $\dfrac{I_s}{I_p} = \dfrac{SF}{1 - SF} = \dfrac{0.9}{1 - 0.9} = 9$

c) $C_s = C(1 - SF) = 0.058\,(1 - 0.9) = 0.0058$

10.6 Scatter reduction

The scatter intensity on a detector can be reduced relative to the case illustrated in Figure 10-9 by decreasing the area of the input beam, as shown in Figure 10-10, or increasing the distance to the detector, as shown in Figure 10-11. Both techniques have disadvantages. The first reduces the size of the image produced at one time and so requires the increased complexity of beam scanning to image a large object. Increasing the detector distance increases the magnification for a divergent beam and so requires a larger detector, and in all cases increases the geometric blur, as given in equation 5-37, resulting in poorer resolution. A thinner object will also produce less scatter, so scatter tends to be much less of an issue for thin microscopy specimens.

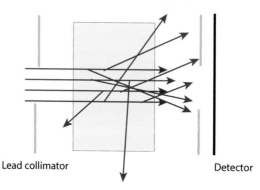

Lead collimator   Detector

FIGURE 10-10. The Compton scatter has been reduced by restricting the area of the beam with fore and aft lead collimators.

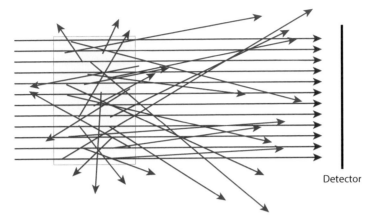

FIGURE 10-11. The Compton scatter in the image has been reduced by increasing the detector distance.

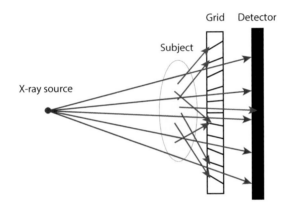

FIGURE 10-12. A scatter-rejection grid shown for a divergent point source. The lead ribbons in the grid extend out of the page. They preferentially stop high-angle scattered radiation.

The most common technique for reducing scatter in medical radiography is to use a scatter rejection grid, as shown in Figure 10-12. The grid consists of a low-density material that is fairly transparent to x rays, such as graphite or aluminum, in which has been embedded a set of lead ribbons aligned with the primary beam. Scattered radiation largely hits at the wrong angle and is absorbed by the lead. A typical grid transmission for primary radiation is about $T_p = 65\%$, and for scatter radiation is about $T_s = 20\%$. Scatter can also be removed by digital techniques, if it is assumed that it is slowly varying compared with the image features.

EXAMPLE 10-5: SCATTER REJECTION

If the scatter fraction is 90% before a typical grid, what is the scatter fraction after the grid? If the subject contrast is $C = 0.058$, what is the contrast in the presence of scatter after the grid?

From the previous example we know that $\dfrac{I_s}{I_p} = 9$.

The scatter intensity after the grid is $I_s' = \mathbf{T}_s I_s$. The primary intensity after the grid is $I_p' = \mathbf{T}_p I_p$. The scatter fraction after the grid is thus

$$SF' = \frac{I_s'}{I_s' + I_p'} = \frac{\mathbf{T}_s I_s}{\mathbf{T}_s I_s + \mathbf{T}_p I_p} = \frac{\mathbf{T}_s \dfrac{SF}{1 - SF}}{\mathbf{T}_s \dfrac{SF}{1 - SF} + \mathbf{T}_p}$$

$$= \frac{\mathbf{T}_s SF}{\mathbf{T}_s SF + \mathbf{T}_p (1 - SF)} \approx \frac{(0.2)(0.9)}{(0.2)(0.9) + (0.65)(0.1)} \approx 0.74.$$

c) $\mathbf{C}_s = \mathbf{C}(1 - SF) = 0.058(1 - 0.74) = 0.015$. The contrast has increased by a factor of 2.7 compared with the no-grid case.

Problems

SECTION 10.1

1. Compute the scattered photon energy for 100 keV x rays scattered at an angle of a) 0°, b) 90°, and c) 180°.

SECTION 10.3

2. What is the backscattered photon energy observed in the lab frame if the initial visible light wavelength was $\lambda_o = 0.5$ μm, and the light was scattered off 25 MeV electrons?

SECTIONS 10.4–10.6

3. Prove equation 10-13.
4. A 10 cm thick patient is imaged at a photon energy with an attenuation length of 10 cm. At that energy, the subject contrast for a tumor in the patient is $\mathbf{C} = 0.1$. The image is taken in 10 s with an incident photon intensity of 2.72×10^6 photons/cm²/s and a detector pixel size of 100 μm × 100 μm. The scatter-to-primary ratio measured after the patient is 9.
   a) What is the scatter fraction?
   b) What is the image contrast?
   c) What are the total and primary photon intensities at the detector (after absorption in the patient)?
   d) What is the number of photons per pixel (in the region of normal tissue, outside the tumor)?
   e) What would the signal-to-noise ratio be if the scatter was removed before it hit the detector?
   f) What is the signal-to-noise ratio with the scatter?

5. An image has an image contrast of 0.05 and a scatter fraction of 80%. A grid is to be designed to increase the contrast by a factor of 3. The primary transmission is 70%.

   a) What is the subject contrast?
   b) What is the maximum scatter fraction that will give three times the original image contrast?
   c) Solve for the maximum allowable scatter transmission for the grid.

Further reading

F. Carroll, Inverse Compton X-ray Sources, chap. 59 in M. Bass, C. DeCusatis, J. Enoch, V. Lakshminarayanan, G. Li, C. MacDonald, V. N. Mahajan, and E. Van Stryland, eds., **Handbook of Optics**, 3rd ed., vol. 5, McGraw-Hill, 2010.

X-ray Data Handbook, http://xdb.lbl.gov/.

Jerrold T. Bushberg, J. Anthony Seibert, Edwin M. Leidholdt Jr, and John M. Boone, **The Essential Physics of Medical Imaging**, 3rd ed., Lippincott Williams & Wilkins, 2012.

11

COHERENT SCATTER I: REFRACTION AND REFLECTION

An electromagnetic wave, such as an x-ray beam, interacts with any object it encounters primarily through the effect of the electric field on the charge carriers in the material. Because electrons are so much lighter than the positive nuclei, electrons are accelerated much more strongly, and the electron response is the primary interaction. The photon can lose all its energy to an electron, as was discussed in chapter 9, on photoelectric absorption. In Compton scattering, as was seen in chapter 10, the photon transfers some of its energy to an electron, and a lower-energy photon is emitted. In this chapter we discuss coherent scattering. The word *coherent* in this context means that the scattered photon has the same energy as the initial photon. Schematic energy diagrams for the three processes are shown in Figure 11-1. For coherent scattering, while the emitted photon has the same energy, it can, as we will see, have a different phase and different direction. The phase is modified by each interaction as photons move through the material. Alternatively, in a wave description, the cumulative change

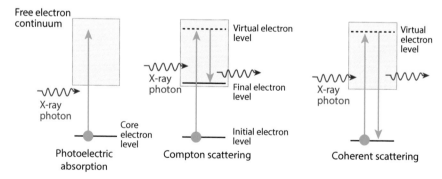

FIGURE 11-1. Absorption (left) or scatter (middle and right) of a photon. In absorption, the electron makes a transition from its initial level to another level or to the free continuum. Scatter can be considered to involve the fluctuation of an electron to a virtual level followed by an immediate downward transition, with the emission of a second photon. For Compton scattering, the final electron level is usually in the continuum, so that the electron is ejected from the atom. For coherent scattering, the electron returns to its original state.

of phase for a plane wave as it propagates through a material is expressed in terms of the index of refraction n, as for equation 9-20,

$$E = E_o e^{i(\kappa z - \omega t)} = E_o e^{i\left(\frac{2\pi}{\lambda} n z - \omega t\right)},$$

(11-1)

where E_o is the electric field amplitude, λ is the wavelength, and z is the distance traveled. The phase change $\Delta\phi$ after the wave moves through a thickness Δz of material is then

$$\Delta\phi = \frac{2\pi}{\lambda} n \, \Delta z.$$

(11-2)

From equation 9-24, the index is related to the atomic scattering factor $f = f_1 - if_2$ by

$$n = 1 - \rho_{atom} \frac{r_e \lambda^2}{2\pi} (f_1 - if_2)$$

(11-3)

The real part of the atomic scattering factor, f_1, is related to the coherent scatter, and the imaginary part, f_2, to the absorption. (The scattering factor does not include Compton scatter because the index of refraction cannot describe changes in the energy of the photon and hence the frequency of the beam.)

11.1 Free-electron theory and the real part of the index of refraction

Finding the index of refraction of a material will allow us to compute the atomic scattering factor. Because the photon energy of x rays is so much larger than valence electron binding energies, most electrons in the material interact with x rays as if they were free electrons, and the semiclassical free electron *Drude model* described below is adequate to begin to estimate the index of refraction. The index of refraction and the dielectric constant are related by Maxwell's equations for a nonmagnetic insulator,

$$\nabla \cdot \vec{E} = 0$$
$$\nabla \cdot \vec{B} = 0$$
$$\nabla \times \vec{E} = -\frac{\partial \vec{B}}{\partial t}$$
$$\nabla \times \vec{B} = \mu_o \varepsilon \frac{\partial \vec{E}}{\partial t},$$

(11-4)

where E and B are the electric and magnetic fields, respectively, μ_o is the vacuum permeability, and ε is the permittivity of the material. Maxwell's equations can be combined to form a wave equation,

$$\frac{\partial^2 \vec{E}}{\partial t^2} = \frac{1}{\mu_o \varepsilon} \nabla^2 \vec{E},$$

(11-5)

from which the phase velocity of the wave is

$$v_p = \frac{1}{\sqrt{\mu_o \varepsilon}} = \frac{c}{n},$$

(11-6)

where c is the speed of light in vacuum, and the index of refraction is

$$n = \sqrt{\frac{\varepsilon}{\varepsilon_o}},$$

(11-7)

where ε_o is the vacuum permittivity. The permittivity ε can be computed from determining the dielectric response of the material to an incident wave. Given an incident plane wave $\vec{E}$, we expect an electric displacement response

$$\vec{D} = \varepsilon \vec{E} = \varepsilon_o \vec{E} + \vec{\Pi},$$

(11-8)

where $\vec{\Pi}$ is the "polarization" field due to the displacement of the electrons with respect to the positive atomic cores. This field is essentially the same as that described for plasmas in section 3.3,

$$\vec{\Pi} = \rho_e (-q_e) \vec{r},$$

(11-9)

in which ρ_e is the electron density, q_e is the magnitude of the electron charge, and $\vec{r}$ is the displacement. For simplicity, choosing $\hat{x}$ as the direction of the electric field, the amplitude of the polarization field is

$$\Pi = \rho_e (-q_e) x,$$

(11-10)

where x is now the electron displacement from equilibrium. To compute the dielectric constant and hence the index of refraction, we need compute only the displacement of the electron due to the electric field, $E = E_o e^{-i\omega t}$. The field exerts a force

$$F = M_e a = M_e \frac{d^2 x}{dt^2} = -q_e E,$$

(11-11)

as shown in Figure 11-2. The force causes the electron to oscillate at the same frequency ω, so its displacement from equilibrium has the form $x = x_o e^{-i\omega t}$, from which

$$M_e \frac{d^2 x}{dt^2} = -M_e \omega^2 x_o e^{-i\omega t} = -q_e E_o e^{-i\omega t} \Rightarrow x_o = \frac{q_e E_o}{M_e \omega^2}.$$

(11-12)

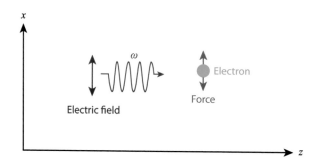

FIGURE 11-2. An x ray with frequency ω traveling in the z direction with its electric field in the x direction exerts a force in the x direction on an electron.

Equation 11-12 is the same as the plasma equation 3-7 except that now we have added a driving force, so that we are looking for the heterogenous rather than the homogeneous solution to the differential equation. The response field is proportional to the electric field and also oscillates at ω, so that it is of the form $\Pi = \Pi_o e^{-i\omega t}$, where the amplitude of the response is

$$\Pi_o = -\rho_e q_e x_o = \frac{-\rho_e q_e^2}{M_e \omega^2} E_o. \tag{11-13}$$

From equation 11-8, $D = D_o e^{-i\omega t} = \varepsilon E_o e^{-i\omega t}$, and

$$\varepsilon E_o = \varepsilon_o E_o + \Pi_o = \varepsilon_o E_o - \frac{\rho_e q_e^2}{M_e \omega^2} E_o. \tag{11-14}$$

Finally, solving for the permittivity gives

$$\varepsilon = \varepsilon_o \left(1 - \frac{\dfrac{\rho_e q_e^2}{M_e \varepsilon_o}}{\omega^2} \right) = \varepsilon_o \left(1 - \frac{\omega_p^2}{\omega^2} \right), \tag{11-15}$$

where the plasma frequency ω_p has been substituted from equation 3-9. This very simple analysis gives a reasonable approximation for the real part of the index of refraction,

$$n_r = \sqrt{\frac{\varepsilon}{\varepsilon_o}} = \sqrt{1 - \frac{\omega_p^2}{\omega^2}} \approx 1 - \frac{1}{2} \frac{\omega_p^2}{\omega^2}. \tag{11-16}$$

The small-quantity approximation is justified because the plasma frequencies of solids are in the range of a few tens of electronvolts, while x-ray energies are about a thousand times higher. The real part of the index of refraction for all materials in the x-ray regime is very slightly less than 1. This justifies the description of equation 9-19 that the real part of n is given by $1 - \delta$ and provides a value for δ,

$$n_r = 1 - \delta \approx 1 - \frac{1}{2} \frac{\omega_p^2}{\omega^2} \Rightarrow \delta = \frac{1}{2} \frac{\omega_p^2}{\omega^2}. \tag{11-17}$$

The index decrement δ falls with the square of the photon frequency, or equivalently, photon energy. It depends only on the electron density in the material.

EXAMPLE 11-1

A fake material has a density of 2 g/cm^3, an atomic number of 10 electrons per atom, and an unphysical molar mass of 10 g/mole.

a) Find the density of electrons per cubic meter.

$$\rho_e = N_A \frac{Z\rho}{M_M} \approx \left(6.2 \times 10^{23} \frac{\text{atoms}}{\text{mole}} \right) \frac{10\,e^-/\text{atom}}{10\,\text{g/mole}} \left(\frac{2\,\text{g}}{10^{-6}\,\text{m}^3} \right) \approx 1.2 \times 10^{30} \frac{1}{\text{m}^3}.$$

b) Find the plasma energy of this material, $U_p = \hbar\omega_p$, in electronvolts.

$$\omega_p^2 = \frac{\rho_e q_e^2}{M_e \varepsilon_o} \approx \frac{\left(1.2\times10^{30}\,\dfrac{1}{m^3}\right)(1.6\times10^{-19}\,\text{Coul})^2}{(9.1\times10^{-31}\text{kg})\left(8.85\times10^{-12}\,\dfrac{\text{Coul}^2}{\text{N}\cdot\text{m}^2}\right)} \approx 6.2\times10^{16}\,\frac{\text{rad}}{\text{s}}.$$

$$U_p = \hbar\,\omega_p \approx \left(\frac{4.1\times10^{-15}\text{eV}\cdot\text{s}}{2\pi}\right)\left(6.3\times10^{16}\,\frac{1}{\text{s}}\right) \approx 41\,\text{eV}.$$

c) Find the index decrement δ for a 10 keV photon.

$$\delta \approx \frac{1}{2}\frac{\omega_p^2}{\omega^2} = \frac{1}{2}\frac{(\hbar\omega_p)^2}{U^2} \approx \frac{1}{2}\left(\frac{41\,\text{eV}}{10\times10^3\,\text{eV}}\right)^2 \approx 8\times10^{-6}.$$

As expected, the index decrement δ is small.

11.2 Atomic scattering factor

Comparison of equation 11-16 with equation 11-3 for the real part of the index gives

$$n_r = 1 - \rho_{atom}\frac{r_e\lambda^2}{2\pi}f_1 \approx 1 - \frac{1}{2}\frac{\omega_p^2}{\omega^2}$$

$$\Rightarrow f_1 \approx \frac{2\pi}{r_e\lambda^2\rho_{atom}}\left(\frac{1}{2}\right)\frac{1}{\omega^2}\left[\omega_p^2\right] = \frac{\pi}{(r_e)\rho_{atom}}\left\{\frac{1}{\lambda^2\omega^2}\right\}\left[\frac{\rho_e q_e^2}{M_e\varepsilon_o}\right]$$

$$\approx \pi\left(\frac{4\pi\varepsilon_o M_e c^2}{q_e^2}\right)\frac{\rho_e}{\rho_{atom}}\left\{\frac{1}{2\pi c}\right\}^2\left[\frac{q_e^2}{M_e\varepsilon_o}\right]$$

$$f_1 \approx \frac{\rho_e}{\rho_{atom}}, \tag{11-18}$$

where the expressions for the plasma frequency ω_p and classical electron radius r_e have been inserted, and the product of the wavelength λ and frequency ω has been replaced by $2\pi c$. Since ρ_e is the electron density and ρ_{atom} is the atomic density, their ratio is just the number of electrons per atom,

$$f_1 \approx Z. \tag{11-19}$$

The atomic scattering factor, in the approximation that all the electrons in the atom are nearly free, is just the atomic number Z. For lower-energy x rays, the binding of electrons to the atomic core becomes more significant. For a combination of free and bound electrons, the coherent scattering factor is written

$$f_1 = f_o + f', \tag{11-20}$$

where f_o corresponds to coherent scatter from (nearly) free electrons, called *Thompson scattering*, and f' corresponds to coherent scattering from bound electrons, called *Rayleigh scattering*. For the atomic scattering factor written in complex form,

$$f = f_1 - if_2 = f_0 + (f' - if_2), \tag{11-21}$$

the term in parentheses represents the contribution, both Rayleigh scattering and absorption, from the bound electrons.

EXAMPLE 11-2

Compute the real part of the atomic scattering factor from the index of refraction in example 11-1.

The atomic density is

$$\rho_{atom} = N_A \frac{\rho}{M_M} \approx \frac{(6.02 \times 10^{23} \text{ atoms/mole})}{10 \text{ g/mole}} \left(\frac{2 \text{ g}}{10^{-6} \text{m}^3} \right) \approx 1.2 \times 10^{29} \frac{1}{\text{m}^3}.$$

The wavelength is

$$\lambda = \frac{hc}{U} = \frac{12.4 \text{ keV} \cdot \text{Å}}{10 \text{ keV}} = 1.24 \times 10^{-10} \text{ m}.$$

$$f_1 = \frac{2\pi}{r_e \lambda^2 \rho_{atom}} \delta$$

$$\approx \frac{2\pi}{(2.84 \times 10^{-15} \text{m})(1.24 \times 10^{-10} \text{m})^2 \left(1.2 \times 10^{29} \frac{1}{\text{m}^3} \right)} (8.4 \times 10^{-6}) \approx 10.$$

This value, as expected, is equal to Z, the number of electrons per atom.

11.3 Phase velocity

An index of refraction less than 1 makes the phase velocity, from equation 11-6, greater than c, the speed of light in vacuum. This is not a violation of relativity, since nothing actually travels at the phase velocity. The phase velocity is simply the velocity one would infer from the spacing Δz of the maxima of the electric field in the material, given by

$$\Delta z = \frac{2\pi}{\kappa} = \frac{2\pi}{\left(\frac{2\pi}{\lambda} n \right)} = \frac{\lambda}{n}, \tag{11-22}$$

where λ is the vacuum wavelength. If the time τ for the oscillation is taken from the angular frequency ω,

$$\tau = \frac{2\pi}{\omega} \Rightarrow v_p = \frac{\left(\frac{\lambda}{n} \right)}{(\tau)} = \frac{c}{n}. \tag{11-23}$$

However, neither signals nor photons travel at this inferred velocity. The photons travel at c always, and signals, produced by changing the intensity of the light beam (modulating the photon density), travel at the group velocity

$$v_g = \frac{\partial \omega}{\partial \kappa} \leq c. \tag{11-24}$$

EXAMPLE 11-3

Compute the phase velocity from the index of refraction in example 11-1.

$$v_p = \frac{c}{1-\delta} \approx c\,(1+\delta) \approx c(1+8.3\times 10^{-6}) \approx 1.000008\,c.$$

11.4 Slightly bound electrons and the phase response

A phase velocity greater than c can also be explained by using a slightly more realistic (although still semiclassical) model of the forces on the electrons in materials, rather than simply free electrons. The force equation 11-11 should include a term for the restoring "spring" force due to the attraction of the electrons to the atomic nuclei, $F=-sx$, so that there is a resonant frequency $\omega_o = \sqrt{\dfrac{s}{M_e}}$. There also should be a velocity-dependent damping force which describes the energy loss due to radiation damping, phonon-electron interaction, and the like, $F=-\varsigma M_e v$, so that

$$F = M_e \frac{d^2 x}{dt^2} = -q_e E - sx - M_e \varsigma v = -q_e E - sx - M_e \varsigma \frac{dx}{dt}. \tag{11-25}$$

The solution to this differential equation is

$$x = x_o e^{i\phi} e^{-i\omega t}, \tag{11-26}$$

where the amplitude is

$$x_o = \frac{\dfrac{q_e}{M_e} E_o}{\sqrt{(\omega_o^2 - \omega^2)^2 + \varsigma^2 \omega^2}}. \tag{11-27}$$

In general, the x-ray frequency ω is much larger than either the resonant frequency ω_o or the damping factor ς. In the limit that both ω_o and ς are zero, the amplitude is the same as for equation 11-12.

The solution to the force equation given by equation 11-26 implies that the electron motion is out of phase with the driving electric field. The phase difference found from substituting that solution into the force equation is

$$\phi = \arctan\left(\frac{-\omega\varsigma}{\omega_o^2 - \omega^2}\right). \qquad (11\text{-}28)$$

As shown in Figure 11-3, the phase difference is zero at low frequencies, changes from $-\pi/2$ to $\pi/2$ at the resonant frequency ω_o (which is much less than the x-ray frequency) and then falls to a small positive value at the very high frequencies associated with x rays.

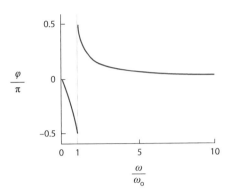

FIGURE 11-3. Phase lag/lead of the electron oscillation with respect to the electric field of the incident x ray.

The small positive phase ϕ corresponds to a slight phase lead relative to the electric field. The coherent scatter from the electron results in an emitted wave which has a slightly larger phase than the incident, so that the apparent phase velocity is larger than c. A phase lead can also be seen in simple mechanical problems, for example, if a mass on a spring is vibrated above its resonant frequency, or a child on a swing is pushed a little too often, at a frequency higher that the natural oscillation.

Equations 11-26 through 11-28 can be combined to express the complex amplitude of the response, using the identity

$$e^{i\phi} = \cos\phi + i\sin\phi = \frac{1}{\sqrt{1+\tan^2\phi}} + i\frac{\tan\phi}{\sqrt{1+\tan^2\phi}}, \qquad (11\text{-}29)$$

so that

$$\tilde{x} = x_o e^{i\phi} = \frac{\frac{q_e}{M_e}E_o}{(\omega_o^2 - \omega^2)^2 + \varsigma^2\omega^2}\left\{(\omega^2 - \omega_o^2) - i\varsigma\omega\right\}. \qquad (11\text{-}30)$$

Equations 11-13 through 11-15 imply that

$$\varepsilon = \varepsilon_o\left(1 - \frac{\rho_e q_e \tilde{x}}{\varepsilon_o E_o}\right), \qquad (11\text{-}31)$$

so that

$$n = \sqrt{\frac{\varepsilon}{\varepsilon_o}} \approx 1 - \frac{1}{2}\frac{\omega_p^2}{(\omega_o^2 - \omega^2)^2 + \varsigma^2\omega^2}\left\{(\omega^2 - \omega_o^2) - i\varsigma\omega\right\}. \qquad (11\text{-}32)$$

The index now has an imaginary part, $\text{Im}\{n\} = \beta$, as well as a real decrement, $\text{Re}\{n\} = 1 - \delta$. X-ray frequencies ω are much larger than the damping factor ς, so that, as seen in chapter 9, β, which depends on ς/ω, is generally small. The imaginary part of the index n, and hence the absorption, peaks at $\omega \sim \omega_o$. The index decrement δ will also dip near an absorption peak when $\omega \sim \omega_o$.

EXAMPLE 11-4

Compute the imaginary part of the index of refraction at 10 keV for the material in example 11-1 and use it to compute the damping factor ζ. From equation 11-5,

$$\sigma_{ab} \approx \sigma_0 \left(\frac{U_o}{U}\right)^3 Z^4 \approx (38.8 \text{ barns}) \left(\frac{1 \text{ keV}}{10 \text{ keV}}\right)^3 (10)^4$$

$$\approx (388 \text{ barns}) \frac{10^{-24} \text{ cm}^2}{\text{barn}} \left(\frac{\text{m}}{10^2 \text{ cm}}\right)^2 = 3.88 \times 10^{-26} \text{ m}^2.$$

From equations 9-22 and 9-11,

$$\beta = \frac{\lambda}{4\pi} \rho_{atom} \sigma_{ab} = \frac{1.24 \times 10^{-10} \text{ m}}{4\pi} \left(1.2 \times 10^{29} \frac{1}{\text{m}^3}\right) 3.88 \times 10^{-26} \text{ m}^2$$

$$= 4.6 \times 10^{-8}.$$

The complex index of refraction gives $\beta \approx \dfrac{1}{2} \dfrac{\omega_p^2 \varsigma \omega}{(\omega_o^2 - \omega^2)^2 + \varsigma^2 \omega^2}$. For x-ray frequencies, $\omega \gg$ both ω_o and ζ, so $\beta \approx \dfrac{1}{2} \dfrac{\omega_p^2 \varsigma \omega}{\omega^4} = \delta \dfrac{\varsigma}{\omega}$, and thus

$$\varsigma \approx \omega \frac{\beta}{\delta} = \frac{U}{\hbar} \frac{\beta}{\delta} \approx \frac{2\pi(10 \times 10^3 \text{ eV})}{4.1 \times 10^{-15} \text{ eV} \cdot \text{s}} \left(\frac{4.6 \times 10^{-8}}{8.4 \times 10^{-6}}\right) \approx 8 \times 10^{16} \frac{1}{\text{s}}. \text{ So, indeed, } \zeta \text{ is more than}$$

two orders of magnitude less than the x-ray frequency.

11.5 Kramers-Kronig relations

As we saw in the previous section, both the real and imaginary parts of the index of refraction (and hence the atomic scattering factor) are affected by the restoring forces, resonances, and damping factors. The real and imaginary components are related to each other, which is useful, because it is often easier to measure absorption than refraction. The real and imaginary parts of any complex analytic function (such as the atomic scattering factor) are connected by the *Kramers-Kronig relationship*, which in this case is

$$f_1(\omega) = Z - \frac{2}{\pi} \int_0^\infty \frac{\omega' f_2}{\omega'^2 - \omega^2} d\omega', \tag{11-33}$$

where the integral is over the primary Cauchy branch in the complex plane. This allows the real part of the atomic scattering factor, and hence the index decrement δ, to be computed numerically from a table of the absorption coefficients as a function of frequency. Care must be taken to ensure that all the factors are complete. The real and imaginary components of the index for nickel are shown in Figure 11-4.

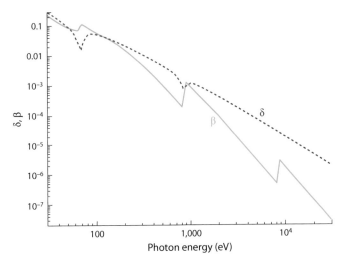

FIGURE 11-4. Real and imaginary parts of the index of refraction for nickel, from http://henke.lbl.gov/optical_constants/.

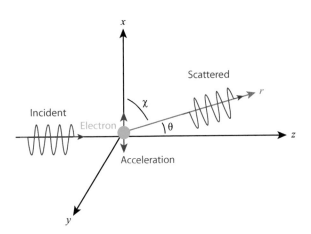

FIGURE 11-5. The geometry for the incident and scattered beam. θ is the angle between the observation direction r and the direction of the incident beam. χ is the angle between the acceleration direction and the observation direction. Because r can be out of the page, the two angles do not necessarily sum to 90°.

11.6 Coherent scatter cross section

The electron is vibrated, and thus accelerated, by the incoming field, and, as we saw in chapters 6 and 7, accelerated charges radiate. This contributes to the loss of energy from the oscillating electron, discussed in section 11.4, and gives rise to the coherent scattered radiation. Hence the intensity of the radiation, and thus the cross section for coherent scatter, can be computed from the acceleration experienced by the electron in the incident electric field. As described by equation 7-5, which was used to discuss synchrotron radiation, the field emitted from an accelerated charge is proportional to the acceleration

$$E_s = \frac{q_e}{4\pi\varepsilon_o c^2 r} a \sin \chi \tag{11-34}$$

where r is the observation distance from the accelerating charge, and χ is the angle between the direction of observation and the direction of the acceleration (which is also the direction of polarization of the incident electric field), as shown in Figure 11-5. Calculating the acceleration from equation 11-11 gives an amplitude

$$a_o = -\omega^2 x_o = -\omega^2 \frac{q_e E_o}{M_e \omega^2} = -\frac{q_e E_o}{M_e}. \tag{11-35}$$

(The minus sign in the "amplitude" is equivalent to a π phase shift, meaning the acceleration is directed in the opposite direction from the field, because the charge is negative.) The scattered field is a spherical wave oscillating at ω with an amplitude

$$\begin{aligned}
E_{so} &= \frac{-q_e}{4\pi\varepsilon_o c^2 r} a \sin \chi \\
&= \frac{-q_e}{4\pi\varepsilon_o c^2 r} \left(-\frac{q_e E_o}{M_e} \right) \sin\chi \\
&= \frac{r_e}{r} E_o \sin \chi,
\end{aligned} \tag{11-36}$$

where r_e is the classical electron radius. The intensity of the scattered radiation is thus

$$\begin{aligned}
I_{so} &= \frac{c\varepsilon_o}{2} |E_{so}|^2 \\
&= \left[\frac{c\varepsilon_o}{2} E_o^2 \right] \left(\frac{r_e}{r} \right)^2 \sin^2\chi \\
&= I_o \left(\frac{r_e}{r} \right)^2 \sin^2\chi,
\end{aligned} \tag{11-37}$$

where I_o is the incident intensity. The power radiated into a solid angle $d\Omega$ at a distance r is

$$P_s = I_s r^2 d\Omega = I_o \left(\frac{r_e}{r} \right)^2 \sin^2\chi r^2 d\Omega = I_o r_e^2 \sin^2\chi \, d\Omega. \tag{11-38}$$

The power can be expressed in terms of the differential cross section,

$$P_s = I_o d\sigma = I_o r_e^2 \sin^2\chi \, d\Omega \Rightarrow \frac{d\sigma}{d\Omega} = r_e^2 \sin^2\chi. \tag{11-39}$$

The differential cross section is r_e^2 if observed in the direction of the incident beam, for which $\chi = 90°$. Hence one meaning of the classical electron radius is that its square represents the "effective area" presented to the beam by a single electron. The differential cross section is zero if observed in the $\chi = 0°$ direction, the direction of the incident electric field polarization, as it is for dipole radiation from decelerating charge, discussed in chapter 6, because the electron creates a field polarized in the direction of its acceleration, and the field polarization must be transverse to the beam direction.

For an incident beam polarized in the x direction, as shown in Figure 11-5, χ is 90° for any observation direction in the yz plane, so that

$$\frac{d\sigma}{d\Omega} = r_e^2. \qquad \perp \text{ polarization} \qquad\qquad (11\text{-}40)$$

If the observation direction is the xz plane, then $\chi = 90° - \theta$, so that

$$\frac{d\sigma}{d\Omega} = r_e^2 \cos^2 \theta. \qquad \parallel \text{ polarization} \qquad\qquad (11\text{-}41)$$

For an unpolarized beam, the differential cross section is

$$\frac{d\sigma}{d\Omega} = r_e^2 \frac{1 + \cos^2 \theta}{2}. \qquad \text{Unpolarized} \qquad\qquad (11\text{-}42)$$

In this case, the differential cross section is still r_e^2 for forward scattering (the $\theta = 0°$ direction). For scattering at $\theta = 90°$, the differential cross section is $1/2\, r_e^2$, because half the time the polarization is in the acceleration direction, and there is no scattered beam. Integrating over all possible scattering angles, the total scattering cross section for a single free electron is

$$\sigma_s = \int_0^{2\pi}\!\!\int_0^{\pi} \frac{d\sigma}{d\Omega} \sin\theta \, d\theta \, d\varphi = r_e^2 \int_0^{2\pi}\!\!\int_0^{\pi} \frac{1 + \cos^2\theta}{2} \sin\theta \, d\theta \, d\varphi$$

$$\Rightarrow \sigma_s = \frac{8}{3}\pi\, r_e^2. \qquad\qquad (11\text{-}43)$$

This is the Thompson scatter cross section, approximately 0.67 barn.

The scatter coefficient depends on the scatter cross section similarly to the way that the absorption coefficient depended on the absorption cross section in equation 9-11,

$$\mu_s = \rho_{atom}\, \sigma_s\, f_1 = \rho_{atom}\, f_1 \frac{8}{3}\pi\, r_e^2, \qquad\qquad (11\text{-}44)$$

where ρ_{atom} is the atomic density, and f_1 is the atomic scattering factor. For photon energies for which all the electrons are nearly free, equation 11-19 implies $f_1 \sim Z$, and

$$\mu_s = \rho_{atom}\, Z \frac{8}{3}\pi\, r_e^2 = \rho_e \frac{8}{3}\pi\, r_e^2. \qquad\qquad (11\text{-}45)$$

At these high photon energies, the electrons are apt to be ejected, and Compton scattering dominates.

For low-energy x rays, the binding energies are important, and the coherent scatter coefficient is

$$\mu_s = \int_0^{2\pi}\!\!\int_0^{\pi} \frac{d\sigma}{d\Omega} f_1(\theta) \sin\theta \, d\theta \, d\varphi$$

$$= \int_0^{2\pi}\!\!\int_0^{\pi} r_e^2 \frac{1 + \cos^2\theta}{2} \big[f_o(\theta) + f'(\theta) \big] \sin\theta \, d\theta \, d\varphi, \qquad\qquad (11\text{-}46)$$

where f_o and f' are the factors for free and bound electrons, which can be a function of scattering angle. As the photon energy increases, the electrons act more nearly free, and f' decreases. A plot of relative attenuation coefficients for absorption and for scattering is shown as a function of energy in Figure 9-2.

EXAMPLE 11-5

---

a) Compute the scatter attenuation coefficient for the material of example 11-1.

$$\mu_s \approx Z\rho_{atom}\,\sigma_s \approx (10)\left(1.2\times10^{29}\,\frac{1}{m^3}\right)0.67\times10^{-28}m^2$$

$$\approx \left(80\frac{1}{m}\right)\left(10^{-2}\,\frac{m}{cm}\right)\approx 0.8\,\frac{1}{cm}$$

b) If the incident photon intensity is 10^{10} photons/s/m^2, what is the count rate of primary, unscattered photons, into a detector immediately behind a cube of the material with side $L = 1$ cm, ignoring absorption and assuming the area of the detector is L^2?

$$\Gamma_{D,\,primary} = IA_D = I_o e^{-\mu_s L}\,L^2 \approx \left(10^{10}\,\frac{photons}{m^2\cdot s}\right)e^{-\left(0.8\frac{1}{cm}\right)(1\,cm)}(0.01\,m)^2$$

$$\approx 4.5\times10^5\,\frac{photons}{s},$$

which is about 45% of the 10^6 photons/s incident on the front face of the cube.

c) Assuming the scatter acts as though it is originating from the center of the block, what is the count rate of scattered photons into the detector?

The angle subtended by the detector is approximately 45°, so the cross section for a single electron to scatter into the detector is roughly

$$\sigma_{s,\,D} = r_e^2 \int_0^{2\pi}\int_0^{\pi/4}\frac{1+\cos^2\theta}{2}\sin\theta\,d\theta\,d\varphi = -\frac{2\pi\,r_e^2}{2}\left[\cos\theta+\frac{\cos^3\theta}{3}\right]_0^{\pi/4}$$

$$= \pi r_e^2\left[\left(1+\frac{1}{3}\right)-\frac{\sqrt{2}}{2}\left(1+\frac{1}{6}\right)\right]\Rightarrow\sigma_{s,\,D}\approx 0.19\,\sigma_s$$

The number of electrons in the beam is

$$N_e \approx Z\rho_{atom}L^3 \approx (10)\left(1.2\times10^{29}\,\frac{1}{m^3}\right)10^{-6}m^3 \approx 1.2\times10^{24}\,\text{electrons.}$$

Because the intensity decreases as the beam passes through the block, we need the average intensity,

$$\langle I\rangle = \frac{1}{L}\int_0^L I_o e^{-\mu_s z}\,dz = I_o\,\frac{1-e^{-\mu_s L}}{\mu_s L}\approx 6.9\times10^9\,\frac{photons}{m^2\cdot s}.$$

Finally,

$$\Gamma_{D,\,scatter} = \langle I \rangle\, N_e\, \sigma_{s,\,D} \approx \left(6.9 \times 10^9\, \frac{\text{photons}}{\text{m}^2 \cdot \text{s}} \right)(1.2 \times 10^{24}\ \text{electrons})\,(0.19)$$

$$\times \left(0.67 \times 10^{-28}\, \frac{\text{m}^2}{\text{electron}} \right) \approx 1 \times 10^5\, \frac{\text{photons}}{\text{s}}.$$

The sum of the scatter and primary beam into the detector is about half the entire 10^6 photons/s incident on the front face. We ignored absorption and multiple scattering (scattered photons themselves being scattered, which would tend to make the scattered beam more isotropic and less forward directed).

11.7 Relativistic cross section

For very high photon energies U, relativistic effects become important in Compton scattering, and the cross section in equation 11-42 must be modified to the Klein-Nishina result,

$$\frac{d\sigma_{KN}}{d\Omega}$$

$$\approx \frac{r_e^2}{2}\, \frac{1 + \cos^2\theta}{\left(1 + \left(\dfrac{U}{M_e c^2} \right)(1 - \cos\theta) \right)^2} \left(1 + \frac{\left(\dfrac{U}{M_e c^2} \right)^2 (1 - \cos\theta)^2}{(1 + \cos^2\theta)\left(1 + \dfrac{U}{M_e c^2}(1 - \cos\theta) \right)} \right) \tag{11-47}$$

For small ratios of the photon energy to the rest mass energy of the electron, the classical result is returned.

11.8 Snell's law

The index of refraction describes the propagation of the field in the medium

$$E = E_o e^{i(\vec{\kappa}\cdot\vec{r} - \omega t)}, \tag{11-48}$$

where the propagation vector $\vec{\kappa}$ is in the direction of propagation and has magnitude

$$\kappa = \frac{2\pi}{\lambda}n. \tag{11-49}$$

Boundary conditions on the field can be used to discover the relation between the directions of the incident, reflected, and transmitted waves when they cross a boundary between two different media, as shown in Figure 11-6.

Consider the case of an interface perpendicular to the z axis with a beam polarized in the y direction, incident at an angle of θ_1 to the interface. It is important to note two facts about the angle θ_1. First, the angle does not have the same meaning as θ had in

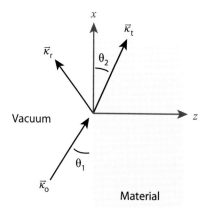

Figure 11-6. Refraction at a vacuum-to-material interface. For a perpendicularly polarized beam, the electric field is out of the plane of the page. The incident ray, with wavevector κ_o, is at an angle of θ_1 to the surface.

the previous sections, as in this case there is a planar boundary rather than an observer at a fixed direction. Second, unlike in visible light conventions, in x-ray applications the angles θ_1 and θ_2 are measured from the surface, not the normal to the surface. The reason for this convention is that, as will become clear when the reflectivity is discussed in section 11.9, one is often considering grazing incidence angles, for which the angle from the normal would be nearly 90°, and it would be awkward to express the angle as 89.99999°. For the geometry of Figure 11-6, the wavevector for the incident beam is

$$\vec{\kappa}_i = \frac{2\pi}{\lambda} n_1 \left(\cos\theta_1 \, \hat{x} + \sin\theta_1 \, \hat{z} \right). \tag{11-50}$$

The incident electric field can then be written

$$\vec{E}_i = E_o \hat{y} \, e^{\, i \frac{2\pi}{\lambda} n_1 (\cos\theta_1 x + \sin\theta_1 z)}, \tag{11-51}$$

where the factor $e^{-i\omega t}$ is unwritten and taken as understood. Similarly, the wavevector for the reflected ray is

$$\vec{\kappa}_r = \frac{2\pi}{\lambda} n_1 \left(\cos\theta_1 \, \hat{x} - \sin\theta_1 \, \hat{z} \right). \tag{11-52}$$

The reflected and transmitted fields can then be written as

$$\vec{E}_r = E_r \, \hat{y} \, e^{\, i \frac{2\pi}{\lambda} n_1 (\cos\theta_r x - \sin\theta_r z)} \quad \text{and} \quad \vec{E}_t = E_t \hat{y} \, e^{\, i \frac{2\pi}{\lambda} n_2 (\cos\theta_2 x + \sin\theta_2 z)}. \tag{11-53}$$

Maxwell's equations 11-4 can be shown to imply that the tangential component of the electric field is continuous at the interface; that is,

$$\left[E_o \, e^{\, i \frac{2\pi}{\lambda} n_1 (\cos\theta_1 x + \sin\theta_1 z)} + E_r \, e^{\, i \frac{2\pi}{\lambda} n_1 (\cos\theta_r x - \sin\theta_r z)} \right]\Bigg|_{z=0}$$

$$= \left[E_t \, e^{\, i \frac{2\pi}{\lambda} n_2 (\cos\theta_2 x + \sin\theta_2 z)} \right]\Bigg|_{z=0}$$

$$\Rightarrow E_o \, e^{\, i \frac{2\pi}{\lambda} n_1 \cos\theta_1 x} + E_r \, e^{\, i \frac{2\pi}{\lambda} n_1 \cos\theta_r x} = E_t \, e^{\, i \frac{2\pi}{\lambda} n_2 \cos\theta_2 x}. \tag{11-54}$$

For this equation to hold not just at $x = 0$, but at all x, it is necessary that the fields not change phase relationships as x increases. That means that the coefficients of the x in the exponents must be equal,

$$n_1 \cos\theta_1 = n_1 \cos\theta_r. \tag{11-55}$$

and

$$n_1 \cos \theta_1 = n_2 \cos \theta_2. \tag{11-56}$$

The first equation states that the angle of reflection equals the angle of incidence. The second equation is *Snell's law* (recall that for x-ray analysis, θ is defined from the interface, not from the normal to the interface).

The index of refraction of all materials is very close to unity in the x-ray regime. Thus θ_1 is very close to θ_2, and there is very little refraction at the interface. This is consistent with our picture of radiographs as simple shadow images of the object, without confusion from refraction or lensing effects due to the surface shape or interfaces within the structure. As we will see in chapters 12 and 15, for low energy x rays or under conditions of very high curvature, refraction can be important, but generally it is safe to assume that x rays travel straight through an object.

Because one is generally dealing with highly transparent objects in visible light refractive optics, it is useful to note another difference in Snell's law as applied to x rays. A complex index of refraction means that the angles can also be complex. For example, consider a ray entering from vacuum, so that $n_1 = 1$ and $n_2 = n$. The incident angle θ_1 is, of course, real, but

$$\cos \theta_1 = (1 - \delta + i\beta) \cos \theta_2 \tag{11-57}$$

implies that $\cos(\theta_2)$ and therefore θ_2 are both complex. Physically, of course, the actual angle is real, but the imaginary part corresponds to absorption. The component of the transmitted wavevector along the surface

$$\kappa_{t,x} = \frac{2\pi}{\lambda} n \cos \theta_2 = \frac{2\pi}{\lambda} \cos \theta_1 \tag{11-58}$$

is purely real, but the component of the wavevector into the material,

$$\kappa_{t,z} = \frac{2\pi}{\lambda} n \sin \theta_2 = \frac{2\pi}{\lambda} n \sqrt{1 - \cos^2 \theta_2}$$

$$= \frac{2\pi}{\lambda} n \sqrt{1 - \frac{\cos^2 \theta_1}{n^2}} = \frac{2\pi}{\lambda} \sqrt{n^2 - \cos^2 \theta_1}, \tag{11-59}$$

can be imaginary even if n is real. Just as for equation 9-20 a complex wavevector implies absorption. Here, the cosine itself is complex,

$$\cos \theta_2 = \frac{\cos \theta_1}{(1 - \delta + i\beta)} = \frac{(1 - \delta - i\beta) \cos \theta_1}{(1 - \delta)^2 + \beta^2} \approx \frac{1 - \delta + i\beta}{(1 - \delta)^2} \cos \theta_1$$

$$= \frac{\cos \theta_1}{1 - \delta} + i \frac{\beta \cos \theta_1}{(1 - \delta)^2}, \tag{11-60}$$

but because $\beta \ll 1$, the imaginary part of $\cos \theta_2$ is small. For most applications it is sufficient to treat $\cos(\theta_2)$ as real.

EXAMPLE 11-6

a) Compute the real part of the cosine of the angle into the material and the resultant angle for a ray incident at an 8 mrad grazing angle onto the material of example 11-1.

$$\cos\theta_2 = \frac{\cos\theta_1}{1-\delta} = \frac{0.999968}{0.999992} = 0.99998 \Rightarrow \theta_2 = 6.9 \text{ mrad.}$$

b) Repeat for a ray at 2 mrad:

$$\cos\theta_2 = \frac{\cos\theta_1}{1-\delta} = \frac{0.999998}{0.999992} = 1.000006 \Rightarrow \theta_2 = 3.6\text{i mrad.}$$

The real part of the angle is zero, which means the transmitted ray travels parallel to the interface.

c) Repeat, including the imaginary part of the index:

For a ray incident at 8 mrad,

$$\cos\theta_2 = \frac{\cos\theta_1}{1-\delta+i\beta} = 0.99998 - 4.6\times10^{-8}\text{i} \Rightarrow \theta_2 = 6.9 \text{ mrad} + 6.7\text{i } \mu\text{rad.}$$

The wavevector into the material now has a small imaginary part, so there will be some loss as the beam travels into the material.

For a ray at 2 mrad:

$$\cos\theta_2 = \frac{\cos\theta_1}{1-\delta+i\beta} = 1.000006 - 4.6\times10^{-8}\text{i} \Rightarrow \theta_2 = 13\,\mu\text{rad} + 3.6\text{i mrad.}$$

The real part of the angle is no longer zero, so the transmitted beam is no longer parallel to the interface but angled slightly into the material.

11.9 Reflectivity

The boundary condition imposed by Maxwell's equations given in equation 11-52, evaluated at $x=0$, gives one relationship between the fields,

$$E_o + E_r = E_t. \tag{11-61}$$

A second relationship is found from the condition that the transverse component of the magnetic field is also continuous (this can also be expressed as noting that the derivative of the electric field is continuous as well),

$$E_o n_1 \sin\theta_1 + E_r(-n_1 \sin\theta_1) = E_t n_2 \sin\theta_2. \tag{11-62}$$

Together these relationships lead to an expression for the reflection coefficient, r, which is defined as the ratio of the reflected to incident amplitudes,

$$r = \frac{E_r}{E_o}. \tag{11-63}$$

For the case that the electric field is polarized out of the plane of incidence (out of the page in Figure 11-6), the *Fresnel reflection coefficient* derived from the two boundary conditions is

$$r = \frac{n_1 \sin \theta_1 - n_2 \sin \theta_2}{n_1 \sin \theta_1 + n_2 \sin \theta_2}. \tag{11-64}$$

For a wave incident from vacuum onto a material of index $n = 1 - \delta + i\beta$,

$$r = \frac{\sin \theta_1 - \sqrt{n^2 - \cos^2 \theta_1}}{\sin \theta_1 + \sqrt{n^2 - \cos^2 \theta_1}}. \tag{11-65}$$

It is common to express this coefficient in terms of the component of the incident wavevector transverse to the interface. The change in wavevector on reflection is

$$\vec{Q}_1 = \vec{\kappa}_r - \vec{\kappa}_1 = \frac{2\pi}{\lambda} n_1 (\cos \theta_1 \, \hat{x} - \sin \theta_1 \, \hat{z}) - \frac{2\pi}{\lambda} n_1 (\cos \theta_1 \, \hat{x} + \sin \theta_1 \, \hat{z})$$

$$= -\frac{4\pi}{\lambda} n_1 \sin \theta_1 \, \hat{z}. \tag{11-66}$$

The magnitude of the change in wavevector is then

$$Q_1 = \left| \vec{\kappa}_r - \vec{\kappa}_1 \right| = \frac{4\pi}{\lambda} n_1 \sin \theta_1 = 2\kappa_{i,z}. \tag{11-67}$$

This quantity is known as the *momentum transfer coefficient*. A similar quantity is defined using the parameters of the second material,

$$Q_2 = 2\kappa_{t,z} = \frac{4\pi}{\lambda} n_2 \sin \theta_2. \tag{11-68}$$

The reflection coefficient can then be written

$$r = \frac{Q_1 - Q_2}{Q_1 + Q_2}. \tag{11-69}$$

For normal incidence (the incident ray is normal, perpendicular, to the surface), θ_1 is 90° and Snell's law gives $\theta_2 = 90°$ as well. In that case, for a ray incident from vacuum to a material with index n,

$$r = \frac{n_1 - n_2}{n_1 + n_2} = \frac{1 - (1 - \delta + i\beta)}{1 + (1 - \delta + i\beta)} \approx \frac{\delta - i\beta}{2}. \tag{11-70}$$

Since the measureable quantity is the intensity, not the field, the relevant quantity is the reflectivity R, which is the ratio of the reflected to incident intensity. The reflectivity is then

$$R = \left| r \right|^2 = rr^* \approx \frac{\delta^2 + \beta^2}{4} \ll 1. \tag{11-71}$$

There is essentially no reflection at normal incidence (at 90° from the surface) for a single interface for any material.

By Snell's law, equation 11-56, the angle inside a material is smaller than the incident angle in vacuum. For grazing incident angles less than a critical angle θ_c, there is no way to have an angle inside the material which satisfies Snell's law. In this case of an incident ray nearly parallel to the surface, no wave can be supported in the medium, and the incident ray is totally externally reflected. The critical angle, for the usual case that $\beta \ll \delta$, is given by setting the angle in the medium equal to 90°,

$$\cos \theta_c \approx (1 - \delta) \cos (0) \tag{11-72}$$

Thus, using small-angle approximations and equation 11-17,

$$\cos \theta_c \approx 1 - \frac{\theta_c^2}{2} = (1 - \delta) \Rightarrow \theta_c \approx \sqrt{2\delta} = \frac{\omega_p}{\omega}. \tag{11-73}$$

Because the plasma frequency is very much less than the photon frequency, total external reflection occurs for very small grazing incidence angles. This phenomenon is used extensively for mirrors for synchrotrons, x-ray microscopes, and x-ray telescopes, as discussed in chapter 12.

EXAMPLE 11-7

What is the critical angle for a 10 keV x ray incident from vacuum onto the material of example 11-1? $\theta_c = \sqrt{2\delta} \approx \sqrt{2(8.3 \times 10^{-6})} \approx 4.1 \, \text{mrad}$

For small incident angles and small β, the component of the wavevector into the material is approximately

$$\kappa_{t,z} = \frac{2\pi}{\lambda} \sqrt{n^2 - \cos^2 \theta_1} \approx \frac{2\pi}{\lambda} \sqrt{(1 - \delta)^2 - \left(1 - \frac{\theta_1^2}{2}\right)^2}$$

$$\approx \frac{2\pi}{\lambda} \sqrt{(1 - 2\delta) - (1 - \theta_1^2)} \approx \frac{2\pi}{\lambda} \sqrt{\theta_1^2 - \theta_c^2}. \tag{11-74}$$

For incident angles less than the critical angle, the transmitted wavevector is imaginary, so there is only a dying exponential "evanescent wave" into the material,

$$E = E_o e^{i\kappa_{t,z} z} = E_o e^{-\frac{2\pi}{\lambda} \sqrt{\theta_c^2 - \theta_1^2} z}. \tag{11-75}$$

For bulk materials there is no transmission, and no energy is carried away from the reflected wave. The penetration depth of the field into the material at $\theta = 0°$ is

$$L_D = \frac{\lambda}{2\pi \theta_c}. \tag{11-76}$$

11.10 Reflection coefficients at grazing incidence

For small angles the Fresnel reflection coefficient, equation 11-64 becomes

$$r = \frac{\sin\theta - \sqrt{n^2 - \cos^2\theta}}{\sin\theta + \sqrt{n^2 - \cos^2\theta}} \approx \frac{\theta - \sqrt{(1 - 2\delta + 2i\beta) - \left(1 - \frac{\theta^2}{2}\right)^2}}{\theta + \sqrt{(1 - \theta_c^2 + 2i\beta) - (1 - \theta^2)}}$$

$$\approx \frac{\theta - \sqrt{\theta^2 - \theta_c^2 + 2i\beta}}{\theta + \sqrt{\theta^2 - \theta_c^2 + 2i\beta}}, \tag{11-77}$$

where θ_1 has been written as θ, and θ_c^2 has been substituted for 2δ.

EXAMPLE 11-8

Compute the reflectivity for a 10 keV beam incident from vacuum onto the material of example 11-1, ignoring absorption (setting $\beta = 0$) for a ray incident at

a) $\theta_1 = 8$ mrad:

$$r \approx \frac{\theta - \sqrt{\theta^2 - \theta_c^2}}{\theta + \sqrt{\theta^2 - \theta_c^2}} \approx 0.075 \Rightarrow R \approx 0.00561948.$$

This angle is above the critical angle, and the reflectivity is very small.

b) $\theta_1 = 2$ mrad:

$$r \approx \frac{\theta - \sqrt{\theta^2 - \theta_c^2}}{\theta + \sqrt{\theta^2 - \theta_c^2}} \approx -0.52 - 0.85i \Rightarrow R = rr^* = 1.$$

The incident angle is below the critical angle, so, since we are ignoring absorption, there is total reflection.

The remainder of this chapter presents mathematical techniques for refining the reflectivity calculation to include absorption and surface roughness effects. In most cases, both those refinements are small. The basic result that the reflectivity is nearly zero above the critical angle, and nearly unity below it, is preserved.

Using equation 11-77 for nonzero angles and nonzero absorption can be tedious without software that easily manipulates complex numbers. To make it easier to perform real-number computations, a series of algebraic substitutions can be made. Writing for convenience,

$$\Delta = \theta_c^2 - \theta^2, \tag{11-78}$$

and noting that it is possible to express

$$\sqrt{-\Delta + 2i\beta} = \theta_A + i\theta_B, \tag{11-79}$$

where

$$\theta_A = \sqrt{-\frac{\Delta}{2} + \frac{1}{2}\sqrt{\Delta^2 + 4\beta^2}} \quad \text{and} \quad \theta_B = \sqrt{\frac{\Delta}{2} + \frac{1}{2}\sqrt{\Delta^2 + 4\beta^2}}, \tag{11-80}$$

then

$$\theta_2 \approx \theta_A + i\theta_B, \tag{11-81}$$

so

$$r \approx \frac{\theta - \theta_A - i\theta_B}{\theta + \theta_A + i\theta_B}, \tag{11-82}$$

and

$$R \approx \frac{(\theta - \theta_A)^2 + \theta_B^2}{(\theta + \theta_A)^2 + \theta_B^2}. \tag{11-83}$$

This approximation is accurate for both polarizations of the incoming x-ray beam. The reflectivity is plotted in Figure 11-7 for two values of the ratio of β to δ. The ratio is smaller for higher-energy photons. For $\theta < \theta_c$ in the limiting case that $\beta = 0$,

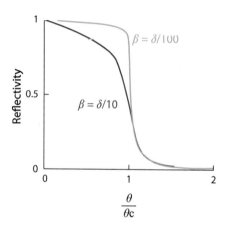

FIGURE 11-7. Plot of the reflectivity as a function of normalized grazing incidence angle for the case that $\beta = \delta/10$ and for $\beta = \delta/100$. At 5 keV, the ratio varies from about 0.001 for Be to 0.2 for Au.

$$r \approx \frac{\theta - \sqrt{\theta^2 - \theta_c^2}}{\theta + \sqrt{\theta^2 - \theta_c^2}} = \frac{\theta - i\sqrt{\Delta}}{\theta + i\sqrt{\Delta}}$$

$$\Rightarrow R \approx \frac{\theta^2 + \Delta}{\theta^2 + \Delta} = 1. \tag{11-84}$$

As before, in the absence of absorption the reflectivity is unity below the critical angle. Above the critical angle, $\Delta = \theta_c^2 - \theta^2 < 0$ and (for θ small enough that $\sin(\theta) \sim \theta$ but large enough that $-\Delta \gg \beta$)

$$r \approx \frac{\theta - \sqrt{\theta^2 - \theta_c^2}}{\theta + \sqrt{\theta^2 - \theta_c^2}} \approx \frac{\theta - \theta\left(1 - \frac{1}{2}\frac{\theta_c^2}{\theta^2}\right)}{\theta + \theta\left(1 - \frac{1}{2}\frac{\theta_c^2}{\theta^2}\right)} \approx \frac{\theta_c^2}{4\theta^2} = \frac{\omega_p^2}{4\theta^2}\frac{1}{\omega^2}. \tag{11-85}$$

The reflection coefficient falls as the photon frequency or energy squared, so the reflectivity at a particular angle drops with the fourth power of energy. As a consequence of this and of the drop in critical angle with energy, reflective surfaces are more effective for lower photon energies.

EXAMPLE 11-9

Compute the reflectivity for a 10 keV beam incident from vacuum onto the material of example 11-1 without ignoring absorption for a ray incident at

a) $\theta_1 = 8$ mrad: $\Delta = \theta_c^2 - \theta^2 \approx (4.1\,\text{mrad})^2 - (8\,\text{mrad})^2 \approx -4.7 \times 10^{-5}$.

$$\theta_A = \sqrt{-\frac{\Delta}{2} + \frac{1}{2}\sqrt{\Delta^2 + 4\beta^2}} \approx \sqrt{\frac{4.7 \times 10^{-5}}{2} + \frac{1}{2}\sqrt{(4.7 \times 10^{-5})^2 + 4(4.6 \times 10^{-8})^2}}$$
$$\approx 6.88\,\text{mrad},$$

$$\theta_B = \sqrt{\frac{\Delta}{2} + \frac{1}{2}\sqrt{\Delta^2 + 4\beta^2}} = \sqrt{-\frac{4.7 \times 10^{-5}}{2} + \frac{1}{2}\sqrt{(4.7 \times 10^{-5})^2 + 4(4.6 \times 10^{-8})^2}}$$
$$= 6.7 \times 10^{-3}\,\text{mrad},$$

so that $\theta_2 \approx \theta_A + i\theta_B \approx 6.9\,\text{mrad} + 6.7i\,\mu\text{rad}$, as was seen in example 11-6, and

$$R \approx \frac{(\theta - \theta_A)^2 + \theta_B^2}{(\theta + \theta_A)^2 + \theta_B^2} = 0.00561964.$$

This result is changed very little from the calculation with $\beta = 0$ because the reflectivity was already so small.

b) $\theta_1 = 2$ mrad: $\Delta = \theta_c^2 - \theta^2 \approx (2\,\text{mrad})^2 - (4.1\,\text{mrad})^2 \approx 1.26 \times 10^{-5}$.

$$\theta_A = \sqrt{-\frac{\Delta}{2} + \frac{1}{2}\sqrt{\Delta^2 + 4\beta^2}} \approx 0.01299\,\text{mrad},$$

$$\theta_B = \sqrt{\frac{\Delta}{2} + \frac{1}{2}\sqrt{\Delta^2 + 4\beta^2}} \approx 3.55\,\text{mrad},$$

so that $\approx \theta_A + i\theta_B \approx 0.013\,\text{mrad} + 3.6i\,\mu\text{rad}$, again in agreement with example 11.6.

$$R \approx \frac{(\theta - \theta_A)^2 + \theta_B^2}{(\theta + \theta_A)^2 + \theta_B^2} = 0.994.$$

The reflectivity dropped from $R = 1$ by 0.6% owing to the inclusion of the absorption constant and the resulting penetration into the material. The change is small because the incident angle is much smaller than the critical angle.

11.11 Surface roughness

The reflectivity calculated in section 11.9 is for an abrupt ideally flat interface. Such surfaces do not exist in reality. A real surface can be regarded from a macroscopic point of view as having a dielectric constant $n^2 = \varepsilon/\varepsilon_o$ that is a continuous function, described in terms of a gradient and interfacial width, rather than an abrupt step function. Depending on the form of the interface function, the electric field can be solved

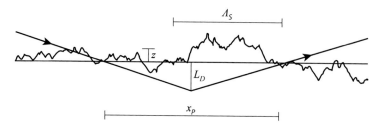

FIGURE 11-8. A surface with rms roughness height z_{rms}, correlation length Λ_s, x-ray penetration depth L_D, and path length x_p. From Kimball and Bittel, *Journal of Applied Physics* 74, no. 2 (1993): 877. Reproduced by permission of AIP Publishing.

analytically, and the reflectivity computed. Alternatively, from a microscopic point of view, the index of refraction can be regarded as changing abruptly, but across an interface which is not ideally flat, as shown in Figure 11-8. As seen in equation 11-75, a wave incident at $0 < \theta_c$ still penetrates some distance into the material, with minimum penetration L_D defined in equation 11-76. The wave then samples a distance x along the surface before reflecting. The minimum path length parameter is (artificially) defined as

$$x_p = \frac{2L_D}{\theta_c}. \tag{11-86}$$

The reflectivity is then not due to a single interaction with an abrupt smooth interface but with the interaction between the wave and the surface sampled over both transverse and longitudinal distances. Generally, the calculations are performed by taking a spatial Fourier transform of the surface and calculating the effect of a sinusoidal surface ripple at various spatial frequencies. Detailed calculations can be complicated.

The simplest model is to approximate the calculation by multiplying the ideal reflectivity by an exponential "Debye-Waller-like" factor (similar to the Debye-Waller factor that will be used in chapter 13 to assess the effect of thermal vibrations on diffraction intensity),

$$\boldsymbol{R}_{DW} \approx \boldsymbol{R}_o e^{-Q_1^2 z_{rms}^2}, \tag{11-87}$$

where $\boldsymbol{R}_o$ is the reflectivity of an ideal smooth surface, Q_1 is the transverse momentum transfer defined in equation 11-67, and z_{rms} is the interfacial width for a gradient surface or the mean square deviation of a rough surface from its ideal location,

$$z_{rms} = \sqrt{\overline{(z - z_o)^2}}. \tag{11-88}$$

The *specular reflectivity*, the fraction of the beam that is reflected at an angle equal to the incidence angle, is reduced. Unlike the reduction of the reflectivity due to the absorption coefficient β, the beam is not necessarily absorbed. Instead, the remainder of the beam is diffusely scattered, both into the glass surface and outward into the

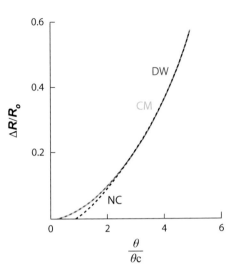

FIGURE 11-9. Plot of the reduction in the reflectivity as a function of normalized grazing incidence angle for the simple Debye-Waller-like model (DW), the Nevot-Croce model (NC), and a correlation model with long correlation length (CM). The three agree for large angles, but the NC model shows less reduction than the other two for angles near and below the critical angle, and so avoids some overcorrection of the reflectivity.

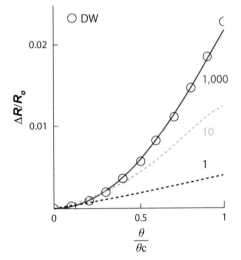

FIGURE 11-10. The effect of the correlation length-to-path length ratios of 1, 10, and 1000 on the reduction in the reflectivity for incident angles less than the critical angle. The open circles show the simple Debye-Waller-like model, which is in good agreement with the long correlation length limit.

vacuum. The roughness causes both a loss in the specularly reflected beam and also a halo of scattered radiation. Because the decrease in reflectivity is small, it can be approximated by

$$\Delta \boldsymbol{R}_{\mathrm{DW}} = \boldsymbol{R}_o - \boldsymbol{R}_{\mathrm{DW}} \approx \boldsymbol{R}_o (1 - e^{-Q_1^2 z_{rms}^2}) \approx \boldsymbol{R}_o Q_1^2 z_{rms}^2 \approx \boldsymbol{R}_o \frac{16\pi^2}{\lambda^2} z_{rms}^2 \theta^2. \quad (11\text{-}89)$$

While this approximation is seen widely because of its ease of use, it tends to overestimate the change to the reflectivity when θ is small. A somewhat better approximation is to use the *Nevot-Croce correction*,

$$\boldsymbol{r}_{\mathrm{NC}} \approx \boldsymbol{r}_o e^{-2\frac{Q_1}{2}\frac{Q_2}{2} z_{rms}^2}, \quad (11\text{-}90)$$

where $Q_2/2$ is the transverse component of the transmitted wavevector. The reflectivity is then

$$\boldsymbol{R}_{NC} = \boldsymbol{r}_{NC}\boldsymbol{r}_{NC}^* = \boldsymbol{r}_o e^{-2\frac{Q_1}{2}\frac{Q_2}{2}z_{rms}^2}\,\boldsymbol{r}_o e^{-2\frac{Q_1}{2}\frac{Q_2^*}{2}z_{rms}^2} = \boldsymbol{R}_o e^{-Q_1\,\mathrm{Re}\{Q_2\}z_{rms}^2}$$

$$\Rightarrow \Delta\boldsymbol{R}_{NC} \approx \boldsymbol{R}_o\,\frac{16\pi^2}{\lambda^2}\,z_{rms}^2\,\theta\theta_A \tag{11-91}$$

computing in the last line the case that $\theta < \theta_c$ and using the definition of θ_A from equation 11-81. The Nevot-Croce equation is in good agreement with the analytical solutions for gradient surfaces. It is plotted with the simpler Debye-Waller-like model in Figure 11-9.

More precise models of the surface take into account the correlations of surface deviations across the surface, often by using a simple model that the correlation between surface spikes has an exponential decay with correlation length Λ_s, computed from the correlation function

$$z \otimes z = \frac{1}{W}\int_0^W z(x-x')z(x')\,dx' = z_{rms}e^{-\frac{x}{\Lambda_s}}. \tag{11-92}$$

The detailed reflectivity then depends on the ratio of the correlation length to the path length, Λ_s/x_p. A plot of the model of Bittel et al. for short and long correlation length ratios is given in Figure 11-10. The reflectivity for long correlation lengths agrees with the simpler Debye-Waller model. At shorter correlation lengths, the reduction to the reflectivity is smaller. The reflectivity is then in better agreement with the Nevot-Croce model below the critical angle.

For long correlation lengths, the diffuse scatter tends to be peaked in the direction of the specular reflection, which can reduce the effective loss (but at the same time can blur the image or focal spot from reflective optics).

EXAMPLE 11-10

Compute the Debye-Waller-like and Nevot-Croce decrement to the reflectivity for a 10 keV beam incident from vacuum onto the material of example 11-1, with a surface with 0.5 nm roughness for a ray incident at

a) $\theta_1 = 8$ mrad:

$$\Delta\boldsymbol{R}_{DW} \approx \frac{16\pi^2}{\lambda^2}\,z_{rms}^2\,\theta^2\boldsymbol{R}_o \approx \frac{16\pi^2}{(1.24\,\text{Å})^2}\,(5\,\text{Å})^2\,(8\times10^{-3})^2\,\boldsymbol{R}_o \approx (0.16)(0.57\%) \approx 0.09\%;$$

$$\boldsymbol{R} = \boldsymbol{R}_o - \Delta\boldsymbol{R}_{DW} \approx 0.47\%.$$

The reflectivity drops by 16% owing to the roughness. Using the Nevot-Croce model gives

$$\Delta \boldsymbol{R}_{NC} \approx \boldsymbol{R}_o \frac{16\pi^2}{\lambda^2} z_{rms}^2 \,\theta\theta_A \approx \boldsymbol{R}_o \frac{16\pi^2}{(1.24 \text{ Å})^2} (5 \text{ Å})^2 (8\times 10^{-3})(6.87\times 10^{-3})$$

$$\approx (0.14)(0.57\%) \approx 0.08\%$$

$$\boldsymbol{R} = \boldsymbol{R}_o - \Delta \boldsymbol{R}_{DW} \approx 0.48\%.$$

The two methods give similar results.

b) $\theta_1 = 2$ mrad: $(\theta < \theta_c)$

$$\Delta \boldsymbol{R}_{DW} \approx \frac{16\pi^2}{\lambda^2} z_{rms}^2 \,\theta^2 \boldsymbol{R}_o \approx \frac{16\pi^2}{(1.24 \text{ Å})^2} (5 \text{ Å})^2 (2\times 10^{-3})^2 \boldsymbol{R}_o \approx (0.01)(99.4\%) \approx 1\%$$

$$\boldsymbol{R} = \boldsymbol{R}_o - \Delta \boldsymbol{R}_{DW} \approx 98\%.$$

The reflectivity drops by 1% owing to the roughness. This model probably overestimates the effect of roughness. Using the Nevot-Croce model, we have

$$\Delta \boldsymbol{R}_{NC} \approx \boldsymbol{R}_o \frac{16\pi^2}{\lambda^2} z_{rms}^2 \,\theta\theta_A \approx \boldsymbol{R}_o \frac{16\pi^2}{(1.24 \text{ Å})^2} (5 \text{ Å})^2 (2\times 10^{-3})(1.29\times 105)$$

$$\approx (6.6\times 10^{-5})(99.4\%) \approx 6.6\times 10^{-5}$$

$$\boldsymbol{R} = \boldsymbol{R}_o - \Delta \boldsymbol{R}_{DW} \approx 99.4\%.$$

Problems

1. Copper has a density of 9 g/cm³, atomic weight of 63 g/mole, and 29 electrons per atom. It is irradiated with a 10 keV x-ray beam.

SECTION 11.1

    a) Compute its plasma energy, in electronvolts.

    b) Compute the index of refraction decrement δ.

SECTION 11.2

    c) Compute the real part of the atomic scattering factor.

SECTION 11.3

    d) Compute the phase velocity.

SECTION 11.6

    e) Compute the cross section for scattering for a single free electron.

    f) Compute the scattering coefficient μ.

    g) Compute the fraction loss to scattering if the beam passes through 1 mm of copper.

SECTION 11.9

    h) Compute the critical angle θ_c.

SECTION 11.10–11.11

    i) Compute the reflectivity for x rays incident at 1.1 θ_c and 0.9 θ_c with and without including β and Debye-Waller-like surface roughness with a height of 5 Å.

Further reading

J. R. Christman, **Fundamentals of Solid State Physics,** John Wiley & Sons, 1987, chap. 10.

David Attwood and A. Sakdinawat, **X-Rays and Extreme Ultraviolet Radiation: Principles and Applications,** Cambridge University Press, 2016.

Jens Als-Nielsen and Des McMorrow, **Elements of Modern X-ray Physics**, 2nd ed., John Wiley & Sons, 2011.

J. Harvey, X-ray Optics, chap. 11 in M. Bass, ed., **Handbook of Optics,** 2nd ed., vol. 2, McGraw Hill, 1994.

J. C. Kimball and D. Bittel, Surface roughness and the scattering of glancing-angle x rays: Application to x-ray lenses, *Journal of Applied Physics* 74, no. 2 (1993): 877.

A. Caticha, Reflection and transmission of x rays by graded interfaces, *Physical Review B* **52**, no.13 (1995): 9214–23.

12

REFRACTIVE AND REFLECTIVE OPTICS

As was seen in chapter 11, the index of refraction for x rays of all materials is very close to unity. That makes optics for x rays considerably more challenging to manufacture than for visible light. Some optics do use refraction, as discussed next, but most x-ray optics employ either reflection at grazing incidence, discussed later in this chapter, or interference effects, as described in chapter 15.

12.1 Refractive optics

Optics for visible light typically employ refraction at two simple interfaces. A double convex lens for visible light is shown in Figure 12-1. The index of refraction of glass in the visible range is about 1.4, so the light is bent significantly towards the surface normal when it enters the glass.

The lens maker's equation,

$$L_f = \frac{R/2}{\mathrm{Re}\,\{n_2\} - 1},$$ (12-1)

gives the relationship between the focal length of a lens L_f, the radius of curvature of the lens R, and the real part of the index of refraction of the glass n_2, for a symmetric lens.

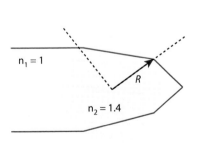

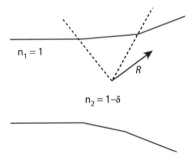

FIGURE 12-1. Visible-light convex focusing optic.

FIGURE 12-2. A convex optic will cause x rays to diverge.

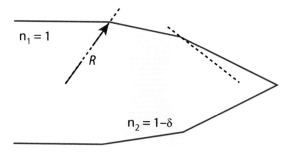

FIGURE 12-3. A concave optic will focus x rays.

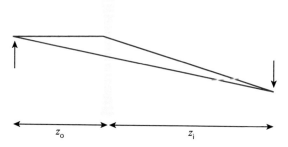

FIGURE 12-4. An inverted image is formed a distance z_i from the lens.

EXAMPLE 12-1

Compute the focal length of a visible light lens with an index of refraction of 1.4, and a radius of 1 cm.

The index of refraction is real, and $L_f = \dfrac{R/2}{n_2 - 1} \approx \dfrac{0.5 \times 10^{-2}\,\text{m}}{1.4 - 1} \approx 1.25\,\text{cm}$.

Refractive optics are more difficult to achieve in the x-ray regime. First, the real part of the index of refraction n_2 is less, not more, than 1, so the lens in Figure 12-1 will have a negative focal length and will diverge the beam, as shown in Figure 12-2. Second, because the real part of n_2 is very slightly less than 1, the focal length produced even by the negative radii (concave) lens in Figure 12-3 is rather long.

EXAMPLE 12-2

Compute the focal length of an x-ray lens with an index of refraction decrement of 10^{-6} and a radius of -1 cm.

Using the real part of the index, the focal length is

$$L_f = \frac{R/2}{1 - \delta - 1} \approx \frac{-0.5 \times 10^{-2}\,\text{m}}{-10^{-6}} \approx 5\,\text{km}.$$

The distance to an image of an object can be determined using the usual thin-lens equation,

$$\frac{1}{L_f} = \frac{1}{z_o} + \frac{1}{z_i} \Rightarrow z_i = \frac{1}{\dfrac{1}{L_f} - \dfrac{1}{z_o}} = \frac{L_f z_o}{z_o - L_f}, \tag{12-2}$$

where z_o is the distance from the object to the lens, and z_i is the distance from the lens to the image, as shown in Figure 12-4. There will not be a real image (with positive z_i) unless $z_o > L_f$. The smallest sum for $z_o + z_i$ occurs for $z_o = z_i = 2L_f$. The distances will be large because the focusing effect of a single lens of moderate radius is negligible.

The problem of the unmanageably large focal length of a single refractive x-ray lens can be overcome by using large numbers of surfaces in a compound lens. A series of two optics is shown in Figure 12-5. The image from the first lens becomes the object for the second. If the lenses are a small distance d apart, the object distance for the second lens is

$$z_{o2} = d - z_{i1} \approx -z_{i1} = -\frac{L_f z_o}{z_o - L_f}. \tag{12-3}$$

The image distance after the second lens is then

$$z_{i2} = \frac{1}{\dfrac{1}{L_f} - \dfrac{1}{z_{o2}}} \approx \frac{1}{\dfrac{1}{L_f} + \dfrac{z_{o1} - L_f}{L_f z_{o1}}} = \frac{L_f z_o}{2z_o - L_f}. \tag{12-4}$$

Considered together, the two lenses have a focal length L_f' given by

$$L_f' = \frac{1}{\dfrac{1}{z_{o1}} + \dfrac{1}{z_{i2}}} = \frac{1}{\dfrac{1}{z_{o1}} + \dfrac{2z_{o1} - L_f}{L_f z_{o1}}} = \frac{L_f}{2}. \tag{12-5}$$

A double lens has half the focal length of a single lens. A series of N lenses has an approximate focal length of

$$L_f' = \frac{L_f}{N}. \tag{12-6}$$

An optic with a useful focal length requires many individual lenses, small radii of curvature, and as large as possible an index decrement δ at the wavelength employed.

EXAMPLE 12-3

Compute the focal length of a compound x-ray lens with an index of refraction decrement of 10^{-5} and a series of 100 individual symmetric lenses, each with a radius of -1 mm.

$$L_f = \frac{1}{N} \frac{1}{n_2 - n_1} \frac{R}{2} \approx \frac{1}{2N} \frac{R}{10^{-5}} \approx \frac{1}{200} \frac{10^{-3} \text{ m}}{10^{-5} \text{ m}} \approx 0.5 \text{ m}.$$

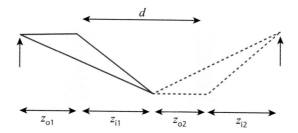

FIGURE 12-5. A double-convex optic. The drawing is not to scale and grossly exaggerates the amount of refraction which will occur in each single lens. In practice, the separation of the lenses, d, is much less than the image distance z_{i1}.

FIGURE 12-6. A compound lens made of bubbles in an epoxy material.

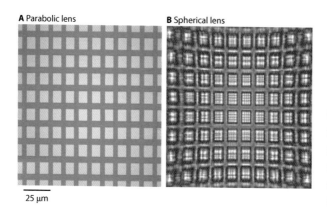

A Parabolic lens **B** Spherical lens

25 μm

FIGURE 12-7. Simulated image from a parabolic and a spherical lens. From B. Lengeler and C. G. Schroer, Refractive X-Ray Lenses, chap. 37, *Handbook of Optics*, 3rd ed., vol. 5, McGraw-Hill, 2010.

Some of the first compound lenses were made of bubbles in epoxy materials, as shown in Figure 12-6. Of course, the thickness of a series of N lenses will not be negligible, so thick-lens corrections must be made to equations 12-2 and 12-6. In addition, because of the nonnegligible thickness, there will be substantial absorption in the lens. The best refractive compound lenses are made of low-Z elements, such as Be, which have low absorption coefficients. Because Be is toxic and difficult to machine, some lenses are made of aluminum or silicon.

Spherical lenses produce good images if the incoming rays are paraxial; that is, they hit the lens near the center of the lens, and nearly on axis. When used off-axis, spherical lenses produce spherical aberration, as shown in Figure 12-7. Because the lens diameter is small, restricting the lens to paraxial use severely limits the useable open area of the lens. Parabolic lenses, as shown in Figure 12-8, produce better images.

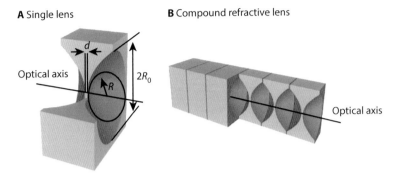

A Single lens

B Compound refractive lens

Optical axis

d

$2R_0$

R

Optical axis

FIGURE 12-8. Parabolic compound lens. From C. Schroer et al., *Proceedings SPIE* 4503 (2002): 62.

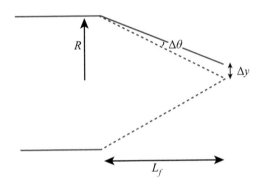

R

$\Delta\theta$

Δy

L_f

FIGURE 12-9. Diffraction-limited spot size due to a finite aperture. The deflection of the ray by $\Delta\theta$ causes widening of the spot by Δy.

Because the lens radii must still be small, even for parabolic lenses, the aperture of the lens cannot be large. The finite aperture size creates a diffraction limit for the smallest possible focal spot. A rough approximation of the effect can be determined from considering Heisenberg uncertainty. If the photon goes through a lens of radius R, the uncertainty in position is $\Delta y = 2R$, so that the uncertainty in the y component of the photon momentum can be estimated from

$$\Delta p_y \Delta y = \Delta p_y \, 2R \sim h, \tag{12-7}$$

where h is Planck's constant.

This corresponds to an uncertainty in the angle of the photon path, $\Delta\theta$, as shown in Figure 12-9, given by

$$\Delta p_y \sim p\Delta\theta \sim \frac{U}{c}\Delta\theta \Rightarrow \Delta\theta \sim \frac{\Delta p_y}{U/c} \sim \frac{h/2R}{U/c} \sim \frac{\lambda}{2R}, \tag{12-8}$$

where the momentum of the photon is given by U/c, and the last substitution is $U = hc/\lambda$. The resulting blur gives a focal spot radius of

$$R_{spot} = \Delta y \sim L_f \Delta \theta \sim \frac{\lambda L_f}{2R}. \tag{12-9}$$

This concept, that confining the beam in diameter results in blurring, is seen again in diffraction analysis, in section 13.12.3, and for the focal spot from zone plates, in section 15.2.

EXAMPLE 12-4

Estimate the focal spot size for a 1 mm diameter lens with a focal length of 1 m, used with 1 Å radiation.

$$R_{spot} \sim \frac{\lambda L_f}{2R_{lens}} \approx \frac{(10^{-10}\,\text{m})(1\,\text{m})}{10^{-3}\,\text{m}} \approx 0.1\,\mu\text{m}.$$

This growth in focal spot size for lenses of small radii can be somewhat mitigated by using lenses with graded, rather than abrupt, apertures. Beams with small focal spots produced with refractive optics can be used to perform analysis with high spatial resolution, for example, to create raster-scan images of elemental composition using x-ray fluorescence. Refractive optics can also be used in x-ray microscopy to magnify small objects. The very small aperture and relatively long focal length of refractive lenses means that refractive optics would capture only a tiny fraction of the output from conventional tube sources. Refractive optics are generally better suited to the narrow, high-intensity beams from synchrotrons.

12.2 Reflective optics

12.2.1 ELLIPTICAL MIRRORS

As seen in chapter 11, the reflectivity of any surface at normal incidence (when the beam is normal to the surface, $\theta = 90°$) is extremely small. However, the reflectivity of all surfaces at grazing incidence approaches unity. All reflective optics for x rays (except for those which use interference effects, as described in chapter 15) employ grazing incidence. A typical geometry is sketched in Figure 12-10 for an elliptical mirror. The major and minor axes of the ellipse are L_e and w, respectively. The foci of the ellipse are located $L_{f,e}$ from the center of the ellipse, where

$$L_{f,e} = \sqrt{L_e^2 - w^2}. \tag{12-10}$$

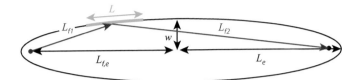

FIGURE 12-10. An elliptical optic of length L, shown in blue. The major and minor axes of the ellipse are L_e and w, respectively. The input focal length of the optic is L_{f1}, which is the distance from the source, placed at one of the foci of the ellipse, to the optic. The output focal length is L_{f2}. The sketch is not to scale. In reality $w \ll L_e$.

An ellipse has the property that the sum of the path lengths from the foci is independent of the point of impact on the ellipse,

$$L_{f1} + L_{f2} = 2L_e. \tag{12-11}$$

For the symmetric case (when the section of the ellipse used for the optic would be the horizontal piece at the top of Figure 12-10) then,

$$L_{f1} = L_{f2} = L_e. \tag{12-12}$$

For the optic to reflect the x rays, the angle of incidence must be less than the critical angle given in equation 11-73. The angle of incidence for the elliptical optic is lowest for the symmetric case, for which

$$\tan \theta \approx \theta \approx \frac{w}{L_{f,e}}. \tag{12-13}$$

Since the critical angle is typically on the order of $\sim 10^{-3}$ radians, the ellipse must be very elongated, on the order of a thousand times longer than it is thick. In that case, L_e, L_{f1}, and $L_{f,e}$ are all nearly equal.

EXAMPLE 12-5

Calculate the major and minor axes for a symmetric elliptical lens with an input focal length (source-to-optic distance) of 10 m, designed to be used with an incidence angle of 1 mrad.

For the symmetric case, the input focal length of the optic, L_{f1}, is approximately equal to the major axis of the ellipse, so $L_e \cong 10$ m. The minor axis is $w \approx L_e \theta \approx (10 \, \text{m})$ $(10^{-3}) = 10^{-2}$ m. The foci are then just $L_e - L_{f,e} = L_e - \sqrt{L_e^2 - w^2} \approx 5 \, \mu\text{m}$ from the edge of the ellipse.

The optic is, of course, only a small section of the ellipse. However, because the incidence angle is small, the length of the optic, L, must be fairly long, as shown in Figure 12-11.

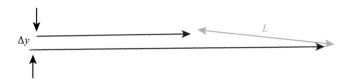

Figure 12-11. Geometry of a beam of height Δy hitting an optic of length L at grazing incidence.

EXAMPLE 12-6

Calculate the required optic length for an optic designed to be used with an incidence angle of 1 mrad, and a synchrotron beam 1 mm in height, as shown in Figure 12-11.

$$L \approx \frac{\Delta y}{\theta} \approx \frac{10^{-3}\,\text{m}}{(10^{-3})} \approx 1\,\text{m}.$$

EXAMPLE 12-7

For a 1 m long elliptical mirror to be used at an incident angle of 1 mrad, with a 10 m focal length, calculate how far the optic bends downward from the center to the edge of the optic.

The equation of the ellipse is

$$\frac{y^2}{w^2} + \frac{z^2}{L_e^2} = 1 \Rightarrow \Delta y = w - w\sqrt{1 - \frac{z^2}{L_e^2}} \approx (10^{-2}\,\text{m})\left(1 - \sqrt{1 - \frac{(0.5\,\text{m})^2}{(10\,\text{m})^2}}\right) \approx 13\,\mu\text{m}.$$

In principle, an ellipsoidal optic produces a two-dimensional focal spot from the input beam. In practice it is difficult to manufacture a 1 m long ellipsoidal optic with profile control to microns, and very low surface roughness. Low surface roughness is required, because, as was discussed in section 11.11, moderate roughness reduces the reflectivity and causes diffuse scatter, which can blur the image. It is simpler to manufacture optics which are curved in only one dimension, for example a piece of an elliptical tube with its axis coming out of the plane of the paper, as illustrated in Figure 12-10.

Such an optic would focus the synchrotron beam of Figure 12-11 to a line coming out of the plane of the figure. A second optic, with its long axis parallel to the first, but the minor axis at 90° to it, is required to change the focus from a line to a point. The combination of two perpendicular optics, as shown in Figure 12-12, is called a *Kirkpatrick-Baez mirror*. These mirrors are commonly used in synchrotron beam lines to condense the spreading beam.

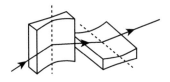

FIGURE 12-12. Kirkpatrick-Baez arrangement of two orthogonal cylindrical mirrors. From M. K. Joy and B. D. Ramsey, Astronomical X-Ray Optics, chap. 47, *Handbook of Optics*, 3rd ed., vol. 5, McGraw-Hill, 2010.

12.2.2 WOLTER OPTICS

Ellipsoidal mirrors work well if the source can be placed at the focal point. However, when using the optic for imaging, the object is necessarily extended beyond a point. In this case the elliptical mirror produces strong aberrations. Good image quality requires a double-bounce system, such as the Wolter optic arrangement of Figure 12-13. The mirror surfaces are generally coated with a high-Z material such as gold or nickel to increase the critical angle for reflection, since the critical angle, from equations 11-73 and 3-9, is proportional to the square root of the electron density. The optics are not very efficient, because in addition to accepting radiation from only a small angle, they have large holes in the center. For that reason, several sets of optics are generally nested together, as shown in Figure 12-14. A false-color x-ray astronomy image taken with a Wolter optic is shown in Figure 12-15.

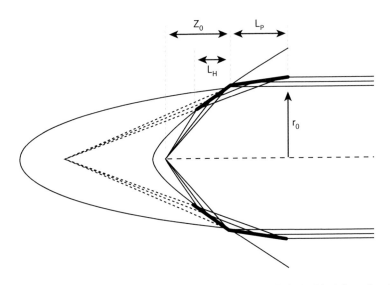

FIGURE 12-13. Geometry of a Wolter I–type x-ray optic. Parallel light incident from the right is reflected at grazing incidence on the interior surfaces of the parabolic and hyperbolic sections; the image plane is at the focus of the hyperboloid. From M. K. Joy and B. D. Ramsey, Astronomical X-Ray Optics, chap. 47, *Handbook of Optics*, 3rd ed., vol. 5, McGraw-Hill, 2010.

FIGURE 12-14. An x-ray optics module for the XMM observatory. 58 electroformed nickel Wolter I optics are nested to increase the effective x-ray collecting area. Photo courtesy of Airbus Ds GmbH and ESA.

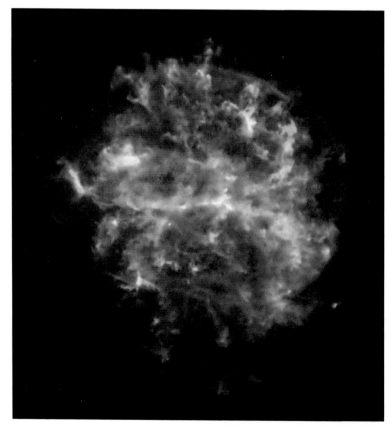

FIGURE 12-15. A Chandra X-ray Space Telescope image of the supernova remnant G292.0+1.8. The colors in the image encode the x-ray energies emitted by the supernova remnant. Photo NASA Chandra X-ray Observatory.

12.2.3 CAPILLARY OPTICS

One of the earliest, and easiest to manufacture, reflective optic is a simple capillary tube, as shown in Figure 12-16. The capillary acts as a cylindrical mirror, requiring multiple reflections to guide the x rays down the pipe. Capillary-tube x-ray pipes were used as early as the 1920s. Glass surfaces are inherently very smooth, so no polishing

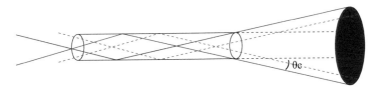

FIGURE 12-16. X rays traveling down a simple hollow glass tube.

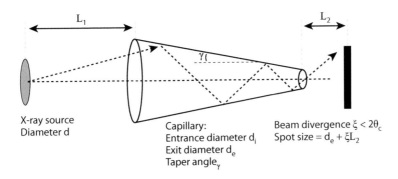

FIGURE 12-17. Condensing capillary optic. From D. H. Bilderback and S. W. Cornaby, Single Capillaries, chap. 52, *Handbook of Optics*, 3rd ed., vol. 5, McGraw-Hill, 2010.

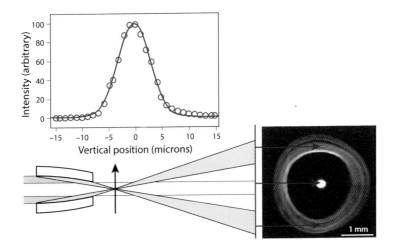

FIGURE 12-18. (Upper panel) Profile of intensity at the focus of a capillary producing a spot size of 5 μm at a distance of 20 mm beyond the tip of the capillary. (Lower panel) The far-field image shows the direct beam (center dot) passing through the capillary and the single-bounce reflected beam forming the outer ring of intensity. From D. H. Bilderback and S. W. Cornaby, Single Capillaries, chap. 52, *Handbook of Optics*, 3rd ed., vol. 5, McGraw-Hill, 2010.

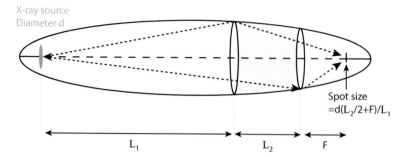

FIGURE 12-19. Focusing ellipse for single-bounce capillary. L_1 is the distance from the x-ray source to the capillary entrance. L_2 is the length of the capillary, and F is the distance from the tip of the capillary to the focus. The shaded area represents the capillary tube. From D. H. Bilderback and S. W. Cornaby, Single Capillaries, chap. 52, *Handbook of Optics*, 3rd ed., vol. 5, McGraw-Hill, 2010.

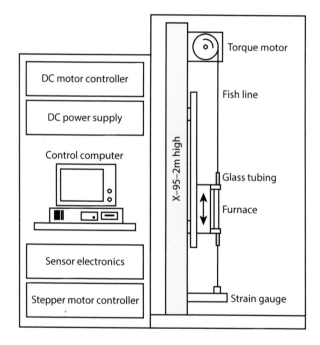

FIGURE 12-20. A glass tube is suspended in an electric furnace. The motor keeps a constant tension as the glass softens and yields during drawing. The furnace is programmed to produce the desired conical, elliptical, or parabolic shape. From D. H. Bilderback and S. W. Cornaby, Single Capillaries, chap. 52, *Handbook of Optics*, 3rd ed., vol. 5, McGraw-Hill, 2010.

is required. The diameter of the pipe must be kept small enough so that the global divergence of the beam subtended by the capillary (the tube diameter divided by the source-to-tube distance) is less than the critical angle for total reflection.

Condensing capillary optics, as shown in Figure 12-17 are used to produce small focused x-ray spots. An example of a focused beam is shown in Figure 12-18. Better focusing is produced if the tapered optic is shaped to an elliptical surface, as shown in Figure 12-19. Shaped optics are produced in computer-controlled drawing towers, as shown in Figure 12-20.

12.2.4 POLYCAPILLARY OPTICS

Polycapillary optics are arrays of a large number of small hollow glass capillary tubes. X rays are guided down these curved and tapered tubes by multiple reflections in a manner analogous to the way fiber optics guide light. They differ from single-bore capillaries and x-ray mirrors in that the focusing or collecting effects come from the overlap of the beams from thousands of channels, rather than from a few surfaces. Generally, this results in relatively efficient collection, especially from large divergent sources such as conventional x-ray tubes, but does not produce submicron beam spot sizes.

X rays can be transmitted down a curved hollow tube as long as the tube is small enough, and bent gently enough, to keep the angles of incidence less than the critical angle for total reflection, θ_c. As shown in Figure 12-21, the angle of incidence for a ray near one edge (toward the center of curvature) increases with tube diameter. The requirement that the incident angles remain less than the critical angle necessitates the use of small channel sizes, typically between 2 and 50 μm. The optics are produced by pulling large-diameter glass tubes to create small-diameter tubes, stacking and pulling them together, and repeating. The final pull is designed to create a section with the desired shape, from which the ends are cut away, as shown in Figure 12-22.

The maximum source size that is captured by the optic is approximately

$$w_{source} \sim L_f \theta_c, \tag{12-14}$$

FIGURE 12-21. X rays traveling in a bent capillary tube. The trajectory of the ray entering at the top at grazing incidence is projected onto the page, but in three dimensions will "toboggan" in a constant-radius spiral. The x ray entering at the bottom (closest to the center of curvature) strikes at a larger angle.

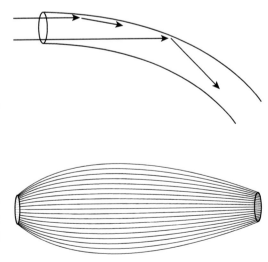

FIGURE 12-22. Sketch of the interior channels of a monolithic polycapillary optic with a short input and longer output focal length.

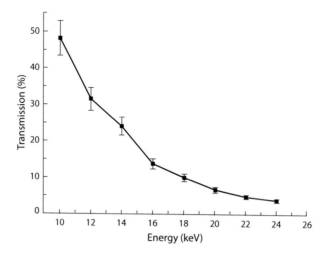

F<small>IGURE</small> 12-23. Transmission versus energy for a focusing optic with a 58 mm input focal length and 119 mm output focal length. Data from Wei Zhou.

where L_f is the input focal length of the optic, and θ_c is the critical angle for reflection. Rays originating from outside this range are incident on the optic channels at too high an angle to be reflected. Smaller sources allow for smaller input focal lengths and hence higher beam intensity.

EXAMPLE 12-8

Calculate the largest usable source size for a polycapillary optic with an input focal length of 50 mm used at 10 keV.

Polycapillary optics are made of borosilicate glass, which has a complex chemical formula but a similar density and hence electron density to simple SiO_2. The plasma frequency for SiO_2 was calculated in the problems at the end of chapter 3 to be $U_p = 30$ eV. From equation 11-73,

$$\theta_c = \frac{U_p}{U} = \frac{30\,\text{eV}}{10\,\text{keV}} = 3\,\text{mrad}.$$

$$w_{source} \sim L_f \theta_c \approx (50\,\text{mm})(3\,\text{mrad}) \approx 150\,\mu\text{m}.$$

A larger source could be used, but x rays from only the central 150 μm would be collected by the optic.

The optic efficiency, also called *transmission*, is the ratio of the number of photons exiting the channels to the number incident on the front face of the optic. Transmission of a focusing lens as a function of photon energy is shown in Figure 12-23.

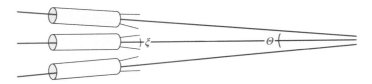

FIGURE 12-24. Global divergence Θ and local divergence ξ from a polycapillary optic.

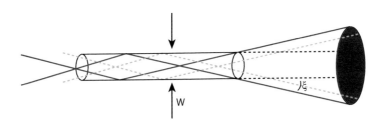

FIGURE 12-25. The spot size is enlarged from the channel diameter w because of the local divergence, which arises from reflections at angles up to the critical angle.

As shown in Figure 12-24, the output from a polycapillary focusing optic has both global divergence Θ and local divergence ξ. Even for a collimating optic, for which the channels are parallel on output ($\Theta = 0$), the output divergence ξ is not zero but is approximately given by twice the critical angle and therefore is dependent on the x-ray energy.

Assuming perfect overlap, the spot size at the focal point is determined by the spot size from each individual capillary channel, which depends on channel size w, output focal length L_f, and local divergence ξ, as shown in Figure 12-25, as

$$w_{spot} \approx w + L_f \cdot \xi. \tag{12-15}$$

Because of the divergence from each channel, optics with smaller focal lengths have smaller spot sizes, as do measurements at higher photon energies.

The transmission losses and the spread of the beam due to the local divergence will result in a beam spot which is less intense than the emission spot on the anode. This result is consistent with the second law of thermodynamics, which requires that the photon state density in phase space cannot increase for any kind of optic. In optics this is generally known as *Liouville's theorem* and is restated as "brilliance is conserved" (for an ideal optical system, with no losses) or that the focal spot cannot be more brilliant that the source. Since brilliance (as defined in section 5.3) is the photon flux per area per solid angle per energy bin, this theorem can also be expressed as saying the product of beam angle × area cannot be decreased, as sketched in Figure 12-26. The total divergence of an optic, from both Θ and ξ, can be estimated by rotating a high-quality crystal in the beam and measuring the angular width of the peak, as shown in Figure 12-27. (A small part of the angular width is due to the energy bandwidth of the x-ray beam. This is discussed in more detail in

FIGURE 12-26. Liouville's theorem requires that $A_o\Omega_o \leq A_f\Omega_f$ for each photon energy transmitted by any optic.

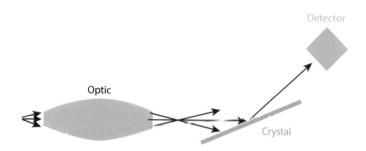

FIGURE 12-27. Rocking curve geometry for focusing optic.

chapter 13, but as noted in chapter 4, the width of a characteristic x-ray emission line is typically just a few electronvolts, which contributes an insignificant amount to the angular width.) The local divergence ξ has been measured using collimating optics, and typically is $\sim 1.3\theta_c$. The factor 1.3 is an experimentally determined parameter that arises because most of the beam has a divergence less than the maximum divergence of $2\theta_c$ produced by reflection at the critical angle (the maximum that could be expected is θ_c on each side, for a full cone width of $2\theta_c$). Unlike the case for pinhole collimation, the local divergence of the beam does not depend on the source size, although it should be remembered, as noted in equation 12-14, that large sources may not be efficiently captured by the optic.

EXAMPLE 12-9

Calculate the expected output focal spot size for a polycapillary optic with a channel size of 3.4 µm and an output focal length of 9 mm used at 20 keV.

First, we need the critical angle at this energy:

$$\theta_c = \frac{U_p}{U} \approx \frac{30\,\text{eV}}{20\,\text{keV}} \approx 1.5\,\text{mrad}.$$

The local divergence is $\xi \sim 1.3\theta_c \approx (1.3)(1.5\,\text{mrad}) \approx 2\,\text{mrad}$. The blur due to the local divergence is $\text{blur} \approx L_f \cdot \xi \approx (9\,\text{mm})(2\,\text{mrad}) \approx 18\,\text{µm}$.

The focal spot width is thus the blur plus the channel size, $w_{spot} \approx w + L_f\xi \approx 21\,\text{µm}$.

EXAMPLE 12-10

Calculate the expected angular width for a 10 mm output diameter polycapillary optic with an output focal length of 50 mm used at 10 keV.

The local divergence is $\xi \sim 1.3\theta_c \approx (1.3)(3 \text{ mrad}) \approx 3.9 \text{ mrad}$.
    The convergence of the beam is given by

$$\Theta \approx 2\arctan\left(\frac{10 \text{ mm}/2}{50 \text{ mm}}\right) \approx 199 \text{ mrad}.$$

The divergences add in quadrature, as $\sqrt{\Theta^2 + \xi^2} \approx 199 \text{ mrad}$.

In this case the total width is dominated by the global divergence.

The global divergence can be found separately from the local divergence by measuring the slope of the beam size versus distance from the optic, as shown in Figure 12-28. In that instance an imaging detector was used. However, near the focal point, the spot size can be small compared with the pixel size of an imaging detector. The spot can still be measured by using a knife-edge technique, as shown in Figure 12-29. The resulting spot size from a typical polycapillary optic as a function of the distance from the optic output is shown in Figure 12-30.

Focusing the beam increases the intensity, for example, onto a small sample, relative to pinhole collimation. (To avoid confusion with the discussion around Figure 12-26, it should be noted that the brightness does not increase, as the beam angle increases when the spot size is decreased, however, increased intensity results in higher

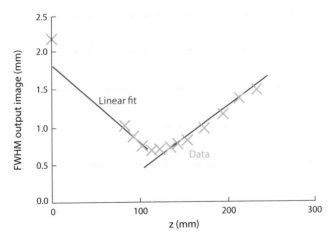

FIGURE 12-28. Global divergence of focusing optic found by a linear fit to the beam size on an imaging detector versus optic-to-detector distance. The slope of the data is the global divergence. From C. MacDonald, *X-Ray Optics and Instrumentation* 2010 (2010): doi:10.1155/2010/867049. Data from Wei Zhou and Dip Mahato.

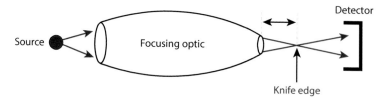

FIGURE 12-29. Geometry for a knife-edge measurement.

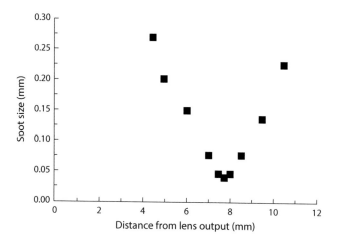

FIGURE 12-30. Measured beam width as a function of distance from the end of the optic. The focal spot size for this lens is about 21 μm. From C. MacDonald, *X-Ray Optics and Instrumentation* 2010 (2010): doi:10.1155/2010/867049. Data from Abrar Hussain.

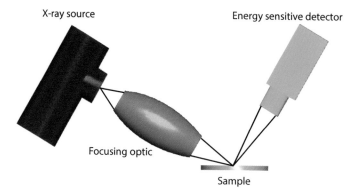

FIGURE 12-31. Geometry for XRF mapping. Courtesy of XOS Inc.

photon flux onto a small sample and is often valuable.) Focusing polycapillary optics are widely used in x-ray fluorescence (XRF) and spectrometry (XRS). A typical geometry for conventional sources is shown in Figure 12-31. Rastering the sample then allows for spatial mapping. An example of the use of micro-x-ray fluorescence (MXRF) to analyze a volcanic inclusion is shown in Figure 12-32.

Instead of using the focusing optic on the excitation side, the focusing optic can be used to collect the fluorescence radiation from the sample. Confocal systems such as

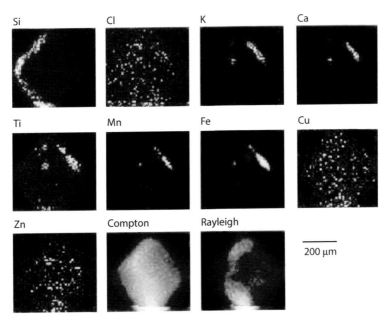

Figure 12-32. MXRF maps of a quartz phenocryst with small volcanic glass inclusions. Courtesy of XOS Inc.

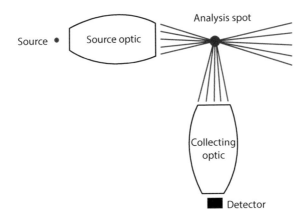

Figure 12-33. Sketch of microfluorescence experiment, showing that overlap of irradiation and collection volumes yields three-dimensional spatial resolution.

the one sketched in Figure 12-33 provide the double benefit of enhanced signal intensity and three-dimensional spatial resolution. Significant reduction in data collection times can be achieved.

12.2.5 ARRAY OPTICS

Micropore (MCP) optics are thin sheets of glass perforated with holes, with length-to-diameter ratios of less than 100. They are designed to depend on a single horizontal and vertical bounce to deflect the ray toward the desired direction and so, in operation, resemble elements of a series of nested mirrors, as in Figure 12-14, only on a smaller scale. They are made from square glass fibers with an inner core and outer glass cladding arranged in an appropriate configuration, for example, a square array of typically 50×50 fibers. These can be arranged in the desired geometry in a block. The assembly is then sliced into thin sheets and the inner core glass is etched away, leaving the hollow channels, as shown in Figure 12-34. The resulting wall thickness can be as small as 1 μm. Because the optics are single-bounce devices, even the relatively rough surface produced by etching yields adequate reflectivity. Measured surface roughnesses are 2.5–5 nm. Because the technology was developed for the multichannel plates used in photomultiplier tubes (as described in section 2.4), the optics are sometimes called *microchannel plate optics*. The holes can also be circular, as shown in Figure 12-35.

The basic principle of operation of micropore optics is shown in Figure 12-36. X rays are reflected from the source to the focal point, so long as the maximum angle at the outer edge of the optic is less than the critical angle. This limits the usable radius R of the optic to $L_f \theta_c$, where L_f is the distance from the optic to the focal point. Similarly, several considerations limit the thickness along the beam axis for practical optics. Although a thicker optic intercepts a larger fraction of the incident beam, multiple reflections, which remove energy from the focal spot, become possible if the thickness L is too great. This requires that

$$L < \frac{w_{channel}}{\theta_c},$$
(12-16)

where $w_{channel}$ is the channel size. Additionally, the spread in the focal spot along the beam axis is roughly twice the optic thickness as shown in Figure 12-37. The optic is optimized for a channel diameter-to-thickness ratio equal to the critical angle. The cross section in Figure 12-36 shows only the vertical reflections from the horizontal faces. A single horizontal reflection also occurs from the vertical faces. Square

FIGURE 12-34. Manufacturing of a microchannel plate optic. A square core is inserted into a square cladding, and fibers are drawn. The fibers are stacked in a block, and multifibers are drawn. From M. Beijersbergen, Pore Optics, chap. 49, *Handbook of Optics*, 3rd ed., vol. 5, McGraw-Hill, 2010.

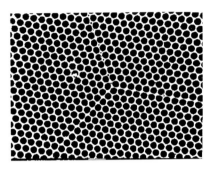

FIGURE 12-35. Circular-pore microchannel plate (most MCPs have square channels). From G.W. Fraser et al., *Nuclear Instruments and Methods in Physics Research Section A* 334, nos. 2–3 (1 October 1993): 579–88. Copyright Elsevier.

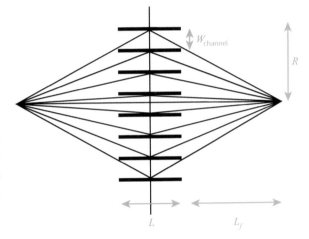

FIGURE 12-36. Focusing from a point source using a multichannel plate optic. The sketch is not to scale: the thickness in the beam direction, L, is very much less than the focal distance L_f.

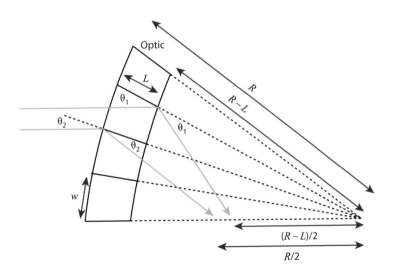

FIGURE 12- 37. Focusing a collimated beam with a slumped multichannel plate.

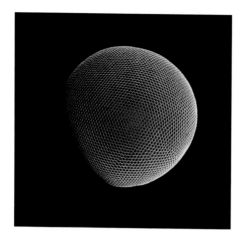

FIGURE 12-38. Krill eye. Compound eye of the Antarctic krill *Euphausia superba*. Photo by Gerd Alberti and Uwe Kils.

pores, as shown in Figure 12-34, are more efficient than circular pores, shown in Figure 12-35.

For the optic to be used to focus a parallel beam, for example, for astronomical applications, or to produce a quasiparallel beam from a point source, the plate must be curved, as shown in Figure 12-37. This curvature can be achieved by thermally "slumping" the optic onto a spherical mandrel. Slumped multichannel plate optics closely

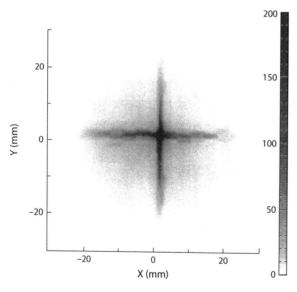

FIGURE 12-39. Output from MCP with 11 μm square pores and thickness-to-diameter ratio of 40:1 illuminated with 1.74 keV Si K x rays. A. N. Brunton, G. W. Fraser, J. E. Lees, and I.C.E. Turcu, Metrology and modeling of microchannel plate x-ray optics, *Applied Optics* 36: 5461–70.

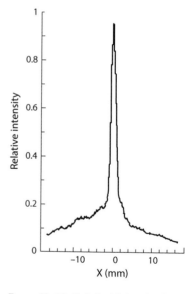

FIGURE 12-40. Detailed intensity from Figure 12-39, A. N. Brunton, G. W. Fraser, J. E. Lees, and I.C.E. Turcu, Metrology and modeling of microchannel plate x-ray optics, in *Applied Optics* 36: 5461–70.

resemble the structure of the eyes of lobsters and krill, which do not have lenses but convey light to the retina by arrays of reflective surfaces, also shown in Figure 12-38. Thus micropore optics are sometimes termed "lobster-eye" optics. The relationship between input and output distances for a curved micropore optic is analogous to the thin-lens equation 12-2,

$$\frac{1}{L_f} = \frac{1}{z_o} + \frac{1}{z_i} = \frac{2}{R}, \tag{12-17}$$

where z_o is the source-to-optic distance, L_f is the optic-to-focal spot distance, and R is the radius of the microchannel plate, as shown in Figure 12-37.

The results of a focusing experiment using low-energy x rays are shown in Figures 12-39 and 12-40. The "cross" pattern seen in Figure 12-39 is typical of square-channel MCP optics, and is predicted by the simulations. In addition to their applications in astronomy, slumped MCP optics may be useful as collimators for x-ray lithography, and to collect scattered radiation.

12.2.6 ENERGY FILTERING

The dependence of the critical angle for reflection on photon energy results in an energy dependent reflectivity, as shown in Figure 11-7, and hence optic performance, for example as shown in Figure 12-23. Thus, reflective optics can be used as a low-pass filter. With this low-pass filter, higher-order harmonics can be removed from the output of a crystal monochromator, or from conventional sources for energy-dispersive x-ray diffractometry and reflectometry. With reflective optics, high anode voltages can be used to increase the intensity of the characteristic lines without increasing the high-energy background. An example of the effect of a polycapillary optic designed to pass 8 keV Cu Kα radiation is shown in Figure 9-6. The optic slightly reduces the Cu Kβ 9 keV peak and suppresses the high-energy bremsstrahlung.

12.2.7 OPTICS METROLOGY

Conventional interferometric techniques normally used for flat or visible light optics are not easily employed in the testing of grazing incidence on aspheric surfaces. *Surface profilometry*, either contact or noncontact, has been the method of choice for measuring

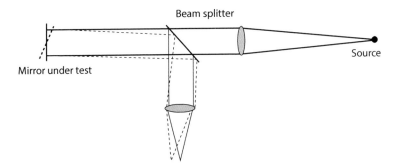

FIGURE 12-41. Autocollimator for measuring x-ray mirrors. From P. Z. Takacs, X-Ray Mirror Metrology, chap. 46, *Handbook of Optics*, 3rd ed., vol. 5, McGraw-Hill, 2010.

the figure of Wolter telescopes and synchrotron radiation mirrors. This has necessitated the development of specialized instruments. A sketch of an autocollimator is shown in Figure 12-41. An example of a surface profile is shown in Figure 12-42. Typically, the shape errors for reflective optics have a $1/v_s$ frequency response, where v_s is the spatial frequency taken from a Fourier transform of the profile error.

12.3 Optics simulations

Numerous applications have been advanced by the development of simulation analyses which allow for increasingly accurate assessment of optics defects. These computer codes, like Shadow, shown in Figure 12-43, are generally based on Monte Carlo simulations of geometric optics trajectories. The simulations provide essential information on performance, design, and potential applications of a variety of optics. Generally, a point is selected on the source, and the ray is propagated until it hits the optic surface, a computation is performed of the angle of incidence and hence reflectivity, and the ray, if reflected, is propagated farther. If the optic has a complex shape, the computation of the point of incidence is usually performed by iteration or approximation, and the computation of the surface normal can be complicated.

Simulations must allow for optics defects, including roughness, surface ripple, and profile error to be taken into account. Because the critical angle θ_c is inversely proportional to the x-ray photon energy, profile error typically has the greatest impact at the highest photon energies. Midrange spatial frequency slope errors, that is, surface oscillations with wavelengths shorter than the optic length and longer than the

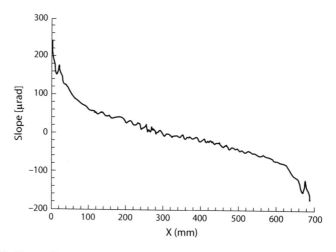

FIGURE 12-42. Slope of a 700 mm long silicon cylinder mirror measured with a long-trace interferometry profilometer. The mean has been subtracted from the data. The profile shows that the surface has an overall convex curvature (profile tilted down to the right) with significant edge roll-off (change in slope at each end). Also, a polishing defect with a 20 mm period is evident in the center of the surface. The slope profile emphasizes high-frequency surface defects. From P. Z. Takacs, X-Ray Mirror Metrology, chap. 46, *Handbook of Optics*, 3rd ed., vol. 5, McGraw-Hill, 2010.

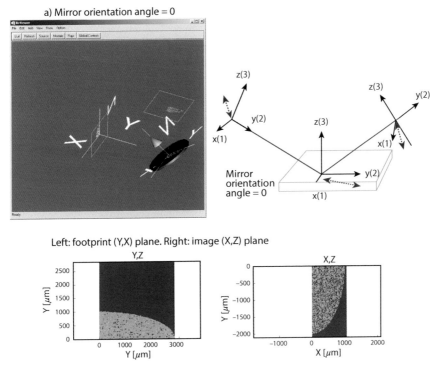

a) Mirror orientation angle = 0

Left: footprint (Y,X) plane. Right: image (X,Z) plane

FIGURE 12-43. Output screen from OASYS simulation. Courtesy Manuel Sanchez del Rio, Elettra.

wavelength of the roughness, are often called *waviness, ripple,* or *surface oscillations.* The detailed shape of the optic surface is generally unknown, but waviness can be modeled as a random tilt of the surface. The distribution of surface angles in the glass is often assumed to be Gaussian. For high-quality glass and photon energies less than 200 keV, the range of surface angles is much smaller than the critical angle. Most borosilicate and lead glass optics have simulation fitting parameters which give a Gaussian width for the waviness of 0.12–0.15 mrad. If the input x-ray beam has small local divergence, for example, from a very small spot source, the waviness increases the average angle of reflection and thus the average angle at which x rays exit the fiber. For instance, for the case of a small source with a local divergence of 2.4 mrad, a simulation of a polycapillary optic at 8 keV with no channel-wall defects produces a divergence of less than the critical angle. For a simulation including a typical waviness of 0.15 mrad, the divergence grows to 3.9 mrad, which matches the measured value. In a geometry in which the number of bounces per photon is small, the output divergence can remain smaller than the critical angle.

Problems

SECTION 12.1

1. a) Compute the focal length for a copper lens with a negative 1 cm radius of curvature at 10 keV (use the parameters for the problems at the end of chapter 11).

b) Compute the focal length for a compound optic with a radius of −0.1 cm, and 100 lenses.

c) Estimate the smallest possible focal spot from this optic.

SECTION 12.2.1

2. An elliptical mirror to be used at 10 keV is made from copper. If the beam to be focused is 2 mm thick, how long must the mirror be?

SECTION 12.2.4

3. What is the expected focal spot size at 10 keV from a polycapillary optic with a focal length of 20 mm and a channel size of 2 μm?

SECTION 12.2.5

4. What is the focal length of a micropore optic with a radius of curvature of 1 m?

Further reading

M. Bass, C. DeCusatis, J. Enoch, V. Lakshminarayanan, G. Li, C. MacDonald, V. N. Mahajan, and E. Van Stryland, eds., **Handbook of Optics**, 3rd ed., vol. 5, McGraw-Hill, 2010

Andong Liu, The x-ray distribution after a focusing polycapillary: A shadow simulation, *Nuclear Instruments and Methods in Physics Research Section B* 243, no. 1 (2006): 223–26.

13

COHERENT SCATTER II: DIFFRACTION

In the discussion in chapter 11 of coherent scatter, refraction, and reflection, materials were regarded as continuous media with a given index of refraction. However, the arrangement of atoms and their associated electrons within the media is neither uniform nor random. The coherent scatter from collections of atoms can cause interference effects, known as *diffraction*.

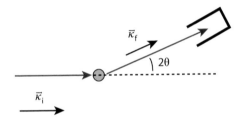

FIGURE 13-1. Scattering from a single electron.

13.1 Scattering from a single electron

Consider the simplest system, a single electron, as shown in Figure 13-1. As before (in equation 11-48, for example), the incident plane wave can be described as

$$E = \mathrm{Re}\left\{ E_o e^{i(\vec{\kappa}_i \cdot \vec{r} - \omega t)} \right\}, \qquad (13\text{-}1)$$

where $\vec{\kappa}_i$ is the incident wavevector, and the variable $\vec{r}$,

$$\vec{r} = x\hat{x} + y\hat{y} + z\hat{z}, \qquad (13\text{-}2)$$

is a location in space. If the incoming beam direction is taken as the z axis, then

$$\vec{\kappa}_i = \kappa_i \hat{z} \Rightarrow \vec{\kappa}_i \cdot \vec{r} = \kappa_i z, \qquad (13\text{-}3)$$

(as it was in equation 11-1). Because we are discussing coherent scatter, ω will not change, so the electric field is usually abbreviated as

$$E_o e^{i(\vec{\kappa}_i \cdot \vec{r})}, \tag{13-4}$$

again, as was done in equation 11-51. From equations 11-36 and 11-42, for an unpolarized input, the average scattered field amplitude is

$$E_1 = r_e \sqrt{\frac{1 + \cos^2(2\theta)}{2}} \frac{E_o}{r}, \tag{13-5}$$

where r_e is the classical electron radius, r is the distance from the electron to the observer, and 2θ is the angle between the incoming and final wavevectors, as shown in Figure 13-1. The scattering angle is defined to be 2θ to be consistent with the familiar notation of Bragg's law, as will be discussed in section 13.7. The scattered beam is observed by placing a detector in a direction defined by the final wavevector $\vec{\kappa}_f$ (or the direction of $\vec{\kappa}_f$ is defined by the location of the detector). The field is then

$$E_s = E_1 e^{i\vec{\kappa}_f \cdot \vec{r}}. \tag{13-6}$$

Because this is coherent scatter, the wavelength does not change, and the magnitudes of the initial and final wavevectors are equal,

$$\kappa_i = \kappa_f = \frac{2\pi}{\lambda} n. \tag{13-7}$$

For the case of a single electron in vacuum, $n = 1$. In general, because the index of refraction is so close to unity, setting $n = 1$ will not introduce much error (compared with sources of uncertainty discussed in sections 13.12 and 13.13), except for very soft x rays, as discussed in section 15.3.2. The scattering wavevector $\vec{Q}$, shown in Figure 13-2 is defined as

$$\vec{Q} = \vec{\kappa}_f - \vec{\kappa}_i. \tag{13-8}$$

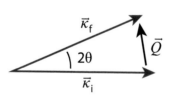

FIGURE 13-2. Scattering vector $\vec{Q}$ and angle 2θ.

Applying the law of cosines to the isosceles triangle of Figure 13-2 gives the amplitude of the scattering vector,

$$Q^2 = \kappa_i^2 + \kappa_f^2 - 2\kappa_i\kappa_f \cos(2\theta) = 2\kappa_i^2 - 2\kappa_i^2(1 - 2\sin^2(\theta))$$

$$= 4\left(\frac{2\pi}{\lambda} n\right)^2 \sin^2(\theta) \Rightarrow Q = \frac{4\pi}{\lambda} n \sin\theta. \tag{13-9}$$

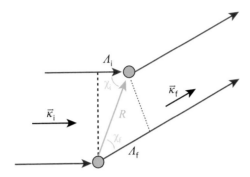

FIGURE 13-3. Scattering from two electrons separated by a distance R.

This result is consistent with the definition of $Q_\perp$ used in the discussion of reflectivity in equation 11-67. It is important to remember that Q depends on both the wavelength and the direction of observation. Changing $\vec{\kappa}_f$ changes the observation direction and hence Q.

13.2 Two electrons

If a second electron is added at a location $\vec{R}$ away from the first, as shown in Figure 13-3, the total scattered field is the sum of the fields from the two electrons. To sum the fields it is necessary to know their relative phase (assuming that R is less than the coherence length of the incident beam, as discussed in chapter 5). The path length difference is $\Lambda_f - \Lambda_i$. Λ_f is the projection of $\vec{R}$ onto $\vec{\kappa}_f$,

$$\Lambda_f = R \cos \chi_f = \hat{\kappa}_f \cdot \vec{R}, \tag{13-10}$$

and Λ_i is its projection onto $\vec{\kappa}_i$, so that the phase difference between the beams scattered off the first or second electron is

$$\Delta\phi = \kappa_f \Lambda_f - \kappa_i \Lambda_i = \vec{\kappa}_f \cdot \vec{R} - \vec{\kappa}_i \cdot \vec{R} = (\vec{\kappa}_f - \vec{\kappa}_i) \cdot \vec{R} = \vec{Q} \cdot \vec{R}. \tag{13-11}$$

Summing the fields from the two electrons, taking care to include the relative phases, as was done in section 5.7, gives

$$E_s \approx E_1 e^{i\vec{\kappa}_f \cdot \vec{r}} + E_1 e^{i(\vec{\kappa}_f \cdot \vec{r} + \Delta\phi)} = E_1 e^{i(\vec{\kappa}_f \cdot \vec{r})} (1 + e^{i\vec{Q} \cdot \vec{R}}). \tag{13-12}$$

The complex amplitude of the field is thus

$$E_{so} \approx E_1 (1 + e^{i\vec{Q} \cdot \vec{R}}), \tag{13-13}$$

where E_1 is the field from a single electron. The physical field amplitude will be the real part of the complex expression

$$\mathrm{Re}\{1 + e^{i\vec{Q} \cdot \vec{R}}\} = 1 + \cos(\vec{Q} \cdot \vec{R}) = 1 + \cos(\Delta\phi). \tag{13-14}$$

As for any interference problem, the amplitude of the combined field can vary from zero to twice the single amplitude depending on the phase difference between the two waves, so that the total scatter intensity from the two electrons varies from zero to four times the intensity from one electron, depending on the direction of observation. The scattered field is maximum when the waves are in phase and adding constructively, that is, when the phase difference is an integral multiple of 2π,

$$\Delta\phi = 2m\pi. \tag{13-15}$$

If we have not two but N_e scatterers, with the jth at a location $\vec{R}_j$ relative to the first, then the scattered field is the sum of the fields from each,

$$E_{so} \approx E_1 \left(1 + \sum_{j=1}^{N_e-1} e^{i\vec{Q}\cdot\vec{R}_j} \right). \tag{13-16}$$

This expression will be used to evaluate scattering from a chain of atoms in section 13.4, but first we need to consider a single atom.

13.3 Scattering from an atom: Fourier transform relationships

In section 11.1 we estimated the index of refraction of materials by treating the material as a loose collection of classical free electrons. Here, however, to develop a picture of the diffraction from atoms, we can no longer treat the electrons as free, or especially not as though they have definite locations within the atom. We can no longer add the electrons classically but must consider the electron probability density $\boldsymbol{P}(\vec{R})$. The field scattered from the atom is the integral over all possible positions $\vec{R}$ of the valence (nearly free) electrons, with the usual exponential factor keeping track of the relative phase from that position,

$$E_{so} \approx E_1 \int \boldsymbol{P}(\vec{R}) e^{i\vec{Q}\cdot\vec{R}} d^3R = f_o E_1, \tag{13-17}$$

where f_o is the contribution to the atomic scattering factor from nearly free electrons. The integral is the Fourier transform of the electron density. The relationship between the amplitude of the scattered electric field and the Fourier transform of the electron density is an important and powerful aspect of diffraction. If the electron field can be estimated from the measured intensities, the electron density can be determined from its inverse Fourier transform. (The difficulty in estimating the phase of the field from intensity measurements is discussed in chapter 14.) Note that f_o is a function of $\vec{Q}$. For forward scattering, $Q \to 0$, and $f_o \to Z$. As Q increases, small distances R give large phase changes. At large Q the phase varies rapidly across the atom, and the field averages to zero, so $f_o \to 0$. The scattering factor as a function of Q (and hence angle) is shown in Figure 13-4.

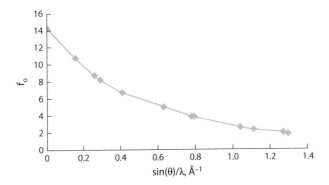

FIGURE 13-4. Atomic scattering factor for silicon, as a function of $Q/4\pi$. Data from J. J. Demarco and R. J. Weiss, *Physical Review A* 137 (1965): 1869–71.

13.4 A chain of atoms

The next step in considering diffraction from a crystal, or from a stack of crystal planes, is to consider a chain of N_a atoms, as shown in Figure 13-5. The scattered field, taking the central atom and then summing over the ones to each side, becomes

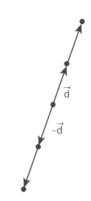

FIGURE 13-5. Chain of atoms, each separated by a distance d.

$$E_s \approx E_1 f_o \left(1 + \left\{ e^{i\vec{Q}\cdot\vec{d}} + e^{i\vec{Q}\cdot 2\vec{d}} + \cdots + e^{i\vec{Q}\cdot\frac{N_a-1}{2}\vec{d}} \right\} \right.$$

$$\left. + \left\{ e^{-i\vec{Q}\cdot\vec{d}} + e^{-i\vec{Q}\cdot 2\vec{d}} + \cdots + e^{-i\vec{Q}\cdot\frac{N_a-1}{2}\vec{d}} \right\} \right)$$

$$= E_1 f_o \left[\sum_{j=-\frac{N_a-1}{2}}^{\frac{N_a-1}{2}} (e^{i\Delta\phi})^j \right], \tag{13-18}$$

where

$$\Delta\phi = \vec{Q}\cdot\vec{d}. \tag{13-19}$$

Finite sums of the form $S = \sum\limits_{j=m}^{n} x^j$ can be solved by subtracting xS from S,

$$xS = \sum_{j=m+1}^{n+1} x^j \Rightarrow S - xS = (1-x)S = x^m - x^{n+1} \Rightarrow S = \frac{x^m - x^{n+1}}{1-x}, \tag{13-20}$$

so that

$$
E_s \approx E_1 f_o \left[\frac{(e^{i\Delta\phi})^{\left(-\frac{N_a-1}{2}\right)} - (e^{i\Delta\phi})^{\left(\frac{N_a-1}{2}+1\right)}}{1 - e^{i\Delta\phi}} \right]
$$

$$
= E_1 f_o \left[\frac{(e^{i\Delta\phi})^{\left(-\frac{N_a-1}{2}\right)} - (e^{i\Delta\phi})^{\left(\frac{N_a+1}{2}\right)}}{1 - e^{i\Delta\phi}} \right].
\tag{13-21}
$$

Multiplying by a common factor in the numerator and denominator gives

$$
E_s \approx E_1 f_o \left[\frac{(e^{i\Delta\phi})^{\left(-\frac{N_a-1}{2}\right)} - (e^{i\Delta\phi})^{\left(\frac{Na+1}{2}\right)}}{1 - e^{i\Delta\phi}} \right] \frac{e^{-i\frac{\Delta\phi}{2}}}{e^{-i\frac{\Delta\phi}{2}}}
$$

$$
= E_1 f_o \left[\frac{(e^{i\Delta\phi})^{\left(-\frac{N_a}{2}\right)} - (e^{i\Delta\phi})^{\left(\frac{N_a}{2}\right)}}{e^{-\frac{i\Delta\phi}{2}} - e^{\frac{i\Delta\phi}{2}}} \right] = E_1 f_o \frac{\sin\left(N_a \frac{\Delta\phi}{2}\right)}{\sin\left(\frac{\Delta\phi}{2}\right)}.
\tag{13-22}
$$

The ratio of sines is a maximum when $\Delta\phi$ is $2\pi m$, in which case, employing l'Hôpital's rule yields

$$
E_s \approx E_1 f_o \lim_{\Delta\phi \to 2\pi m} \frac{\sin\left(N_a \frac{\Delta\phi}{2}\right)}{\sin\left(\frac{\Delta\phi}{2}\right)} = E_1 f_o \lim_{\Delta\phi \to 2\pi m} \frac{\frac{N_a}{2} \cos\left(N_a \frac{\Delta\phi}{2}\right)}{\frac{1}{2} \cos\left(\frac{\Delta\phi}{2}\right)}
$$

$$
= E_1 f_o N_a,
\tag{13-23}
$$

and the scattered field for constructive interference from N_a atoms is N_a times the field from one atom.

The ratio is a minimum when $\Delta\phi$ is $2\pi/N_a$. The width of a plot of field versus $\Delta\phi$ is taken as half the distance between the zeros at $\pm 2\pi/N_a$, or also $2\pi/N_a$. For large N_a, the peaks are very high and very narrow. If N_a is a macroscopic number, like 6×10^{23}, the peaks are essentially delta functions, and the scattering is nonzero only if

$$
\Delta\phi = \vec{Q} \cdot \vec{d} = 2\pi m.
\tag{13-24}
$$

A delta function is given by

$$
\delta(x) = \begin{cases} \infty & x = 0 \\ 0 & \text{else} \end{cases},
\tag{13-25}
$$

such that

$$\int_{-\infty}^{\infty} \delta(x)\,dx = 1. \tag{13-26}$$

The delta function has the property

$$\int_{-\infty}^{\infty} \delta(x - x_o)\,f(x)\,dx = f(x_o), \tag{13-27}$$

since it is nonzero only at the one point $x = x_o$.

Because $\vec{Q}$ is determined from the observation direction $\vec{\kappa}_f$, the result that the scattering is nonzero only for certain values of $\vec{Q}$ means that there are only certain observation directions which give strong intensity. This is the phenomenon observed as diffraction.

13.5 Lattices and reciprocal lattices

To predict the directions in which diffraction will be observed, it is necessary to extend the model of a crystal from the one-dimensional chain of Figure 13-5 to a three-dimensional structure, sketched in Figure 13-6. A crystal lattice has translational symmetry: every lattice point in the ideal crystal is displaced from the one at the origin by a translation vector

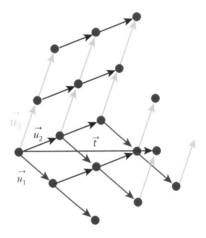

FIGURE 13-6. A crystal lattice consists of a regular array of lattice points.

$$\vec{t} = j_1 \vec{u}_1 + j_2 \vec{u}_2 + j_3 \vec{u}_3, \tag{13-28}$$

where j_1, j_2, and j_3 are integers, and $\vec{u}_1$, $\vec{u}_2$ and $\vec{u}_3$ are the lattice vectors which define the crystal geometry (note that these vectors do not have to be orthogonal or of the same length). The crystal can then be constructed by placing an atom at each lattice point. Extending equation 13-18 to a three-dimensional array requires replacing the vector representing the location of each atom with respect to the origin with the new vector $\vec{t}$,

$$E_s \approx E_1 f_o \sum_{\text{all } \vec{t}} e^{i\vec{Q}\cdot\vec{t}} = E_1 f_o \sum_{j_1} \sum_{j_2} \sum_{j_3} e^{i\vec{Q}\cdot(j_1\vec{u}_1 + j_2\vec{u}_2 + j_3\vec{u}_3)}. \tag{13-29}$$

As before, to have constructive interference,

$$\Delta\phi = \vec{Q}\cdot\vec{t} = 2\pi m. \tag{13-30}$$

Now, the problem of finding the directions which give strong diffraction has become one of finding the scattering vectors $\vec{Q}$ which satisfy the diffraction condition for all possible translation vectors $\vec{t}$ in the crystal. These vectors are surprisingly straightforward to construct, and are called *reciprocal lattice vectors* (since, like the wavevector κ and like Q, they have dimensions of reciprocal length). A reciprocal lattice vector $\vec{G}$ is defined as

$$\vec{G} = h\vec{g}_1 + k\vec{g}_2 + l\vec{g}_3,\tag{13-31}$$

where h, k, and l are any integers, and

$$\vec{g}_1 = \frac{2\pi}{V_c}\vec{u}_2 \times \vec{u}_3, \quad \vec{g}_2 = \frac{2\pi}{V_c}\vec{u}_3 \times \vec{u}_1, \quad \vec{g}_3 = \frac{2\pi}{V_c}\vec{u}_1 \times \vec{u}_2.\tag{13-32}$$

V_c is the volume of the polyhedron, called the *unit cell*, bounded by $\vec{u}_1, \vec{u}_2$ and $\vec{u}_3$, and

$$V_c = \left|\vec{u}_1 \cdot (\vec{u}_2 \times \vec{u}_3)\right|.\tag{13-33}$$

The reciprocal vectors are chosen so that $\vec{g}_j \cdot \vec{u}_j = 2\pi$, for example,

$$\vec{g}_1 \cdot \vec{u}_1 = 2\pi \frac{\vec{u}_2 \times \vec{u}_3}{\vec{u}_1 \cdot (\vec{u}_2 \times \vec{u}_3)} \cdot \vec{u}_1 = 2\pi, \text{ and so that } \vec{g}_j \cdot \vec{u}_m = 0 \text{ if } j \neq m, \text{ for example,}$$

$\vec{g}_1 \cdot \vec{u}_2 = \dfrac{2\pi}{V_c}(\vec{u}_2 \times \vec{u}_3) \cdot \vec{u}_2 = 0$. Thus the diffraction condition will be satisfied if $\vec{Q} = \vec{G}$, since

$$\vec{G} \cdot \vec{t} = (h\vec{g}_1 + k\vec{g}_2 + l\vec{g}_3) \cdot (j_1\vec{u}_1 + j_2\vec{u}_2 + j_3\vec{u}_3) = 2\pi(hj_1 + kj_2 + lj_3),\tag{13-34}$$

which is 2π times an integer. The condition

$$\vec{Q} = \vec{G}\tag{13-35}$$

is the diffraction condition. The reciprocal lattice is the Fourier transform of the real-space lattice, so that, as discussed for equation 13-17, the diffraction from a lattice will give a field proportional to the reciprocal lattice, and hence the directions of $\vec{Q}$ and the diffracted wavevector $\vec{\kappa}_f$ in three-dimensional space. When $\vec{Q} = \vec{G}$, the field is again $E_s \approx E_1 f_o N_a$, where N_a is the number of atoms in the beam.

EXAMPLE 13-1: RECIPROCAL LATTICE

Give the reciprocal lattice vectors for a cubic lattice.

Setting the coordinate axes to the cube edges, the real-space lattice vectors for a cube are the three cube edges, so that the three vectors are $\vec{u}_1 = u\hat{x}, \vec{u}_2 = u\hat{y}, \vec{u}_3 = u\hat{z}$, and the unit cell is a cube of volume u^3. A general translation vector is then $\vec{t} = j_1\vec{u}_1 + j_2\vec{u}_2 + j_3\vec{u}_3$, where the j's are integers. The possible endpoints of this vector form a cubic array.

The reciprocal lattice vectors are then

$$\vec{g}_1 = 2\pi \frac{\vec{u}_2 \times \vec{u}_3}{\vec{u}_1 \cdot (\vec{u}_2 \times \vec{u}_3)} = 2\pi \frac{u^2 \hat{x}}{u^3} = \frac{2\pi}{u} \hat{x}, \quad \vec{g}_2 = \frac{2\pi}{u} \hat{y}, \quad \vec{g}_3 = \frac{2\pi}{u} \hat{z}.$$

These form a cubic array as well, but the "length" of the cube edge is $2\pi/u$.

13.6 Planes

The reciprocal lattice vectors are also useful for enumerating the possible planes of the crystal, since, as will shortly be shown, each vector $\vec{G} = h\vec{g}_1 + k\vec{g}_2 + l\vec{g}_3$ is perpendicular to a particular set of real space planes, denoted (hkl). Note that no commas are used between the indices h, k, and l in the plane notation to distinguish it from the notation for a point. The (hkl) plane closest to the origin intercepts the $\vec{u}_1$ axis at u_1/h, the $\vec{u}_2$ axis at u_2/k, and the $\vec{u}_3$ axis at u_3/l. Points are described in terms of the the lattice vectors, so that the intercept point on the $\vec{u}_1$ axis is written as $(1/h, 0, 0)$, meaning the point at the location $\frac{1}{h}\vec{u}_1 + 0\vec{u}_2 + 0\vec{u}_3$.

EXAMPLE 13-2: PLANES

Sketch the (100) and (111) planes of a cubic crystal, and the direction of the corresponding $\vec{G}$ vectors.

a) The (100) plane has intercepts u, ∞, ∞, meaning it intercepts the x axis at u and does not intercept the y and z axes (it is parallel to them). This is the cube face closest to the observer in Figure 13-7. The reciprocal lattice vector is

$$\vec{G} = (1\vec{g}_1 + 0\vec{g}_2 + 0\vec{g}_3) = (1)\frac{2\pi}{u}\hat{x} + 0\frac{2\pi}{u}\hat{y}$$

$$+ 0\frac{2\pi}{u}\hat{z} = \frac{2\pi}{u}\hat{x},$$

so it is parallel to the x axis and thus perpendicular to the (100) plane.

b) The (111) plane has intercepts all equal to u, as shown. The reciprocal lattice vector is $\vec{G} = \frac{2\pi}{u}\hat{x} + \frac{2\pi}{u}\hat{y} + \frac{2\pi}{u}\hat{z}$, so it is parallel to the body diagonal.

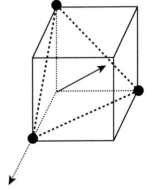

FIGURE 13-7. The (111) plane in a cubic system and the direction of the reciprocal lattice vector representing this plane.

The somewhat awkward description of the real-space (hkl) planes in terms of inverse intercepts on the axes is chosen so that the corresponding $\vec{G}$ vector is perpendicular to the (hkl) plane. The equation of the (hkl) plane through the three intercepts is

$$hx_1 + kx_2 + lx_3 = 1, \tag{13-36}$$

and $\vec{d} = x_1\vec{u}_1 + x_2\vec{u}_2 + x_3\vec{u}_3$ is a vector from the origin to the point (x_1, x_2, x_3) on the plane. The equation of the plane 13-36 is satisfied for the intercept on the u_1 axis, which is the point $(1/h, 0, 0)$, and for the other two intercepts. One vector on the (hkl) plane is the line between its intercepts on the u_1 and u_2 axes,

$$\vec{r}_1 = (1/h)\vec{u}_1 - (1/k)\vec{u}_2. \tag{13-37}$$

The angle between $\vec{G}$ and $\vec{r}_1$ is given by

$$\cos(\varphi) = \frac{\vec{G} \cdot \vec{r}_1}{Gr_1} = \frac{(h\vec{g}_1 + k\vec{g}_2 + l\vec{g}_3) \cdot ((1/h)\vec{u}_1 - (1/k)\vec{u}_2)}{Gr_1}$$
$$= \frac{h(1/h)2\pi + k(1/h)0 + \cdots - k(1/k)2\pi}{Gr_1} = 0 \Rightarrow \varphi = \frac{\pi}{2}. \tag{13-38}$$

The same result is found for any other two vectors on the plane, so $\vec{G}$ is perpendicular to the plane.

To find the distance between the planes, $\vec{d}$ should be chosen so that it is parallel to $\vec{G}$. Then, the dot product gives

$$\vec{G} \cdot \vec{d} = Gd\cos(0) = (h\vec{g}_1 + k\vec{g}_2 + l\vec{b}_3) \cdot (x_1\vec{u}_1 + x_2\vec{u}_2 + x_3\vec{u}_3)$$
$$= 2\pi(hx_1 + kx_2 + lx_3) = 2\pi, \tag{13-39}$$

where the last step used the equation for the plane, 13-36. Thus the magnitude of $\vec{G}$ is equal to the spacing d between (hkl) planes,

$$|\vec{G}| = \frac{2\pi}{d}. \tag{13-40}$$

This makes it possible to routinely calculate plane spacings using vector operations even in crystal systems for which the lattice vectors $\vec{u}_1$, $\vec{u}_2$, and $\vec{u}_3$ are not orthogonal. For the cubic system, the d spacing is straightforward, since

$$|\vec{G}| = \left| h\frac{2\pi}{u}\hat{x} + k\frac{2\pi}{u}\hat{y} + l\frac{2\pi}{u}\hat{z} \right| = \frac{2\pi}{u}\sqrt{h^2 + k^2 + l^2}, \tag{13-41}$$

so

$$d = \frac{2\pi}{|\vec{G}|} = \frac{u}{\sqrt{h^2 + k^2 + l^2}}. \tag{13-42}$$

EXAMPLE 13-3: PLANE SPACINGS

a) Compute the distance between (100) planes of a cubic crystal.

b) Compute the distance between (111) planes of a cubic crystal.

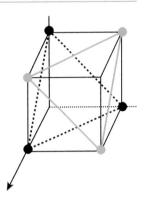

a) For (100), $d = \dfrac{u}{\sqrt{1^2 + 0^2 + 0^2}} = u$, which is clearly the spacing between cube faces.

b) For (111), $d = \dfrac{u}{\sqrt{1^2 + 1^2 + 1^2}} = \dfrac{u}{\sqrt{3}}$, which is the spacing between the two planes shown in Figure 13-8.

FIGURE 13-8. Two neighboring (111) planes in a cubic crystal are represented by the dotted and blue triangles.

13.7 Bragg's law

Understanding the reciprocal relationship between the plane spacing and G, equation 13-40, allows us to reformulate the diffraction condition equation 13-35 in terms of wavelength and d spacing,

$$\vec{G} = \vec{Q} \Rightarrow G = \frac{2\pi}{d} = Q. \tag{13-43}$$

Substituting the length of Q from equation 13-9 gives

$$\frac{2\pi}{d} = G = Q = \frac{4\pi}{\lambda} n \sin \theta_B, \tag{13-44}$$

where the angle θ which satisfies the equation is denoted by θ_B. Setting $n = 1$ gives Bragg's law,

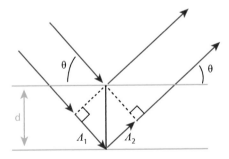

FIGURE 13-9. Mnemonic for remembering Bragg diffraction from planes with spacing d. The extra path length, $(\Lambda_1 + \Lambda_2)$, must be an integral multiple of λ.

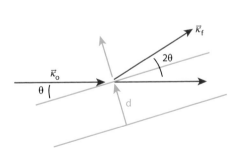

FIGURE 13-10. Diffraction from a set of planes.

$$\lambda = 2d \sin \theta_B. \tag{13-45}$$

The angle θ_B is called the Bragg angle for the set of planes with spacing d and an x-ray beam at wavelength λ. A sketch justifying Bragg's law in terms of the path lengths for x rays "reflecting" off a set of planes with a plane spacing d is shown in Figure 13-9. The symmetry condition that the angle θ between the plane and the incoming ray be equal to the angle between the plane and the outgoing ray arises from the vector relationship $\vec{Q} = \vec{G}$. The vector $\vec{Q}$ is the difference between the outgoing and incoming wavevectors, which must be equal to $\vec{G}$, which is perpendicular to the planes. The geometry is sketched in Figure 13-10. Bragg's law is extremely powerful for diffraction geometries in which it is only necessary only to know the magnitude of the angle of diffraction from the incident ray. To predict the directions for single-crystal diffraction, it is necessary to understand the plane orientations in three dimensions, as will be discussed in chapter 14.

13.8 θ-2θ diffractometer

A sketch of a θ-2θ diffractometer, used to measure diffraction from planes parallel to the surface of a sample, is shown in Figure 13-11. Ideally, the incident beam is a monochromatic plane wave (with a single wavelength λ and single incident angle θ). As the sample is rotated, the angle of incidence θ changes. The detector is moved to always be at θ to the surface, 2θ to the incoming ray. At the particular incident angle θ_B at which Bragg's law is satisfied, a beam is diffracted into the detector, and a peak in intensity is seen.

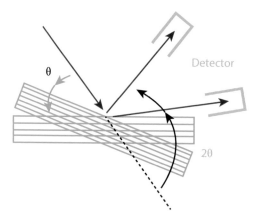

Figure 13-11. θ-2θ diffractometer: as the sample is moved to be at angle θ from the incident beam, the detector is moved to 2θ.

13.9 Powder diffraction

In general, unless it is known that the crystal planes are parallel to the sample surface, the geometry of the θ-2θ diffractometer will not lead to finding a diffraction

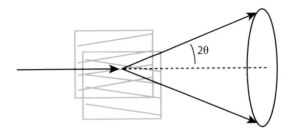

FIGURE 13-12. Powder diffraction. Two small crystallites with different plane orientation are shown. The sample is not drawn to scale; the plane spacings are of atomic scale, while the sample size is macroscopic.

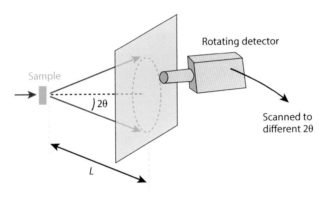

FIGURE 13-13. Powder diffraction data can be collected with an imaging detector or, in its absence, with a scanned detector.

peak. For example, if the crystal planes in Figure 13-11 are tilted out of the plane of the page, the detector will never be aligned with the diffracted beam.

In powder diffraction, the sample consists of very many small crystals, randomly oriented to all possible directions, ensuring that some planes will be aligned at a Bragg condition. The geometry is sketched in Figure 13-12. The diffracted beam becomes a cone including all possible directions at an angle 2θ from the incident beam. The data can be collected with an imaging detector, or a small detector scanned through the beam, as shown in Figure 13-13. The measured intensity from a detector scan is shown in Figure 13-14. The diffraction pattern on an imaging detector is shown in Figure 13-15. The image is processed by doing a radial integration (summing the intensity from all the pixels at the same radius from the center of the circle), computing the diffraction angle 2θ from the ring radius, and then plotting the total intensity versus 2θ, in the same manner as in Figure 13-14.

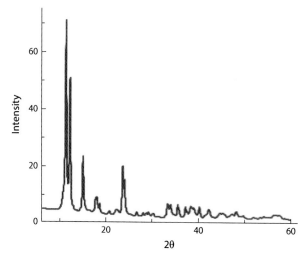

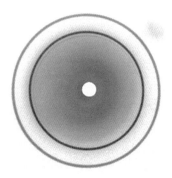

FIGURE 13-14. Output from a detector scan for a powder $Rh_2O_2CH_3$ sample. Data from Alexi Verchinine, Organic powder diffraction with polycapillary X-ray optics, MS thesis, 2006.

FIGURE 13-15. Powder diffraction image from an fcc metal sample. From W. Zhou et al., *Thin Solid Films* 518 (2010): 5047–56. Reproduced by permission of Elsevier.

EXAMPLE 13-4: POWDER DIFFRACTION FROM A CUBIC CRYSTAL

Compute the radii of all the diffraction rings incident on a 200 cm wide detector placed $L = 50$ cm from a powder sample of a cubic crystal with unit cell cube edge $u = 2$ Å, irradiated with a parallel x-ray beam with wavelength $\lambda = 1$ Å.

To find all possible planes, we need all possible (hkl), which we enumerate in table 13-1 in order of increasing $h^2 + k^2 + l^2$. We list only one of each set that have the same $h^2 + k^2 + l^2$, for example of the six cube faces (100), (010), (001), (−100) (0−1 0) and (00−1), we list only the one plane (100). *Note*: The last three planes in the list reveal a problem with the notation for negative numbers. As a result, negative numbers are usually written with a bar above, so that the planes become $(\bar{1}00), (0\bar{1}0)$, and $(00\bar{1})$.

Proceeding with the computation, from the (hkl) we can compute d, and from d and λ and Bragg's law we can compute θ. Finally from 2θ, we can compute the radii of the diffraction rings using the geometry of Figure 13-12.

TABLE 13-1

h	k	l	$h^2 + k^2 + l^2$	$d = \dfrac{2\,\text{Å}}{\sqrt{h^2 + k^2 + l^2}}$	$\sin\theta_B = \dfrac{1\,\text{Å}}{2d}$	θ_B	$2\theta_B$	$R_{ring} = L\tan(2\theta)$	
0	0	0	0	∞	0	0	0	0	This is the undiffracted, straight-through beam.
1	0	0	1	2	1/4	14°	29°	27 cm	
1	1	0	2	$\sqrt{2}$	$\sqrt{2}/4$	21	41	44	
1	1	1	3	$2/\sqrt{3}$	$\sqrt{3}/4$	26	51	62	
2	0	0	4	1	1/2	30	60	87	The (200) plane is equivalent to the second-order diffraction from the (100) plane, i.e., what you would get if you wrote $m\lambda = 2d\sin(\theta)$, with $m = 2$ and $d = 2$ Å.
2	1	0	5	$2/\sqrt{5}$	$\sqrt{5}/4$	34	68	123 (not recorded)	This ring has a diameter larger than the detector and so will not be recorded unless the detector is moved closer.
2	1	1	6						Not recorded.
2	2	0	8						Not recorded.
2	2	1	9	2/3	3/4	49	97	none	This diffraction is at more than 90°, which means it is directed back toward the source. To see the diffraction, the detector would have to be moved on the other side of the sample (possibly with a hole in it to allow the incident beam to pass).
...									
4	1	0	17	$2/\sqrt{17}$	$\sin(\theta) = \sqrt{17}/4 > 1$				Diffraction from this set of planes (and any with higher $h^2 + k^2 + l^2$) is not possible. The table of possible (hkl) is finite.

13.10 Structure factor

The examples have considered only the simplest crystal, one which has atoms only at the corners of the unit cell. In reality the unit cells (the little cubes in Figures 13-7 and 13-8) can contain very many atoms, for example, all the atoms in the α helices and β sheets of a protein crystal. The location $\vec{R}$ of each atom in the crystal is described in terms of the location of the origin of the unit cell $\vec{t}$ and a vector $\vec{r}$ which gives the location of the atom within the unit cell,

$$\vec{R} = \vec{t} + \vec{r}, \tag{13-46}$$

as shown in Figure 13-16. The list of locations $\vec{r}_j$ of the atoms within the unit cell is called the *basis*. Summing the contributions due to scattering off each atom for a direction described by the scattering vector $\vec{Q}$ as in equation 13-18, becomes

$$E_s = E_o \sum_{\vec{R}} f_o e^{i\vec{Q}\cdot\vec{R}} = E_o \sum_{t,\vec{r}} f_o e^{i\vec{Q}\cdot(\vec{t}+\vec{r})} = E_o \sum_j \left\{ \sum_t e^{i\vec{Q}\cdot\vec{t}} \right\} f_j e^{i\vec{Q}\cdot\vec{r}_j}, \tag{13-47}$$

where f_o has been replaced with f_j to allow for different types of atoms (different elements) within the unit cell. The term in curly brackets is simply the sum over all translation vectors, which, as noted in equations 13-24 and 13-35, is nonzero only if the scattering vector is a reciprocal lattice vector, $\vec{Q} = \vec{G}$. Given that constraint, the scattered field amplitude for a single unit cell becomes

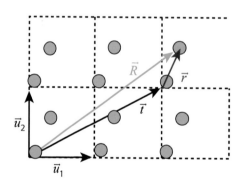

$$E_s = E_o \boldsymbol{F}, \quad \boldsymbol{F} = \sum_j f_j e^{i\vec{G}\cdot\vec{r}_j}. \tag{13-48}$$

FIGURE 13-16. A two-dimensional sketch of a crystal with multiple atoms per unit cell.

The sum $\boldsymbol{F}$, called the *structure factor*, is the Fourier transform of the structure within a single unit cell.

EXAMPLE 13-5: STRUCTURE FACTOR

Compute the structure factor for a body-centered cubic (bcc) crystal, that is, a crystal which has an atom at each corner of a cubic unit cell, plus one atom in the center.

The lattice is cubic, so, as before, $\vec{u}_1 = u\hat{x}$, $\vec{u}_2 = u\hat{y}$, $\vec{u}_3 = u\hat{z}$, and $\vec{G} = \dfrac{2\pi}{u}(h\hat{x} + k\hat{y} + l\hat{z})$.

A body-centered cubic crystal has two identical atoms per cell, one at the corner, $\vec{r}_1 = (0\vec{u}_1 + 0\vec{u}_2 + 0\vec{u}_3) = 0$, and one in the middle, at $\left(\dfrac{1}{2}, \dfrac{1}{2}, \dfrac{1}{2}\right)$, so that

$$\vec{r}_2 = \left(\frac{1}{2}\vec{u}_1 + \frac{1}{2}\vec{u}_2 + \frac{1}{2}\vec{u}_3\right) = \frac{u}{2}(\hat{x} + \hat{y} + \hat{z}).$$

(Note that there are not eight corner atoms per cell. There are two ways to see this. First, each of the eight corner atoms is in eight unit cells, so 1/8th of each atom is in this cell, which adds to 1. Alternatively, the full crystal is created if a single atom is placed in the corner of each cell (so that this one and seven neighboring cells complete the cube.) Using this basis, $\vec{G} \cdot \vec{r}_1 = 0$, and

$$\vec{G} \cdot \vec{r}_2 = \frac{2\pi}{u}(h\hat{x} + k\hat{y} + l\hat{z}) \cdot \frac{u}{2}(\hat{x} + \hat{y} + \hat{z}) = \pi(h + k + l).$$

The structure factor is (using the fact that $e^{i\pi} = -1$)

$$\mathbf{F} = \sum_{\vec{r}} f_o e^{i\vec{G} \cdot \vec{r}} = f_o(e^0 + e^{i\pi(h+k+l)}) = f_o(1 + (-1)^{h+k+l})$$

$$= \begin{cases} 2f_o & h+k+l = \text{even} \\ 0 & \text{else} \end{cases}.$$

EXAMPLE 13-6: POWDER DIFFRACTION FROM A BODY-CENTERED CUBIC CRYSTAL

Compute the radii of all the diffraction rings incident on a 200 cm wide detector placed $L = 50$ cm from a powder sample of a body-centered cubic crystal with cube edge $u = 2$ Å, irradiated with a parallel x-ray beam with wavelength $\lambda = 1$ Å.

This problem is identical with example 13-4, except for the structure factor, which is zero whenever $h + k + l$ is odd. Since the amplitude of the diffracted beam is proportional to the structure factor, a zero means no diffraction is seen. So the (100) ring disappears, as does the (111). Only the two rings, the (110) at 44 cm and the (200) at 87 cm are seen.

EXAMPLE 13-7: INDEXING A DIFFRACTION PATTERN

A powder sample is irradiated with a parallel monochromatic x-ray beam, with a wavelength of 1.5 Å. The detector is 30 mm from the sample. The sample is known to be cubic, and believed to be bcc, but it is uncertain. The radii of the diffraction rings on the detector are 17.9 mm, 29.0 mm, 42.3 mm, 61.7 mm, 96.5 mm, and 189.6 mm.

a) Can the sample be bcc?
b) What is the lattice constant (cube edge), u?

First, we use $R_{ring} = L \tan(2\theta)$, with $L = 30$ mm to get θ, and Bragg's law with $\lambda = 1.5$ Å to get the d values in table 13-2. Next, we make a list of possible (hkl) in order of increasing $h^2 + k^2 + l^2$, as shown in table 13-3 and compute the structure factor for each ring. If it is nonzero, then compute $u = d\left(\sqrt{h^2 + k^2 + l^2}\right)$.

The lattice constant values u are consistent, so the conclusion is the sample could be bcc, with a lattice constant of 3.999 ± 0.004 Å.

TABLE 13-2

Ring radius (mm)	2θ (deg)	d (Å)
17.9	30.82	2.822
29.0	44.03	2.001
42.3	54.65	1.634
61.7	64.07	1.414
96.5	72.73	1.265
189.6	81.01	1.155

TABLE 13-3

(hkl)	$h^2 + k^2 + l^2$	F_{bcc}	bcc ring no.	d	u, bcc
(100)	1	0			
(110)	2	$2f_o$	1	2.822	3.991
(111)	3	0			
(200)	4	$2f_o$	2	2.001	4.002
(210)	5	0			
(211)	6	$2f_o$	3	1.634	4.002
(220)	8	$2f_o$	4	1.414	3.999
(221)	9	0			
(300)	9	0			
(310)	10	$2f_o$	5	1.265	4.000
(311)	11	0			
(222)	12	$2f_o$	6	1.155	4.000

13.11 Intensity

To compute the intensity in the diffraction cone, it is helpful to start with the scatter from the single chain of atoms, discussed in section 13.4, which from equations 11-36–11-42 and 13-22 is

$$I = \frac{I_o}{L^2} r_e^2 f_o^2 \frac{1 + \cos^2(2\theta)}{2} \frac{\sin\left(N_a \frac{\phi}{2}\right)}{\sin\left(\frac{\phi}{2}\right)}, \tag{13-49}$$

where L is the distance of the detector from the sample, and N_a is the number of atoms. Substituting u_1 for the one-dimensional chain spacing gives

$$\frac{\sin\left(N_a \dfrac{\phi}{2}\right)}{\sin\left(\dfrac{\phi}{2}\right)} = \frac{\sin\left(N_a \dfrac{Q_1 u_1}{2}\right)}{\sin\left(\dfrac{Q_1 u_1}{2}\right)}, \tag{13-50}$$

where Q_1 is the one-dimensional Q in the u_1 direction. Using

$$\lim_{\substack{\text{large } N_a}} \frac{\sin(N_a \pi x)}{\sin(\pi x)} = N_a \delta(x - h), \tag{13-51}$$

where h is an integer, then

$$\frac{\sin\left(N_a \dfrac{Q_1 u_1}{2}\right)}{\sin\left(\dfrac{Q_1 u_1}{2}\right)} = N_a \delta\left(\frac{u_1}{2\pi} Q_1 - h\right). \tag{13-52}$$

Finally,

$$I = \frac{I_o}{L^2} r_e^2 f_o^2 \frac{1 + \cos^2(2\theta)}{2} N_a \delta\left(\frac{u_1}{2\pi} Q_1 - h\right). \tag{13-53}$$

Because it is necessary to integrate over Q_1, a change of variable is necessary,

$$\int \delta\left(\frac{u_1}{2\pi} Q_1 - h\right) dQ_1 = \frac{2\pi}{u_1} \int \delta\left(\frac{u_1}{2\pi} Q_1 - h\right) d\left(\frac{u_1}{2\pi} Q_1\right)$$

$$= \frac{2\pi}{u_1} \int \delta(x) dx = \frac{2\pi}{u_1}(1). \tag{13-54}$$

The intensity calculation is then extended to three dimensions and integrated over all possible detector directions and crystal orientations. The integration must be performed over the geometry of the actual measurement, for example, for a fixed or rotating crystal or for a powder geometry. For a powder diffraction experiment, the total power in the ring is

$$P = I_o r_e^2 \mathbf{F}_{hkl}^2 m_{hkl} \frac{1 + \cos^2(2\theta)}{2} \frac{1}{8 \sin^2 \theta \cos \theta} \frac{V_{eff}}{V_c}\left\{\frac{1}{V_c}\right\} \lambda^3 f_{DW} \tag{13-55}$$

where V_c is the volume per unit cell. In comparing equation 13-55 for the powder with equation 13-49 for a single chain, first, the sample-to-detector distance L has canceled out after integration over the area of the diffraction ring on the detector, so if the full ring is not captured, the expected power must be adjusted. Second, the atomic scattering factor has been absorbed into the structure factor, which also takes into account the number of atoms per unit cell. A multiplicity factor m_{hkl} has been added to take into account that the (100) and (010) planes, for example, are indistinguishable after integration over all possible crystalline orientations. The $\sin^2\theta\cos\theta$ term is due to geometric factors in the powder diffraction experiment, for the case that the sample is in the form of a cylinder. The number of atoms has been replaced by the number of unit cells, computed as the ratio of the effective sample volume to the cell volume. The term V_{eff} is the effective sample volume, taking into consideration absorption in the sample. The additional factor of V_c in the curly brackets comes from the same change of variable as for equation 13-54. A factor of λ^3 arises from the $1/\kappa^3$ term that is due to changes of variables in integrating over all final directions. The last factor is the Debye-Waller factor of equation 13-67, which takes into account the reduction of intensity due to thermal vibrations, and is discussed in section 13.12.2.

13.12 Defects

13.12.1 MOSAICITY

In the idealized case of a monochromatic parallel beam incident on a single crystal placed in a θ-2θ diffractometer with planes parallel to the surface, a plot of the diffracted intensity versus 2θ will be a delta function at $2\theta_B$, as shown in Figure 13-17. Such a plot is called a *rocking curve*. There are a number of deviations from ideality which affect the rocking curve. If the crystal is not perfect but has some mosaicity, as shown in Figure 13-18, the rocking curve will have some finite width, since different parts of the crystal will diffract at slightly different incident angles. In fact, as will be discussed in chapter 14, even a "perfect" crystal with zero mosaicity has a finite-width rocking curve.

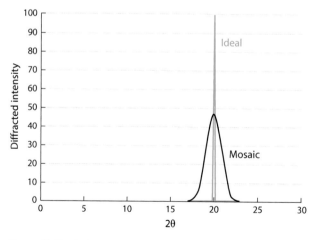

FIGURE 13-17. Ideal intensity from a "perfect" or mosaic crystal with a Bragg angle of 20°.

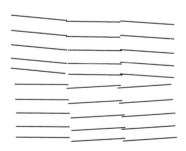

FIGURE 13-18. A model of a "mosaic" crystal, which has areas ("tiles") of nearly perfect crystal, with each "tile" at a slight angle to the next.

13.12.2 THERMAL VIBRATIONS

The most pervasive defect in the ideal crystal described in section 13.10 is thermal vibrations. No atom is expected to actually be at its specified location at any moment in time. Instead, the atoms experience thermal vibrations about the ideal location, as shown in Figure 13-19. To see the effect of thermal vibrations, first reconsider the ideal case of a fixed chain of atoms, of section 13.4, with no thermal vibrations, but this time using a chain of infinite length. The scatter from an ideal chain can be found by writing the locations of the atoms in terms of their probability density,

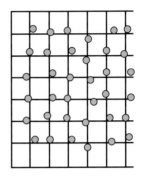

FIGURE 13-19. A snapshot of a two-dimensional crystal undergoing thermal vibrations.

$$\boldsymbol{P}_a(\vec{R}) = \sum_j \delta(\vec{R} - j\vec{d}). \tag{13-56}$$

Choosing the x direction to be along the chain, we have

$$\boldsymbol{P}_a(x) = \sum_j \delta(x - jd). \tag{13-57}$$

Using the Fourier result of equation 13-17, we obtain the scattered field,

$$E_{so} \approx E_1 f_o \int \boldsymbol{P}_a(x) e^{iQ_x x} dx = E_1 f_o \sum_j \int \delta(x - jd) e^{iQ_x x} dx$$

$$= E_1 f_o \sum_j e^{iQ_x jd} = E_1 f_o \sum_j e^{i\vec{Q} \cdot j\vec{d}}. \tag{13-58}$$

The final sum is the same as the one in equation 13-18, except with $N_a \to \infty$. The result is a series of delta functions,

$$E_{s_o}(Q_x) = \sum_m \delta\left(Q_x - m\frac{2\pi}{d}\right). \tag{13-59}$$

This is the expected result, that the ideal chain of atoms results in diffraction in only those particular directions in which the Bragg condition is satisfied.

Thermal vibrations are included by changing the probability density to a Gaussian with width $\overline{\Delta r}$,

$$\boldsymbol{P}_a(x) = \frac{1}{\sqrt{2\pi\,\overline{\Delta r}}} \sum_j e^{-\frac{(x - jd)^2}{2\overline{\Delta r}^2}}. \tag{13-60}$$

The value of the constant $\overline{\Delta r}$, which describes the average deviation of an atom from its ideal location, can be estimated by setting the strain energy per atom equal to the thermal energy per atom,

$$\frac{1}{2} C_Y \left(\frac{\overline{\Delta r}}{r_0}\right)^2 \frac{V_c}{N_{a,c}} \sim \frac{3}{2} k_B T, \tag{13-61}$$

where C_Y is the elastic (Young's) modulus, r_o is the nearest-neighbor distance between two atoms, $\overline{\Delta r}/r_o$ is the strain, and $N_{a,c}$ is the number of atoms per unit cell of the crystal, so that $V_c/N_{a,c}$ is the volume per atom.

The diffracted field is

$$E_{so} \approx E_1 f_o \int P_a(x) e^{iQ_x x} \, dx = E_1 f_o \frac{1}{\sqrt{2\pi}\,\overline{\Delta r}} \sum_j \int e^{-\frac{(x-jd)^2}{2\overline{\Delta r}^2}} e^{iQ_x x} \, d\jmath \qquad (13\text{-}62)$$

The integral is solved by completing the square,

$$\frac{(x-jd)^2}{2\overline{\Delta r}^2} + iQ_x x = \frac{1}{2\overline{\Delta r}^2}\Big[(x + (-jd + i\overline{\Delta r}^2 Q_x))^2$$

$$+ 2ijd\overline{\Delta r}^2 Q_x + \overline{\Delta r}^4 Q_x^2 \Big] = s^2 + ijdQ_x + \frac{1}{2}\overline{\Delta r}^2 Q_x^2, \qquad (13\text{-}63)$$

where

$$s = \frac{x + (-jd + i\overline{\Delta r}^2 Q_x)}{\sqrt{2}\,\overline{\Delta r}}. \qquad (13\text{-}64)$$

Then the integral becomes

$$\int e^{-\frac{(x-jd)^2}{2\overline{\Delta r}^2}} e^{iQ_x x} \, dx = \left(\int e^{-s^2} \sqrt{2}\,\overline{\Delta r}\, ds \right) e^{\left(-ijdQ_x - \frac{\overline{\Delta r}^2 Q_x^2}{2} \right)}$$

$$= \sqrt{2\pi}\,\overline{\Delta r}\, e^{-idQ_x j}\, e^{-\frac{\overline{\Delta r}^2 Q_x^2}{2}}, \qquad (13\text{-}65)$$

since $\int e^{-s^2} \, ds = \sqrt{\pi}$, and the field becomes

$$E_{so} \approx E_1 f_o \left(\sum_j e^{-ijdQ_x} \right) e^{-\frac{\overline{\Delta r}^2 Q_x^2}{2}}. \qquad (13\text{-}66)$$

The sum in parentheses is the same as in equation 13-58, which results in the condition that the diffracted field is nonzero only for $Q_x = m\dfrac{2\pi}{d}$. Thus the field is once again a series of delta functions. Thermal vibrations do not broaden the peak even though the atom locations are smeared out. However, the intensity of the peak is reduced by the Debye-Waller factor,

$$f_{DW} = \left(e^{-\frac{\overline{\Delta r}^2 Q^2}{2}} \right)^2 = e^{-\overline{\Delta r}^2 Q^2}. \qquad (13\text{-}67)$$

At room temperature f_{DW} is of order unity for most materials.

The thermal vibrations described by equation 13-60 ignore correlations between atomic vibrations. Such vibrations tend to be correlated only for close neighbors, that is, over short distances. Because small distances in real space correspond to large wavevectors in Fourier space, the scatter from the correlated vibrations shows up as low-intensity background scatter with broad peaks, known as *thermal diffuse scattering*. This scattering creates tails on Bragg peaks, which either can be a nuisance or can be analyzed to study the elastic wave structure of the material.

13.12.3 CRYSTAL SIZE

Another omnipresent "defect" is that the sample is finite. The effect of the finite size can be considered in terms of the Heisenberg uncertainty principle, as was done for the effect of finite aperture sizes in optics in section 12.1. If the electrons are confined to a sample of size W, there is an uncertainty in their momentum,

$$\Delta p_e \, \Delta x_e = \Delta p_e \, W \sim \text{h},\qquad(13\text{-}68)$$

which creates an equal uncertainty in the momentum of the x-ray photons scattered off the electrons,

$$\Delta p = \Delta(\hbar\kappa_f) = \Delta p_e \sim \frac{\text{h}}{W} \Rightarrow \Delta\kappa_f \sim \frac{2\pi}{W}.\qquad(13\text{-}69)$$

Narrowing the crystal in one dimension creates an uncertainty in the component of the diffracted wavevector in that direction.

Alternatively, the effect can be calculated by expressing the finite crystal as an infinite crystal multiplied by an aperture function and using Fourier transform techniques. That is not necessary here because a calculation of the effect of finite sample size was already seen for the finite chain, in section 13.4, which we will now use as a model for a stack of planes. The size of the sample is the length of the stack,

$$W = N_a d,\qquad(13\text{-}70)$$

where now d is the distance between the planes, and N_a is the number of atomic planes. The resulting diffraction peaks were given by equation 13-22,

$$E_s \approx E_1 f_o \frac{\sin\left(N_a \frac{Qd}{2}\right)}{\sin\left(\frac{Qd}{2}\right)},\qquad(13\text{-}71)$$

where Qd has been substituted for $\Delta\phi = \vec{Q}\cdot\vec{d}$, because $\vec{Q}$ must be parallel to $\vec{d}$. As before, the width of the peak is given by half the distance between zeros of the numerator,

$$\Delta Q = \frac{2\pi}{N_a d} = \frac{2\pi}{W}.\qquad(13\text{-}72)$$

This is the same result as that from the Heisenberg uncertainty relation. Differentiating the definition of Q, equation 13-9, gives

$$Q = \frac{4\pi}{\lambda} \sin\theta \Rightarrow \Delta Q = \left(\frac{4\pi}{\lambda} \cos\theta\right)\Delta\theta, \tag{13-73}$$

so that

$$\Delta\theta = \frac{1}{\left(\dfrac{4\pi}{\{\lambda\}} \cos\theta\right)}[\Delta Q] = \frac{1}{\dfrac{4\pi}{\{2d\sin\theta\}}\cos\theta}\left[\frac{2\pi}{W}\right], \tag{13-74}$$

where λ has been replaced by using Bragg's law. Finally,

$$\frac{\Delta\theta}{\tan\theta} = \frac{d}{W}. \tag{13-75}$$

FIGURE 13-20. A polycrystalline sample: a solid formed with very many small crystallites of different orientation.

Unless the angle is near 90°, the broadening in angle is approximately the ratio of the plane spacing to the sample size. The effective size of the sample for a powder is the size of an individual crystal grain. The grains can be very small if the sample is polycrystalline, that is, a solid made up of very many single crystal grains, as shown in Figure 13-20. If other contributions to the angular broadening of the diffraction peak are kept small, the width of the peak can be used to estimate the grain size of the sample.

EXAMPLE 13-8: GRAIN-SIZE BROADENING

A polycrystalline sample has a grain size of $W_g = 0.03$ μm and is irradiated with 12.4 keV x rays. How much is the Bragg peak at $2\theta = 40°$ broadened?

First, we need the wavelength and plane spacing. The wavelength is 1 Å.

The plane spacing is $d = \dfrac{\lambda}{2\sin\left(\dfrac{2\theta}{2}\right)} \approx \dfrac{1\,\text{Å}}{2\sin(20°)} \approx 1.46\,\text{Å}.$

Then $\Delta\theta = \dfrac{d}{W_g}\tan\left(\dfrac{2\theta}{2}\right) \approx \dfrac{1.46\,\text{Å}}{0.03\times10^4\,\text{Å}}\tan(20°) \approx 1.8\,\text{mrad} \approx 0.1°.$

13.12.4 AMORPHOUS MATERIALS

It is also possible for the sample not to be crystalline at all. Glasses and many organic materials are amorphous; that is, they do not possess the translational symmetry assumed in section 13.5. However, the Fourier transform relationship between the electron density and the diffracted field of equation 13-17 is still true, so measuring the diffraction pattern yields information about the radial distribution function $g(\vec{R})$, the probability that there is an atom at $\vec{R}$. Because the structure is more random, the distances between neighbors vary somewhat, and the diffracted rings tend to be broader than for a crystalline powder. An example of the diffraction from water is shown as a wide ring on the protein crystal diffraction pattern of figure 14.28.

13.13 Resolution

13.13.1 THE EFFECT OF ANGULAR BROADENING

The point of a diffraction measurement is generally to determine the d spacing of a crystal (or nearest-neighbor distances for an amorphous material). Given a measured diffraction angle 2θ and known wavelength λ, the d spacing is calculated from Bragg's law. The effect of a broadened diffraction peak is generally to create an uncertainty in the Bragg angle determined from the peak. If the uncertainty in Bragg angle is $\Delta\theta$, the uncertainty in plane spacing d is easily calculated from Bragg's law,

$$\lambda = 2d \sin \theta \Rightarrow d = \frac{\lambda}{2 \sin \theta} \Rightarrow |\Delta d| = \left| \frac{-\lambda}{2 \sin^2 \theta} \cos \theta \, \Delta\theta \right| = \left| d \, \frac{\Delta\theta}{\tan \theta} \right|$$

$$\Rightarrow \left| \frac{\Delta d}{d} \right| = \left| \frac{\Delta\theta}{\tan\theta} \right|. \tag{13-76}$$

EXAMPLE 13-9: PEAK BROADENING

Consider the powder sample of example 13-7. Compute the uncertainty in lattice constant computed from the first and last rings if the sample a) has a mosaicity of 1°, b) has grain size broadening of 1°.

a) Mosaicity simply changes the orientation of the planes to the beam. While it broadens the rocking curve from a single crystal, it has no effect on a powder sample (except possibly to reduce the effective grain size, since each individual crystal is now broken up into mosaic blocks).

b) The peaks are no longer delta functions, but are 1° wide.

The uncertainty in 2θ is 1°, so the uncertainty in θ is $\Delta\theta \approx \dfrac{1° \dfrac{\pi \, \text{rad}}{180°}}{2} \approx 8.7 \ \text{mrad}.$

For the first ring, $\Delta d_1 = d_1 \left| \dfrac{\Delta\theta}{\tan\theta} \right| \approx (2.82 \ \text{Å}) \dfrac{8.7 \times 10^{-3} \ \text{rad}}{\tan\left(\dfrac{30.82}{2} \right)} \approx 0.09 \ \text{Å}$

$$\Rightarrow \Delta u = \sqrt{2} \, \Delta d \approx 0.13 \ \text{Å}.$$

For the sixth ring, $\Delta d_6 = d_6 \left| \dfrac{\Delta \theta}{\tan \theta} \right| \approx (1.155 \text{ Å}) \dfrac{8.7 \times 10^{-3} \text{ rad}}{\tan\left(\dfrac{81.01}{2}\right)} \approx 0.012 \text{ Å}$

$$\Rightarrow \Delta u = \sqrt{12} \Delta d \approx 0.04 \text{ Å}.$$

There is less uncertainty in lattice constants found for high-angle rings.

Here the uncertainty in peak position has been taken as the peak width. In some cases it is possible to accurately estimate the location of the center of the peak even if the peak is wide, at least if the peak is smooth. For example, polycapillary and other optics tend to create smooth peaks, which allow for greater ease of peak center estimation.

13.13.2 ENERGY SPREAD

The diffraction peak is also broadened by nonidealities in the x-ray beam. A real x-ray beam is not monochromatic but has some energy spread. The effect of the energy width of the incoming beam is again easily calculated from Bragg's law,

$$\lambda = 2d \sin\theta \Rightarrow \Delta\lambda = (2d \cos\theta) \Delta\theta \Rightarrow \Delta\theta = \frac{\Delta\lambda}{2d \cos\theta}$$

$$= \frac{\Delta\lambda}{2\left(\dfrac{\lambda}{2\sin\theta}\right)\cos\theta} = \tan\theta \frac{\Delta\lambda}{\lambda} \Rightarrow \frac{\Delta\theta}{\tan\theta} = \frac{\Delta\lambda}{\lambda} = \frac{\Delta U}{U}, \tag{13-77}$$

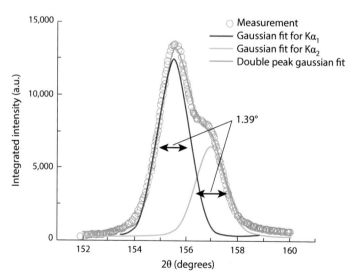

FIGURE 13-21. Diffraction peak showing the effect of the Kα doublet, which gives a shoulder to the peak. The peaks were broadened by a highly convergent beam to 1.4°, so the doublet is barely resolvable, but the peak location could still be found with high accuracy. From W. Zhou et al., *Thin Solid Films* 518 (2010): 5047–56. Reproduced by permission of Elsevier.

where Bragg's law was used again to substitute for the plane spacing d. If a characteristic line is used for the source, the line width is typically a few electronvolts, compared with kiloelectronvolts for the photon energy, so the uncertainty in angle contributed by the energy spread is a few tenths of a percent. However, a doublet line can also increase the broadening, as shown for the barely resolved $K\alpha$ doublet in Figure 13-21.

13.13.3 GLOBAL DIVERGENCE AND APERTURE SIZE

If the x-ray beam is not collimated, but has some global divergence Θ, as discussed in section 5.4 and shown in Figure 13-22, then the apparent radius of the ring on the detector is

$$r_{ring} = L \tan\left(2\theta_B \pm \frac{\Theta}{2}\right) \pm \frac{W}{2}. \tag{13-78}$$

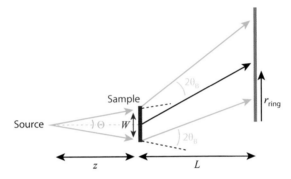

FIGURE 13-22. The effect of global divergence Θ on the width of the diffraction ring.

If the ring radii are used to calculate the diffraction angle, the apparent angle becomes

$$2\theta = \arctan\left(\frac{r_{ring}}{L}\right) = \arctan\left(\tan\left(2\theta_B \pm \frac{\Theta}{2}\right) \pm \frac{W}{2L}\right)$$
$$\approx 2\theta_B \pm \frac{\Theta}{2} \pm \frac{W}{2L}\cos^2(2\theta_B). \tag{13-79}$$

The width of the peak is then

$$\Delta 2\theta \approx \frac{W}{L}\cos^2(2\theta_B) + \Theta. \tag{13-80}$$

For a collimated beam, Θ is zero, but there is still blur due to the finite size, W, of the beam on the sample.

13.13.4 LOCAL DIVERGENCE

A real x-ray beam, even if it is collimated, is not completely parallel but has a local divergence ξ, as discussed in section 5.5 and shown in Figure 13-23. The difference between Figures 13-22 and 13-23 is that in the former, where we were considering global

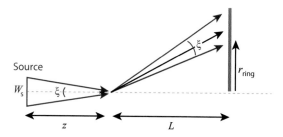

FIGURE 13-23. The effect of local divergence ξ on the width of the diffraction ring.

divergence, the source is small, but the sample is large. In the latter, to consider local divergence, the sample is small, but the source is large. (Under real conditions, of course, neither source nor sample is ideally small, and both effects must be considered.) The apparent radius of the ring on the detector due to the local divergence (finite source size) alone is

$$r_{ring} = L \tan\left(2\theta_B \pm \frac{\xi}{2} \right),$$

(13-81)

so that the angular width of the peak is

$$\Delta 2\theta \approx \xi.$$

(13-82)

Problems

SECTION 13.2

1. A face-centered cubic (fcc) crystal has four atoms per unit cell: at a corner and at the center of the three nearest faces: $(0,0,0)$, $(0, 1/2, 1/2)$, $(1/2, 0, 1/2)$, and $(1/2, 1/2, 0)$. Show that the structure factor for an fcc crystal is

$$F = f\left(1 + (-1)^{h+k} + (-1)^{h+l} + (-1)^{k+l}\right),$$

when h, k, and l are all even or all odd.

2. A polycrystalline sample of an fcc crystal with a unit cell cube edge of 0.3 nm is irradiated with 10 keV photons.

a) What is the reciprocal lattice constant (the length of the edge of the reciprocal lattice cube)?

b) Find the diameters of all the diffraction rings on a 300 mm diameter detector placed 100 mm on the source side of the sample (the detector has a hole to allow the incident beam to pass through).

c) The grain size is 50 nm. What is the width of the rings due to grain-size broadening?

d) What is the largest sample size that will create a ring width small enough to observe the grain size broadening in part c if the source is 200 mm from the sample?

3. Silicon is diamond cubic. A diamond cubic crystal has eight atoms per unit cell in two groups of four: the first group includes a corner and the center of the three nearest faces: $(0, 0, 0)$, $(0, 1/2, 1/2)$, $(1/2, 0, 1/2)$, and $(1/2, 1/2, 0)$. The second group is displaced $1/4, 1/4, 1/4$ from those four: $(1/4, 1/4, 1/4)$, $(1/4, 3/4, 3/4)$, $(3/4, 1/4, 3/4)$, and $(3/4, 3/4, 1/4)$. Show that the structure factor for the (100), (200), and (300) planes is zero.

Further reading

B. D. Cullity, **Elements of X-Ray Diffraction**, Addison-Wesley, 1956.

Jens Als-Nielsen and Des McMorrow, **Elements of Modern X-ray Physics**, John Wiley & Sons, 2001.

Eric Lifshin, **X-ray Characterization of Materials**, John Wiley & Sons, 1999.

David Attwood and A. Sakdinawat, **X-Rays and Extreme Ultraviolet Radiation: Principles and Applications**, Cambridge University Press, 20016

Christman, **Fundamentals of Solid State Physics,** John Wiley & Sons, 1987

D. Halliday, R. Resnick, and J. Walker, **Fundamentals of Physics**, 10th ed., John Wiley & Sons, 2013, sec. 36.7.

14

SINGLE-CRYSTAL AND THREE-DIMENSIONAL DIFFRACTION

14.1 The Ewald sphere

In chapter 13, we saw that the diffraction condition, $\vec{Q} = \vec{G}$, equation 13-35, gives rise to Bragg's law, $\lambda = 2d \sin\theta_B$, equation 13-45. Bragg's law provides a straightforward technique for analyzing and predicting some diffraction geometries, especially powder diffraction. However, equation 13-35 is a vector equation, and some of the directional information is lost in the scalar Bragg equation. This is especially the case in assessing diffraction from single crystals rather than polycrystalline or powder samples. An alternative view of the diffraction problem is gained by the wavevector, "k-space" view of Ewald sphere analysis. Because diffraction arises from coherent scatter, the incoming and diffracted wavevectors must be equal. For a given incident wavevector, the set of possible outgoing wavevectors forms a hollow sphere, as shown in Figure 14-1. The diffraction condition requires that the scattering vector $\vec{Q}$ is equal to a reciprocal lattice vector $\vec{G}$. One way to assess this is to sketch the reciprocal lattice with its origin at the endpoint of the incident wavevector, as shown in Figure 14-2. In general, for an arbitrary x-ray wavelength and orientation of the crystal, the diffraction condition is not satisfied, and there is no diffracted beam.

An intersection between the reciprocal lattice point and the Ewald sphere can be contrived by using a powder sample, as was done in section 13.9. For powders, all possible orientations of the reciprocal lattice are present, so each reciprocal lattice point turns into a reciprocal lattice sphere (each vector $\vec{G}$ represents the radius of a sphere centered at the reciprocal lattice origin), as shown in Figure 14-3. Alternatively, a single crystal sample can

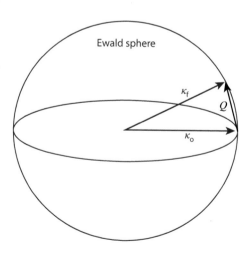

FIGURE 14-1. The incoming wavevector, κ_o, diffracted wavevector, κ_f, and scattering vector Q. For a given incoming wavevector, the locus of all possible diffracted wavevectors is a sphere.

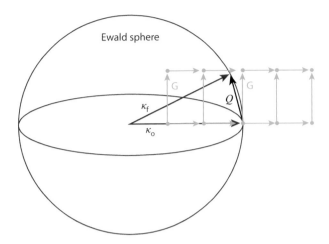

FIGURE 14-2. The scattering vector Q must be equal to a reciprocal lattice vector G. For an arbitrary case, the two vectors are not equal.

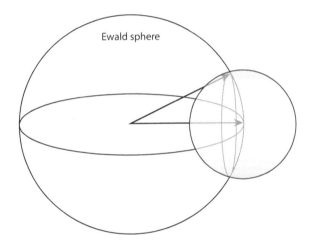

FIGURE 14-3. For a powder sample, the reciprocal lattice points become reciprocal lattice spheres, and the two spheres intersect in a cone of directions. Note that the origin of the reciprocal lattice sphere is the endpoint (not the origin) of the incident wavevector, as required to satisfy the diffraction condition.

be rotated in a θ-2θ diffractometer, as discussed in section 13.8 and illustrated in Figure 13-11. In reciprocal wavevector space, this causes the reciprocal lattice points to be rotated into a position in which they can intersect the Ewald sphere, as shown in Figure 14-4.

14.2 The θ-2θ diffractometer and the Rowland circle

In a θ-2θ diffractometer with a typical conventional source, the x-ray beam is divergent, so that the ray hitting the part of the sample close to the source is incident at a

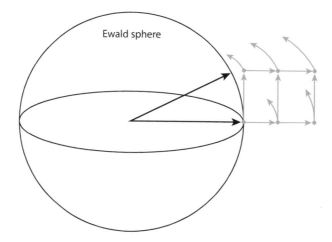

FIGURE 14-4. In a θ-2θ diffractometer, the reciprocal lattice points are rotated until they intersect with the Ewald sphere.

larger angle than the ray hitting farther away, as shown in Figure 14-5. The allowed length of the sample in the beam direction is therefore limited by the angular acceptance of the sample, for example by its mosaicity as discussed in section 13.12.1. Because each incident ray must hit at the Bragg angle to diffract, the ray hitting closer to the source must be diffracted off those regions of the mosaic sample whose planes are tilted slightly toward the source, and the ray hitting farther from the source must be diffracted

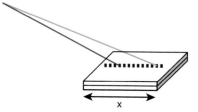

FIGURE 14-5. Two rays striking a sample. The ray hitting nearer the source hits at a larger angle.

from planes tilted slightly away. Thus the ray closer to the source also leaves the sample at a larger angle, so that the beams from the two parts of the sample will converge, as shown in Figure 14-6. This provides a convenient place to situate the detector, as shown in Figure 14-7. Viewed from above, the source, sample, and detector are in a single line, as shown in Figure 14-8. In Figure 14-7 the source and detector are placed symmetrically with respect to the sample. However, the sample can be placed anywhere on this circle, called the *Rowland circle*, containing both the source and detector, as shown in Figure 14-9 and described in more detail in section 14.3.

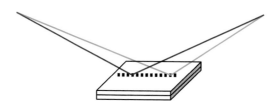

FIGURE 14-6. Convergent beam diffracted from a divergent source.

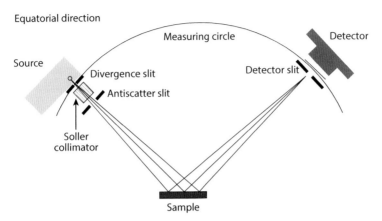

FIGURE 14-7. Setup for a Bragg-Brentano diffractometer. If the source is fixed, the sample moves θ when the detector moves 2θ. If the sample is fixed, both the source and detector move θ in opposite directions. This is a side view of the setup; the top view is shown in Figure 14-8. From S. T. Misture, Requirements for X-Ray Diffraction, chap. 28, *Handbook of Optics*, 3rd ed., vol. 5, McGraw-Hill, 2010.

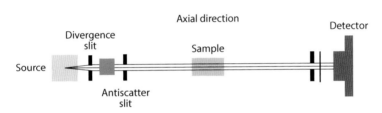

FIGURE 14-8. Top view of the setup in Figure 14-7. The dark square between the divergence and antiscatter slits is a soller slit. From S. T. Misture, Requirements for X-Ray Diffraction, chap. 28, *Handbook of Optics*, 3rd ed., vol. 5, McGraw-Hill, 2010.

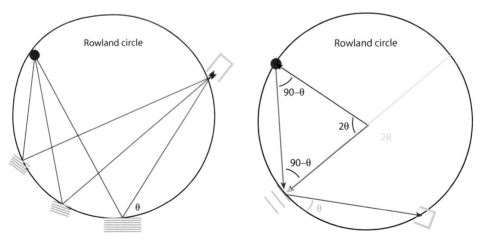

FIGURE 14-9. The Rowland "measuring" circle.   FIGURE 14-10. Geometry of the Rowland circle.

14.3 Aside: Proof that the angle of incidence is always θ_B on the Rowland circle

To show that the angle of incidence is always the Bragg angle on the Rowland circle, we start with a source and detector equidistant from the sample, both at an angle θ, as shown in Figure 14-10. This angle is chosen so that $\theta = \theta_B$, the Bragg angle for the crystal planes at the employed x-ray energy. The Rowland circle has as its diameter the normal to the crystal and is drawn through both the source and detector. The center of the circle is thus a distance R from the crystal and the source. The radius R can be found from the source-to-detector distance L using the law of cosines,

$$L^2 = R^2 + R^2 - 2R^2 \cos(2\theta) \Rightarrow L = 2R\sin\theta. \tag{14-1}$$

Now consider a sample placed on the Rowland circle at angle α with respect to the original, as shown in Figure 14-11. The sample is also tilted at the angle α, so that the crystal normal is parallel to the chord drawn from the outer edge of the circle, and the planes are no longer tangential to the circle. The triangle formed by the sample, circle center, and outer edge is isosceles, so both angles are α, and the third angle is then $180 - 2\alpha$. That means the angle between the original radius to the first sample and the new radius to the second sample is 2α. As shown in Figure 14-12, another isosceles triangle is formed by the source, sample, and circle center. Since its interior angle is $2\theta - 2\alpha$, the other two angles are both $90 - \theta + \alpha$. Thus a ray from the source will be incident on the crystal at the original Bragg angle θ, and will diffract. A bit more geometry will show that this ray also converges onto the detector.

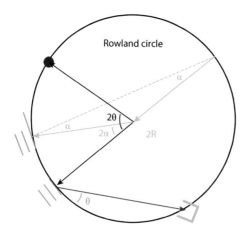

FIGURE 14-11. A second sample placed on the Rowland circle, with its normal tilted at α from the first (the crystal is perpendicular to the green dashed line; it is not tangent to the circle). The green triangle is isosceles, so the interior angle is $180 - 2\alpha$.

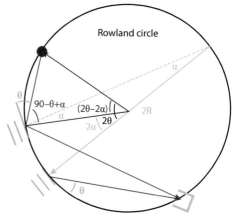

FIGURE 14-12. The interior angle of the purple isosceles triangle is $2\theta - 2\alpha$. This makes each of the other angles equal to $90 - \theta + \alpha$. The angle between the green dashed crystal normal and the source ray is then $90 - \theta$, and the ray hits the crystal at the Bragg angle θ.

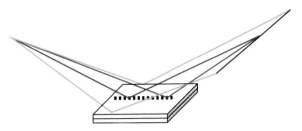

FIGURE 14-13. The blue divergent rays hit the sample in directions into and out of the plane of Figures 14-7 and 14-9. These rays continue to diverge.

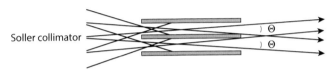

FIGURE 14-14. A soller slit is a series of parallel plates used to limit the global divergence of the beam in the out-of-plane direction. From S. T. Misture, Requirements for X-Ray Diffraction, chap. 28, *Handbook of Optics*, 3rd ed., vol. 5, McGraw-Hill, 2010.

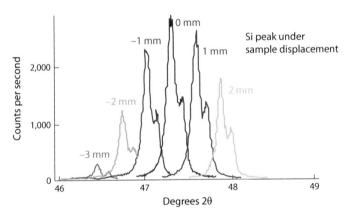

FIGURE 14-15. The diffracted intensity, and the apparent angle of the Bragg diffraction, changes if the sample is displaced from the correct location on the Rowland circle. From Brian York, Denver X-ray Conference.

14.4 Beam divergence

The convergence on the Rowland circle shown in Figures 14-6 and 14-9 occurs only in the diffraction plane. In the out-of-plane direction the beam is still divergent, as shown in Figure 14-13. For that reason, a series of parallel plates, called a *soller slit*, is used to regulate the beam divergence in the out-of-plane direction, as shown in in Figures 14-7 and 14-14. One consequence of the in-plane beam divergence is that the angle between the source and the sample changes if the sample is moved in the

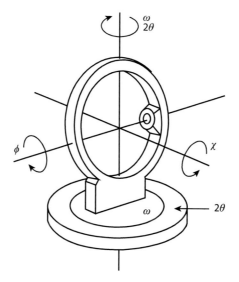

FIGURE 14-16 A "four circle" goniometer, which can rotate the sample by θ as the detector is rotated by 2θ, but can also separately rotate the sample by the three angles ω, χ, and ϕ.

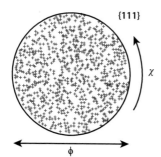

FIGURE 14-17. Measured (random) orientation of grains in a polycrystalline sample. Courtesy of Mark F. Horstemeyer, Mississippi State University.

direction perpendicular to the crystal planes, as shown in Figure 14-15. Care must be taken in placing the sample to avoid this artifact. Alternatively, the peak shift is avoided if the incident beam is made parallel, for example by using a collimating x-ray optic like a polycapillary optic, described in section 12.2.4.

14.5 Texture and strain measurements

A θ-2θ diffractometer, like that shown in Figures 13-11 and 14-7, can be used to measure polycrystalline samples. Often, these are films deposited on a substrate, and the sample is mounted with the substrate at angle θ to the source. If the sample is polycrystalline, only a small fraction of the grains in the sample will be aligned so that the particular set of planes (the ones with the necessary d spacing) are at the given angle θ to the beam. If the sample is placed on a goniometer, as shown in Figure 14-16, it can be tilted so that other grains can be brought into the diffraction condition. For a sample with relatively large grains, it may be that only one grain at a time is in alignment. In that case, by recording the goniometer angles for which diffraction is seen, the orientation of each grain can be mapped out, as shown in Figure 14-17. If the sample is a fairly uniform polycrystalline sample, the pattern is random. If the sample is, for example, a thin film grown on a substrate, there may be a preferred orientation of one set of planes to be parallel to the substrate. The sample is not a single crystal but consists of small grains with the other two crystal axes at

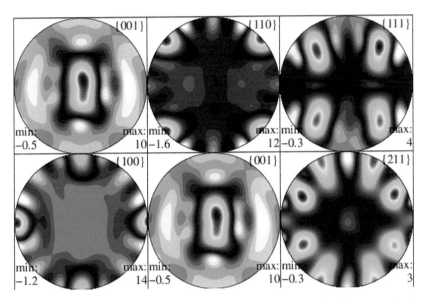

FIGURE 14-18. A map of the orientation of the TiAl grains in an Al alloy. For example, for a sample aligned for diffraction from (001) planes, the peak falls off rapidly if the sample is tilted in ϕ. For the (111) orientation, after tilting in ϕ, there is strong diffraction at 90° intervals in χ. Liss et al., *Textures and Microstructures* 35, nos. 3/4 (2003): 219–52.

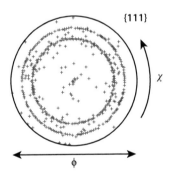

FIGURE 14-19. Orientation of the grains shown in Figure 14-17 after the sample was strained along the χ axis. Courtesy of Mark F. Horstemeyer, Mississippi State University.

random rotations. From the direction perpendicular to the substrate, the crystal will still look polycrystalline, as in Figure 13-20. Once the crystal is rotated to the correct θ (by changing ω, as shown in Figure 14-16), rotation by χ about an axis perpendicular to the diffracting planes will not change the output. However, tilt about an axis perpendicular to both the ω and χ axes will cause the diffracted beam to deflect out of the plane of the detector, and not be captured. The result is an intensity map similar to that of Figure 14-18. The signal is very strong only near $\phi = 0$, at the center of the pole figure. This property of nonrandom orientations is known as *texture*. Texture can also be created if an external stress causes the grains to rotate so that the selected planes have a preferred orientation, as shown in Figure 14-19.

In general, strain also causes the plane spacings to change. Diffraction measurements can only detect changes in d spacing in the direction of the scattering vector, $\vec{Q}$ (which is also the plane normal direction). The measured d spacing will vary with $\sin^2\phi$, as shown in Figure 14-20, where ϕ is the angle between the plane normal and the measurement Q vector.

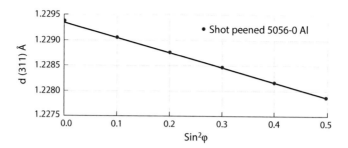

FIGURE 14-20. From P. S. Prevey, *Metals Handbook*, 9th ed., vol. 10. Copyright ASM.

14.6 Single-crystal diffraction

In the previous sections we investigated creating an intersection between the Ewald sphere and the reciprocal lattice points by manipulating the reciprocal lattice, by either rotating the crystal or using a powder or polycrystalline sample. Another technique, called *Laue diffraction*, is to broaden the spherical shell that is the Ewald sphere by using broadband rather than monochromatic radiation, as shown in Figure 14-21. The larger sphere is given by the largest available wavevector (and hence shortest wavelength), generally determined by the tube voltage for a conventional source; and the smaller sphere is

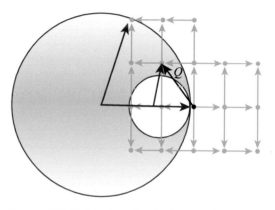

FIGURE 14-21. Solid sphere for single-crystal Laue diffraction measurements. The radius of the larger sphere is determined by the maximum available photon energy, and the radius of the smaller sphere, by the minimum energy required to be registered in the detector. Any reciprocal lattice vector between the two spheres will result in a possible diffraction.

given by the smallest available wavevector (and hence longest wavelength), generally limited by absorption in the detector window. Any reciprocal lattice point which falls between the two spheres corresponds to a possible Q vector for diffraction. Each reciprocal lattice point corresponds to a unique outgoing wavevector, and so a single point on the detector, as shown in Figure 14-22. As for the analysis of the ring pattern of Figure 13-15, larger distances of the spot from the center of the pattern correspond

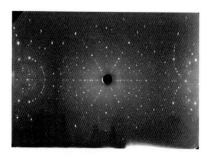

FIGURE 14-22. Laue diffraction pattern of silicon. The film was wrapped in a cylinder to cover nearly 360° around the crystal. From S. C. Jones, Union College. The lines of spots indicate high symmetry directions.

to larger 2θ angles, and the expected 2θ angle can be calculated from Bragg's law in the same manner as was done in section 13.9, except for two issues. First, the wavelength of the radiation that created that particular spot is unknown. Second, the azimuthal angle is necessary to describe the pattern but is not addressed by Bragg's law. A more complete vector description is required. To start, it is necessary to know the direction of the incident beam with respect to the crystal axes.

EXAMPLE 14-1

Consider a body-centered cubic (bcc) single crystal with lattice constant $u = 2$ Å irradiated with a "white" source producing x rays of energy ranging from 0 to 20 keV. The x-ray beam is incident parallel to the z direction for the crystal. Calculate the x, y locations of all possible diffraction spots on a 300 mm wide detector 100 mm from the sample.

The first step is to compute the relationship between the planes and the photon energy. The incident beam is given by the wavevector $\vec{\kappa}_o = \dfrac{2\pi}{\lambda}\hat{z}$. The reciprocal lattice vector for a cubic lattice (hkl) plane is $\vec{G} = \dfrac{2\pi}{u}(h\hat{x} + k\hat{y} + l\hat{z})$. Thus the outgoing wavevector after diffraction from the (hkl) planes is

$$\vec{\kappa}_f = \vec{\kappa}_o + \vec{G} = \frac{2\pi}{\lambda}\hat{z} + \frac{2\pi}{u}(h\hat{x} + k\hat{y} + l\hat{z}). \tag{14-2}$$

The length of the incident wavevector is unknown, but because this is coherent scatter, the photon energy must be unchanged, so the length of the outgoing wavevector must equal that of the incident wavevector,

$$\vec{\kappa}_f \cdot \vec{\kappa}_f = \frac{4\pi^2}{\lambda^2} + \frac{8\pi^2}{\lambda u}l + \frac{4\pi^2}{u^2}(h^2 + k^2 + l^2) = \vec{\kappa}_o \cdot \vec{\kappa}_o = \frac{4\pi^2}{\lambda^2}. \tag{14-3}$$

Thus the wavelength associated with the particular (hkl) plane can be computed:

$$-\frac{8\pi^2}{\lambda u}l = \frac{4\pi^2}{u^2}(h^2 + k^2 + l^2) \Rightarrow \lambda = 2u\left(\frac{-l}{h^2 + k^2 + l^2}\right). \tag{14-4}$$

Because the wavelength must be positive, the index l must be negative. This is consistent with Figure 14-21, which shows that all the allowed Q vectors have a component antiparallel to the incident beam; that is, if the incident beam is in the z direction, the Q vector will have a negative z component. Now for each (hkl) plane, the wavelength can be computed, and the d spacing is given by the usual formula,

$$d = \frac{u}{\sqrt{h^2 + k^2 + l^2}}, \tag{14-5}$$

so that the Bragg angle can be computed:

$$\theta = \sin^{-1}\left(\frac{\lambda}{2d}\right) = \sin^{-1}\left(\frac{\dfrac{-2lu}{h^2 + k^2 + l^2}}{2\dfrac{u}{\sqrt{h^2 + k^2 + l^2}}}\right) = \sin^{-1}\left(\frac{-l}{\sqrt{h^2 + k^2 + l^2}}\right). \tag{14-6}$$

Note that the angle is independent of u. Crystals with the same structure but different lattice spacings will give identical diffraction patterns, although they will require different x-ray wavelengths. Once the Bragg angle is determined, it remains to determine the azimuthal angle, as shown in Figure 14-23. This can easily be done by realizing that the diffracted ray, although it exists in real space, is parallel to the diffracted wavevector $\vec{\kappa}_f$. Thus the azimuthal angle is

$$\varphi = \tan^{-1}\left(\frac{y}{x}\right) = \tan^{-1}\left(\frac{\kappa_{f,y}}{\kappa_{f,x}}\right) = \tan^{-1}\left(\frac{k}{h}\right). \tag{14-7}$$

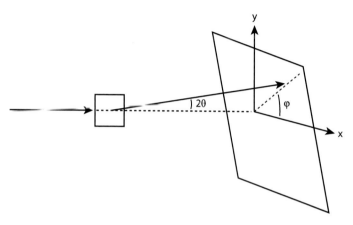

FIGURE 14-23. Diffracted x-ray beam incident on the detector.

Now all that remains is to tally all possible (hkl) planes.

a) As was calculated for example 13-6, the structure factor and hence intensity is zero for all (hkl) for which the sum $h+k+l$ is odd. This eliminates the (100) plane. The (110) plane won't work because l is not negative.

b) However, unlike the case for powder diffraction, the different "equivalent" planes give rise to different diffraction spots. So we must consider the $(10\bar{1})$ plane (for which $h=1$, $k=0$, and $l=-1$). The Bragg angle for the $(10\bar{1})$ plane is determined from equation 14-6 as 45°, so 2θ is 90°, and no ray will hit the detector. This is also the Bragg angle for the $(01\bar{1})$ and $(0\bar{1}\bar{1})$ planes. It is also the angle for the integer multiples of these planes, for example, the $(20\bar{2})$ and $(30\bar{3})$ planes, so none of these give diffraction spots.

c) The $(11\bar{1})$ plane has a structure factor of zero.

Tabulating in order of increasing $-l$, then increasing $h^2+k^2+l^2$:

d) The $(21\bar{1})$ plane gives a 2θ angle of 48°. The radius on the detector is given by the distance L to the detector as

$$R = L \tan(2\theta) \approx (100\,\text{mm}) \tan(48°) \approx 112\,\text{mm}, \tag{14-8}$$

which gives a diameter of 224 mm. This is less than the 300 mm diameter of the detector and so will hit the detector if the required photon energy is available. The

Table 14-1

h	k	l	$h^2+k^2+l^2$	F	$\theta = \sin^{-1}\left(\dfrac{-l}{\sqrt{h^2+k^2+l^2}}\right)$	$2\theta_B$	$R=L$ $\tan(2\theta)$	$\lambda = -ul/$ $(h^2+k^2+l^2)$	U	
0	0	0	0	2f	0	0	0			This is the undiffracted, straight-through beam
1	0	0	1	0						No intensity as $F=0$; see (a)
1	0	1	2	2f					<0	Not possible; see (a)
1	0	−1	2	2f	45	90	∞			Does not hit detector; see (b)
1	1	−1	3	0						No intensity; see (c)
0	0	−2	4	2f	90	180				Does not hit detector; see (f)
2	0	−1	5	0						No intensity
2	1	−1	6	2f	24	48	112	0.67	19	8 spots are seen for the various combinations of (−2,1,−1), (1,2,−1), etc.; see (d)
1	1	−2	6	2f	55	109				Does not hit detector, see (h)
2	0	−2	8	2f	45	90				Does not hit detector
2	2	−1	9	0						No intensity
3	0	−1	10	2f	18	37	75	0.4	31	The required photon energy is higher than available, so no diffraction is seen, see (e)
2	2	−2	12	2f	35	70	282			Does not hit detector, see (i)
2	3	−1	14	2f	15.5	31	60	0.29	43	Required energy too high
1	3	−2	14	2f	32	65	211			Does not hit detector
2	4	−2	24	2f	24	48	112	0.33	37	Required energy too high
2	5	−1	30	2f	10.5	21	38	0.13	93	Required energy too high
1	5	−2	30	2f	21	43	93	0.27	47	Required energy too high

wavelength from equation 14-4 is 0.67 Å, which corresponds to an energy of 18.6 keV, which is available from the source. So this set of planes has an allowed structure factor, an angle which makes it possible to hit the detector at a radius which is smaller than the detector, and uses an available energy. The azimuthal angle from equation 14-7 is 26.6°, so that $x = R \cos (\varphi) = 100$ mm; $y = R \sin (\varphi) = 50$ mm. The other permutations, $(12\overline{1})$, $(\overline{2}1\overline{1})$, and so on, are also possible, with $(x, y) = (50, 100)$, $(-100, 50)$, ...

e) The $(30\overline{1})$ plane gives an acceptable 2θ angle of 37°, but the required energy, calculated from equation 14-4 is 31 keV, higher than provided by the 20 kV x-ray tube. Since the energy increases with $h^2 + k^2 + l^2$, higher values of h and k require photon energies which are not available with this l.

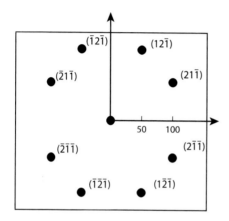

FIGURE 14-24. Diffraction pattern produced in example 14-1.

f) The $(00\overline{2})$ plane gives a 2θ angle of 180°, back at the source, and so cannot be detected.

g) The $(10\overline{2})$ plane has zero structure factor.

h) The $(11\overline{2})$ plane gives a 2θ angle of 109°, which does not hit the detector. In general, from equation 14-6, to keep the 2θ angle less than 90°, it is necessary to have $l^2 < h^2 + k^2$. This also eliminates, for example, the $(20\overline{2})$ and $(10\overline{3})$ planes. As l^2 increases, h and k must increase.

i) The $(22\overline{2})$ plane gives a 2θ angle of 71° and an energy of 18.6 keV, but the radius is 282 mm, so the diameter is larger than the detector. Again, increasing h and k would cause the energy to be too high

j) The conflicting requirements for a low enough angle to hit the detector and low enough energy to be possible means that no other diffraction spots are possible. The only diffraction spots are from the (211)-type planes, as shown in Figure 14-24.

The reverse problem, using a Laue diffraction pattern to determine information about the crystal structure, is more complex, as it may be that the direction of the crystal axes relative to the beam direction is unknown. In that case, it may be necessary to rotate the crystal in the beam to locate a high symmetry pattern to use symmetry clues to assign (hkl) indices to individual diffraction spots. With complex structures such as proteins, discussed in section 14.8, iterative techniques are generally employed, with model structures developed to compare measured intensity with simulated results.

14.7 Laue geometry

In discussing white-beam diffraction geometry, it is important to point out another commonly used meaning of *Laue geometry*. In this case, *Laue* simply means transmission geometry, as shown in Figure 14-25, as opposed to reflection, or *Bragg* geometry, as shown in Figure 14-26. This distinction is commonly made in discussing the use of crystals as monochromators, but is also used when the incoming beam is itself monochromatic.

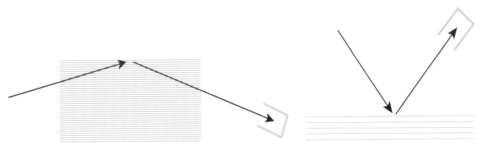

FIGURE 14-25. Laue (transmission) geometry for diffraction.   FIGURE 14-26. Bragg (reflection) geometry for diffraction.

14.8 Protein crystallography

Protein crystallography is sometimes referred to as using Laue diffraction, but it is a different technique, using monochromatic radiation. The necessary coincidence between the reciprocal lattice and Ewald sphere, as in Figure 14-2, is created by very small oscillations of the sample, of no more than a couple of degrees. For a simple inorganic material, this would not create any intersections, and no diffraction pattern would be seen. The difference is that protein crystals have very large unit cells, with very many atoms in each unit cell. The size of the reciprocal lattice spacing, $G = 2\pi/u$, is thus very small compared with the radius of the Ewald sphere, $2\pi/\lambda$, as sketched in Figure 14-27. The diffraction pattern resulting from a 2° oscillation is shown in Figure 14-28. A further complication with protein crystallography is that it is notoriously difficult to obtain large protein crystals. Thus the samples are small, and, because the proteins are made up of low-Z elements, weakly diffracting, resulting in low diffracted intensities. The image of Figure 14-28 was taken using a focusing optic. Some types of optics are typically used with lab sources, and higher-quality data are generally obtained using synchrotron sources. The figure shows the pattern from a single

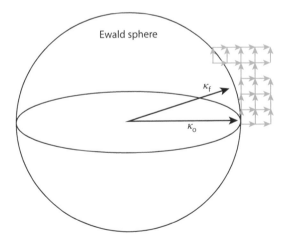

FIGURE 14-27. In protein crystallography, the reciprocal lattice spacing is small, so that only a small rotation is required to move several reciprocal lattice points through the surface of the Ewald sphere.

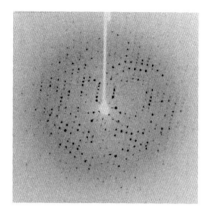

FIGURE 14-28. Part of a lysozyme pattern taken in 8 s with a 2.8 kW rotating anode and a polycapillary focusing optic. The dark spots are high-intensity diffraction spots (the image is a negative). The white bar coming in from upper right is a beam stop blocking the undiffracted beam. The dark ring is due to amorphous scatter from water. S. M. Owens et al., *Proceedings of the 46th Annual Denver X-ray Conference, 1997.*

orientation of the sample. To analyze the crystal, a large dataset of images is produced by rotating the crystal to a new location and repeating the measurement.

14.9 The phase problem

The purpose of obtaining a dataset of images such as the one shown in Figure 14-28 is to obtain a three-dimensional map of the location of every atom in the protein crystal. As noted in equation 13-17, the diffracted field is proportional to the Fourier transform of the electron density. If the field is known, the transform can be inverted to give a precise map of the location of the electrons. However, what is measured is not the field, but the intensity. The phase information is lost. Thus for most measurements an iterative technique is required to deduce the structure. A reasonable structure is proposed based on knowledge of the system, and the diffraction pattern is calculated in a manner similar to that in example 14-1 and compared with the measured pattern. The structure is then modified and the quality of the fit recalculated. This process is relatively straightforward if the sample in question is similar to a known structure. It can be problematic if the structure is unknown.

One technique for reconstructing the phase of the field relies on the fact that the phase is strongly dependent on the wavelength. Multiple measurements are performed at different x-ray wavelengths, and making use of relationships like the Kramers-Kronig calculation of section 11.5, the dependence of the intensity on wavelength is used to deduce the phase.

Alternatively, some of the atoms within the protein can be replaced with heavier-Z elements while, it is hoped, retaining the shape of the original protein. This technique, of *multiple isomorphic replacements*, allows the position of the heavy ion, and hence its contribution to the phase of the structure factor, to be deduced. The *Patterson difference map* compares the Fourier transform of the measured intensities before and after the replacement. Another technique using substitution requires three wavelength measurements: near, below, and above the K absorption edge of the target atom. Because the scattered phase is strongly affected by the absorption edge, it can be used to deduce the location of the atom. This technique is known as *multiple anomalous diffraction*. Under some conditions it can be accomplished with a single wavelength measurement.

Three-beam techniques produce phase information directly, creating a holographic reconstruction of the crystal.

14.10 Coherent diffraction imaging

The Fourier transform relationship, equation 13-17, between the real-space electron density of a sample and the diffracted field holds for any object, not just crystals. In principle, if two images are taken of an object—one, the "absorption" image, with the detector placed close to the sample, and a second, "far field" image, with the detector as far as practical downstream—it should be possible to compute an electron density map of the sample. For the method to work, considerable spatial coherence of the beam is required, and so the technique is limited to synchrotron and free-electron sources. Iterative techniques are employed, first, by assuming the sample is completely homogenous, comparing the result with the Fourier transform of the actual image, and using Bayesian statistical techniques to modify the model of the sample. An example of phase-contrast imaging was shown in Figure 5-17.

14.11 Dynamical diffraction

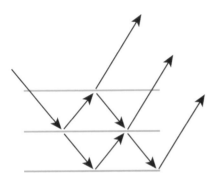

FIGURE 14-29. Multiple reflections must be considered in computing dynamical diffraction.

As was discussed in chapter 13, the energy spread and divergence of the beam incident on the sample determines the resolution with which measurements can be performed. Thus, if a crystal is to be used as a monochromator, it is important to know the inherent rocking curve width of the crystal. As was discussed in section 13.12.1, this characteristic can be limited by defects in the crystal. However, even for a perfect crystal, the inherent rocking curve is not infinitely narrow. In changing from a poor crystal to a nearly perfect crystal, it is necessary to include the effects of interference within the crystal between the beams that have been multiply reflected over many layers, as shown in Figure 14-29. Each layer acts in a manner similar to a Fabry-Perot etalon (an optical device consisting of two facing mirrors like a laser cavity, used to select a narrowband, resonant beam, for example, for spectroscopy). A large number of layers must be considered, facilitated by matrix techniques. The geometry is sketched in Figure 14-30. The problem is to find the complex amplitude E_{ro} of the reflected beam given that the incident beam has amplitude E_o. The amplitude E_{ro} depends on both the incident amplitude and the amplitude of the beam E_{r1} rising through the crystal,

$$E_{ro} = \boldsymbol{r}_{o1} E_o + \boldsymbol{t}_{1o} e^{i\phi_1} E_{r1}, \tag{14-9}$$

where $\boldsymbol{r}_{o1}$ is the reflection coefficient of the first layer, $\boldsymbol{t}_{o1}$ is its transmission coefficient, and ϕ_1 is the phase change on crossing the layer,

$$\phi = \frac{4\pi}{\lambda} nd \sin\theta. \tag{14-10}$$

Each amplitude is then similarly related to two other amplitudes. This calculation produces a sparse matrix which can be solved to give the result shown in Figure 14-31. The width of the rocking curve for the reflected beam in the symmetric case (the planes are parallel to the crystal surface) for unpolarized incident radiation is

$$\Delta\theta_{Darwin} = 2r_e\lambda^2\left(\frac{1+\cos^2\theta}{2}\right)\frac{F}{\pi V_c \sin 2\theta},\tag{14-11}$$

where r_e is the classical electron radius, λ is the x-ray wavelength, θ is the Bragg angle, F is the amplitude of the structure factor for the reflection, and V_c is the volume of the unit cell, which is u^3 for a cubic lattice. The rocking curve width for a perfect crystal is known as the *Darwin width*.

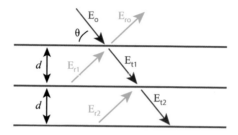

FIGURE 14-30. Matrix elements for the computation of dynamical diffraction intensity.

EXAMPLE 14-2

Compute the Darwin width for silicon (220) reflection at 12.4 keV.

The classical electron radius is $r_e \approx 2.8\times10^{-15}$ m $\approx 2.8\times10^{-5}$ Å.

The wavelength is $\lambda = \dfrac{hc}{U} \approx \dfrac{12.4\,\text{keV}\cdot\text{Å}}{12.4\,\text{keV}} \approx 1$ Å.

The lattice constant for silicon is $u \approx 5.4$ Å, so that $d = \dfrac{u}{\sqrt{2^2+2^2}} \approx 1.9$ Å.

The Bragg angle is $\theta = \sin^{-1}\left(\dfrac{\lambda}{2d}\right) \approx 15°$.

Silicon is diamond cubic with eight atoms per unit cell. The structure factor is

$$F(h,k,l) = F_{FCC}\left(1+e^{i\frac{\pi}{2}(h+k+l)}\right) = 8f_o.$$

The atomic scattering factor would be about 14 (the atomic number Z of silicon) for forward scattering, but f_o is a function of angle, as shown in Figure 13-4. In this case $\dfrac{\sin(\theta)}{\lambda} \approx 0.26$ Å^{-1}, and, from Figure 13-4, the scattering factor is about 8.

$$\Delta\theta_{Darwin} = 2r_e\lambda^2\left(\frac{1+\cos^2\theta}{2}\right)\frac{F}{\pi V_c \sin 2\theta}$$

$$\approx \frac{2(2.8\times10^{-5}\text{ Å})(1\text{ Å})^2}{\pi(5.4\text{ Å})^3}\left(\frac{1+\cos^2(15°)}{2}\right)\frac{8(8)}{\sin(30°)} \approx 14\times10^{-6}\text{ rad.}$$

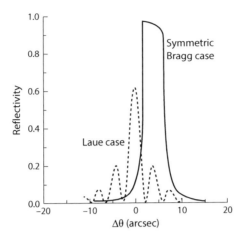

Figure 14-31. Transmission and reflection from a nearly perfect crystal silicon crystal at 12.4 keV. The solid curve is for the case of reflection geometry; the dashed curve is for transmission geometry for a 20 μm thick crystal. From P. Siddons, Crystal Monochromators and Bent Crystals, chap. 39, *Handbook of Optics*, 3rd ed., vol. 5, McGraw-Hill, 2010.

The Darwin width depends on the crystal, the plane selected, and the photon energy, but is typically in the range of a few to a few tens of microradians.

Problems

1. A powder diffractometer uses 10 keV x rays. What is the radius of the Ewald sphere?
2. A face-centered cubic (fcc) single crystal with lattice constant $u = 3$ Å is irradiated with a bremsstrahlung source producing x rays of energy ranging from 0 to 40 keV. The x-ray beam is incident in the $-y$ direction. The detector is 30 mm from the sample (on the side away from the source). Find the wavelength for the diffraction from the (111) plane and the (x, y) coordinate of the spots on the detector.
3. A bcc crystal in a θ-2θ diffractometer run with Cu Kα radiation has a peak for the (111) planes at $2\theta = 40°$.
   a) What is the d spacing of the crystal?
   b) The sample is strained by 1% (the lattice constant increases by 1%). What is the new peak angle if the strain is parallel to the (111) planes?
   c) If the strain is perpendicular to the planes?
4. Compute the Darwin width for diffraction from a (111) silicon crystal at 8 keV.

Further reading

B. D. Cullity, **Elements of X-Ray Diffraction**, Addison-Wesley, 1956.
P. Siddons, Crystal Monochromators, chap. 39 in M. Bass, C. DeCusatis, J. Enoch, V. Lakshminarayanan, G. Li, C. MacDonald, V. N. Mahajan, and E. Van Stryland, eds., **Handbook of Optics**, 3rd ed., vol. 5, McGraw-Hill, 2010
Eric Lifshin, **X-ray Characterization of Materials**, John Wiley & Sons, 1999.

15

DIFFRACTION OPTICS

The previous chapters concentrated on the use of diffraction to measure the properties of materials. Alternatively, natural or artificial crystals and gratings can be used to shape x-ray beams (in both space and spectrum), for a variety of applications.

15.1 Gratings

A simple transmission grating is shown in Figure 15-1. Compared with the lowest aperture in the figure, the "extra" distance traveled on the incident side of the grating in passing through the adjacent aperture is $\Delta \Lambda_{inc} = d \sin (\theta_{inc})$. The extra distance on the exit side of the grating is $\Delta \Lambda_{exit} = d \sin (\chi)$. For constructive interference, the sum of these two distances must be an integral multiple of a wavelength, which gives rise to the grating equation,

$$d(\sin (\theta_{inc}) + \sin (\chi)) = m \lambda. \tag{15-1}$$

Gratings can also be reflective surfaces with artificial periodicity d created by blazing the surface, as shown in Figure 15-2. The grating equation is derived the same way for reflection gratings, as shown in Figure 15-2. The blaze angle φ, the angle at which the surface is tilted relative to the substrate, does not enter into the equation, just the blaze spacing d. However, for mirror reflections off the blazed surface, the angle of incidence must equal the angle of reflection expressed relative to the blazed surface, so that

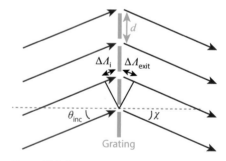

FIGURE 15-1. Transmission grating with spacing d and incident angle θ_{inc}.

$$\theta_{inc} - \varphi = \varphi - \chi. \tag{15-2}$$

If the blaze angle and incident angle are fixed, then to describe the dispersion of wavelength with exit angle, the incident angle θ_{inc} should be replaced in equation 15-1 with $2\varphi - \chi$, yielding, after a little trigonometry,

$$m\lambda = 2d \sin\varphi \cos(\varphi - \chi). \tag{15-3}$$

Grating substrates can be flat, as shown in Figure 15-2, or curved to produce a focused beam, as shown in Figure 15-3. Gratings are useful for performing spectroscopy in the extreme ultraviolet (EUV) and very soft x rays, but as the wavelengths become shorter, the spacing d becomes impractical.

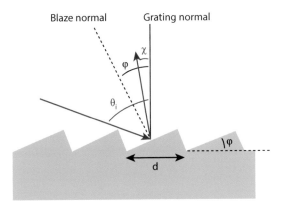

FIGURE 15-2. A blazed grating with incident angle (relative to grating normal) of θ and exit angle χ. The blaze angle is ϕ.

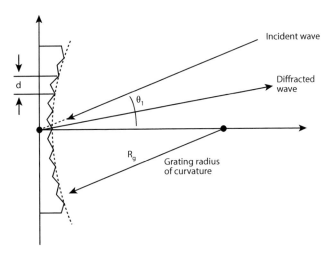

FIGURE 15-3. A curved ruled grating produces a focused beam after reflection. From A. E. Siegman, *Journal of the Optical Society of America A* 2 (1985): 1793–93.

EXAMPLE 15-1

A transmission grating has a slit spacing of 0.5 μm. The incident beam is 0.3 keV. The slit-to-detector distance is 1 m.

a) If the incident beam is at normal incidence, what is the position of the third peak ($m = 3$), and what is its angle with respect to the incident beam?

$$\lambda = \frac{hc}{U} \approx \frac{12.4 \times 10^{-10} \text{ m} \cdot \text{keV}}{0.3 \text{ keV}} \approx 4.1 \times 10^{-9} \text{ m}.$$

$$d\left(\sin\left(\theta_{inc}\right) + \sin\left(\chi\right)\right) = m\lambda \Rightarrow \chi = \sin^{-1}\left(\frac{m\lambda}{d} - \sin\left(\theta_{inc}\right)\right)$$

$$\approx \sin^{-1}\left(\frac{3\left(4.1 \times 10^{-9} \text{ m}\right)}{0.5 \times 10^{-9} \text{ m}} - 0\right) \approx \sin^{-1}\left(0.025\right) \approx 1.4°.$$

$$y = L\tan\left(\chi\right) \approx \left(1 \text{ m}\right)\tan\left(1.4°\right) \approx 25 \text{ mm}.$$

b) If the incident beam hits the grating at 70° from normal incidence (20° from the grating surface), at what angle does the third-order diffraction appear?

$$\chi = \sin^{-1}\left(\frac{m\lambda}{d} - \sin\left(\theta_{inc}\right)\right) \approx \sin^{-1}\left(0.025 - \sin\left(70°\right)\right) \approx -66°.$$

A negative angle in Figure 15-1 corresponds to the same side of the normal as the incident beam, so the angle of the third-order diffraction is 4° from the incident beam. This is an improvement compared with only 1.4°, as it was in the case of normal incidence.

c) If the incident energy is changed to 0.4 keV, how much do the angles change?

$$\lambda = \frac{hc}{U} = 3.1 \times 10^{-9} \text{m}; \; \chi_o = \tan^{-1}\left(m\frac{\lambda}{d}\right) \approx \tan^{-1}\left(3\frac{\left(3.1 \times 10^{-9} \text{ m}\right)}{0.5 \times 10^{-6} \text{ m}}\right)$$

$$\approx \tan^{-1}\left(0.019\right) \approx 1.1°.$$

Changing the energy by 0.1 keV changed the angle by only 0.3°.

$\chi_{70} \approx \sin^{-1}\left(0.019 - \sin\left(70°\right)\right) \approx -67°$. At 70° incidence, changing the energy by the same 0.1 keV changed the outgoing ray by about 1°. This difference makes the change in photon energy easier to see.

Another use of flat transmission gratings is in x-ray phase imaging (which was also described in section 5.10). As noted in section 9.3, the real part of the index of refraction of the object, which gives rise to the phase delay, is very much larger than the imaginary part of the index, which gives rise to absorption. Thus the phase image has inherently much higher contrast than the normal radiographic absorption image. Consider a plane wave with field $E_o e^{ikz}$ normally incident ($\theta_{inc} = 0$) on a grating like that shown in Figure 15-1. If the grating transmission is sinusoidal, it can be expressed as

$$\boldsymbol{T}(Y) = e^{i\frac{2\pi}{d}Y}, \tag{15-4}$$

where d is the grating spacing. Using analysis similar to that of the double-slit experiment of section 5.8, or of the diffraction effects of section 13.3, the intensity can be

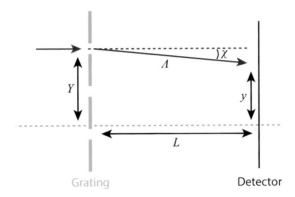

Grating   Detector

FIGURE 15-4. Detector downstream from a grating. The sketch is not to scale; the ray is near the axis, so that $y \ll L$.

computed at a point y on the detector screen, shown in Figure 15-4. The wave travels a distance

$$\Lambda = \sqrt{L^2 + (Y - y)^2}. \tag{15-5}$$

In the paraxial approximation the Y component is small compared with the z component (the ray stays near the axis), so that

$$\Lambda \approx L + \frac{(Y - y)^2}{2L} \approx L - \frac{y}{L} Y \approx L - Y \sin \chi, \tag{15-6}$$

where χ is now the angle between the incident and deflected beams.

Integrating over all possible paths through the grating Y to find the field at the detector point y, the field is

$$E(y) = E_o \int \boldsymbol{T} \frac{e^{i(\kappa_o \Lambda)}}{\Lambda} dY \approx E_o e \frac{e^{i(\kappa_o L)}}{L} \int \boldsymbol{T} e^{-i(\kappa_o Y \sin \chi)} dY, \tag{15-7}$$

where terms were kept to first order in Y in phase but only zeroth order in amplitude $\left(\frac{1}{\Lambda} \sim \frac{1}{L} \right)$. The integral is the Fourier transform of the grating in the variable, $\kappa_y = \kappa_o \sin \chi$, and, just as for the chain of atoms in equation 13-18, is given by

$$\int \boldsymbol{T} e^{-i(\kappa_y Y)} dY = \delta \left(\kappa_y - m \frac{2\pi}{d} \right). \tag{15-8}$$

The final step is to reexpress the z component of the wavevector at the detector. Considering only in-plane propagation ($\kappa_x = 0$), the wavevector at the detector is given by

$$\kappa = \kappa_o = \sqrt{\kappa_y^2 + \kappa_z^2} \Rightarrow \kappa_z = \sqrt{\kappa_o^2 - \kappa_y^2}. \tag{15-9}$$

In the paraxial approximation the y component is small compared with the z component, so that

$$\kappa_z \approx \kappa_o - \frac{\kappa_y^2}{2\kappa_o}. \tag{15-10}$$

The first-order ($m = 1$) field at the detector is thus

$$E = E_o' e^{i\left(\kappa_o L - \frac{\kappa_y^2}{2\kappa_o}L\right)} \delta\left(\kappa_y - \frac{2\pi}{d}\right) = E_o'' e^{-i\left(\frac{\kappa_y^2}{2\kappa_o}L\right)} \delta\left(\kappa_y - \frac{2\pi}{d}\right), \tag{15-11}$$

where E_o' includes both the factor $1/L$ and a factor proportional to the size of the grating, and E_o'' includes the overall phase factor. Evaluating κ_y given the delta function, the field is

$$E = E_o'' e^{i\left(\frac{(2\pi/d)^2}{2\kappa_o}\right)L}. \tag{15-12}$$

The field oscillates as the distance L to the detector is increased. At the special *Talbot distance* L_T at which the argument is 2π, the field is proportional to the incident field,

$$\frac{(2\pi/d)^2}{2\kappa_o} L_T = 2\pi \Rightarrow L_T = \frac{\kappa_o d^2}{\pi} = \frac{2d^2}{\lambda}. \tag{15-13}$$

At this distance the field at the grating is re-created on the detector. This same phenomenon is observed whether the grating is an amplitude grating or a phase grating, for which the transmission is given by

$$\mathbf{T}(Y) = e^{i\varphi_o \sin\left(\frac{2\pi}{d}Y\right)}. \tag{15-14}$$

Phase imaging can then be performed by placing a nearly transparent object in front of a phase grating, as shown in Figure 15-5. The object causes the resultant fringes to shift, and the amount of the shift can be analyzed to determine the phase imparted by the object. A complication in practical application is that the fringe spacing is typically much smaller than the pixel size of a large-area x-ray detector, so that a second analyzer grating is required to analyze the fringes. This process can require precise stepping and multiple images.

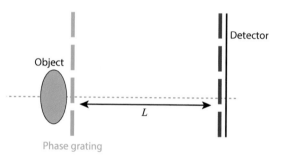

FIGURE 15-5. Object placed before a phase grating, with an analyzer grating placed before the detector.

EXAMPLE 15-2

---

a) What phase grating spacing d is necessary to have a Talbot distance of less than 1 m for 1 Å radiation?

The Talbot distance is $L_T = \frac{2d^2}{\lambda} \Rightarrow d = \sqrt{\frac{\lambda L_T}{2}} < \sqrt{\frac{(1 \times 10^{-10}\,\text{m})(1\,\text{m})}{2}} \approx 7\,\mu\text{m}$. Shorter wavelengths or more compact systems require even finer spacings, which can be difficult to produce with the precision-controlled thickness required for the correct phase modulation over a large area.

b) What is the fringe spacing at the Talbot distance with that grating?

Fringes occur when $\kappa_y = m \dfrac{2\pi}{d}$. The fringe spacing is the difference between consecutive maxima, $\Delta y = y_{m+1} - y_m$, and $\kappa_y = \kappa_o \sin \chi = \dfrac{2\pi}{\lambda} \dfrac{y_m}{L_T}$, so $\Delta y = \dfrac{2\pi}{d} \dfrac{L_T}{\kappa_o} = \dfrac{\lambda L_T}{d}$

$= \dfrac{2d^2}{d} = 2d \approx 14 \, \mu\text{m}.$

15.2 Zone plates

Zone plates are a special case of gratings with a circular geometry. A circular grating with uniform spacing is shown in Figure 15-6. To analyze the field after a zone plate, consider the geometry of Figure 15-7. A ray leaves the axis from a source at a distance z from the aperture, hits the aperture at a distance R from the axis, then travels back to the axis to a detector a distance L away. As for the Talbot grating example shown in Figure 15-4, the path distance Λ for the paraxial approximation $R \ll z$, is

$$\Lambda = \Lambda_s + \Lambda_D = \sqrt{z^2 + R^2} + \sqrt{L^2 + R^2}$$
$$\approx z + L + \frac{R^2}{2}\left(\frac{1}{z} + \frac{1}{L}\right) \approx z + L + \frac{R^2}{2L_f}, \tag{15-15}$$

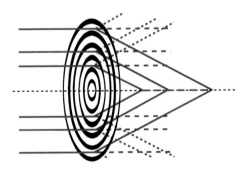

FIGURE 15-6. An (unusual) circular grating of uniform spacing. From A. Michette, Zone Plates, chap. 40, *Handbook of Optics*, 3rd ed., vol. 5, McGraw-Hill, 2010.

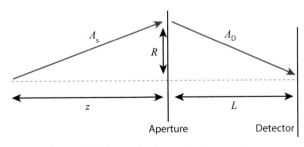

FIGURE 15-7. Geometry for a circular aperture.

where L_f is defined as

$$\frac{1}{L_f} \equiv \frac{1}{z} + \frac{1}{L}. \tag{15-16}$$

If the aperture is a circular annulus open from radius R_a to R_b, then the field at the detector arises from integrating over the aperture,

$$E = \int\limits_{aperture} \frac{E_A e^{i\kappa\Lambda}}{\Lambda}, \tag{15-17}$$

where E_A is proportional to the field incident on the aperture. Substituting to zeroth order in intensity ($\Lambda \sim z+L$), and first order in phase, gives

$$E = \int\limits_0^{2\pi} \left(\int\limits_{R_a}^{R_b} \frac{E_A e^{i\kappa\left(z+L+\frac{R^2}{2L_f}\right)}}{z+L} R\,dR \right) d\varphi$$

$$= \left[2\pi \frac{E_A e^{i\kappa(z+L)}}{z+L} \frac{L_f}{\kappa} \right] \int\limits_{R_a}^{R_b} e^{i\kappa\frac{R^2}{2L_f}} d\left(\frac{\kappa R^2}{2L_f} \right) = E_A' \left(e^{i\kappa\frac{R_b^2}{2L_f}} - e^{i\kappa\frac{R_a^2}{2L_f}} \right), \tag{15-18}$$

where the constants have been absorbed into E_A'. Substituting for κ gives

$$E = E_A' \left(e^{i\frac{2\pi}{\lambda}\frac{R_b^2}{2L_f}} - e^{i\frac{\pi}{\lambda}\frac{R_a^2}{L_f}} \right) = E_A' (e^{i\pi m_b} - e^{i\pi m_a})$$

$$= E_A'' (1 - e^{i\pi(m_b - m_a)}), \tag{15-19}$$

where $E_A'' = -E_A' e^{i\pi m_a}$, and the two variables m_a and m_b are defined as

$$m_{a,b} \equiv \frac{R_{a,b}^2}{\lambda L_f}. \tag{15-20}$$

If the radius R_m is chosen so that m is an integer, so that

$$R_m = \sqrt{m\lambda L_f}, \tag{15-21}$$

then it defines the edge of a Fresnel zone. The mth Fresnel zone is the ring-shaped area in the aperture plane between the radius $R_{m-1} = \sqrt{(m-1)\lambda L_f}$ and the radius $R_m = \sqrt{m\lambda L_f}$. The zone radii are defined so that the beam from one edge of the annulus is exactly out of phase with the beam at the other edge. Thus the beam from any point in one zone is exactly out of phase with the beam from the same point in an adjacent zone. The area of each zone is

$$A_m = \pi R_m^2 - \pi R_{m-1}^2 = \pi(m\lambda L_f - (m-1)\lambda L_f) = \pi\lambda L_f, \tag{15-22}$$

independent of zone number. Hence the field passed by each zone has equal amplitude, and adjacent zones cancel.

The field after passing through an aperture which is open from R_a to R_b, for which m_a and m_b are integers, measured on axis at the distance L as shown in Figure 15-7 is

$$E = E_A' \left(1 - e^{i\pi(m_b - m_a)}\right) = \begin{cases} 0 & m_b - m_a \text{ even} \\ 2E_A' & m_b - m_a \text{ odd} \end{cases}. \tag{15-23}$$

The highest field that can be obtained at this point with a single aperture is twice the incident field.

EXAMPLE 15-3

What simple pinhole radius produces the highest intensity for a source 50 cm away and a detector 33 cm away when irradiated by coherent 1 Å x rays?

$$\frac{1}{L_f} \equiv \frac{1}{z} + \frac{1}{L} \Rightarrow L_f = \frac{1}{\dfrac{1}{L} + \dfrac{1}{z}} \approx \frac{1}{\dfrac{1}{33\,\text{cm}} + \dfrac{1}{50\,\text{cm}}} \approx 20\,\text{cm}.$$

The first Fresnel zone has radius $R_1 = \sqrt{\lambda L_f} \approx \sqrt{(1 \times 10^{-10}\,\text{m})(0.2\,\text{m})} \approx 4.5\,\mu\text{m}$. Using a pinhole any larger than that would cause destructive interference and would not increase the intensity at the image.

To avoid the destructive interference, a Fresnel zone plate is created by blocking every other Fresnel zone, as shown in Figure 15-8. If the first, central, zone is blocked, then the first open annulus covers the second zone, the third zone is blocked, and the second open annulus exposes the fourth zone, and so forth. The amplitude at the focal point of the field from a zone plate with N_z open zones is N_z times the amplitude from a single zone,

$$E_N = N_Z E_1. \tag{15-24}$$

The intensity at the focal point is thus

$$I_N = N_Z^2 I_1. \tag{15-25}$$

It is clearly desirable to have as many zones as possible in the zone plate. However, as can be seen in Figure 15-8 and equation 15-21, the width of each zone becomes smaller toward the outer edge of the optic. The width of the mth zone is

$$\begin{aligned}
w_m = R_m - R_{m-1} &= \sqrt{m\lambda L_f} - \sqrt{(m-1)\lambda L_f} \\
&= \sqrt{m\lambda L_f}\left(1 - \sqrt{1 - \frac{1}{m}}\right) \approx \sqrt{m\lambda L_f}\left(\frac{1}{2m}\right) = \frac{R_1}{2\sqrt{m}}.
\end{aligned} \tag{15-26}$$

Often, one is given the number of open apertures rather than the number of Fresnel zones. In that case, the width of the Nth open zone (which is the $m = 2N$th Fresnel zone, since every other one is blocked) is

$$\begin{aligned}
w_N = R_{2N_{open}} - R_{2N_{open}-1} &= \sqrt{2N_{open}\,\lambda L_f} - \sqrt{(2N_{open}-1)\lambda L_f} \\
&\approx \frac{R_1}{\sqrt{8N_{open}}}.
\end{aligned} \tag{15-27}$$

EXAMPLE 15-4

A zone plate designed for 1 Å x rays has an L_f of 20 cm. The manufacturing process allows for a minimum feature size (blocked zone width) of 0.5 μm. What is the outer diameter of the zone plate? What is the intensity relative to the intensity through a single zone?

As in example 15-3, $R_1 = \sqrt{\lambda L_f} \approx 4.5\,\mu m$. The outer zone width then gives us the number of open zones, $w_N = \dfrac{R_1}{\sqrt{8N_{open}}} \Rightarrow N_{open} = \dfrac{R_1^2}{8w_N^2} = 10$. From this result we can calculate the outer radius, using $m = 2N_{open} = 20$: $R_{20} = \sqrt{20\lambda L_f} \approx \sqrt{20\,(1\times10^{-10}\,\text{m})(0.2\,\text{m})} \approx 20\,\mu m$, and the diameter is 40 μm. (To check, the 19th Fresnel zone has radius $\sqrt{19\lambda L_f} = 19.5\,\mu m$, which confirms a feature size of 20 μm − 19.5 μm = 0.5 μm.)

Because there are 10 open zones, the intensity is 100 times the single zone intensity.

The small size of zone plates makes them a reasonable fit for a synchrotron source but tends to result in very poor intensity from a point source (and the transverse coherence length of the point source would have to be longer than the optic diameter to achieve constructive interference). While the constructive addition from multiple zones results in a concentration of intensity, half the zone plate area is blocked, and of the remaining beam, half is undiffracted, so only one-quarter of the incident energy is available to be diffracted into the focal point. Further, some of the remaining energy is diffracted to higher-order beams.

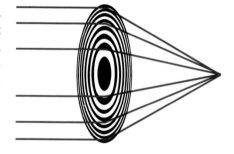

FIGURE 15-8. Diffraction from a zone plate. Every other Fresnel zone, starting with the central one, is blocked. From A. Michette, Zone Plates, chap. 40, *Handbook of Optics*, 3rd ed., vol. 5, McGraw-Hill, 2010.

The concentration of the field onto the axis means that a parallel input beam (with $z = \infty$ in equation 15-16), as shown in Figure 15-8, focuses at a distance L_f. Equation 15-16 acts just like a thin-lens equation, describing the relationship between the source and image locations given the focal length L_f of the zone plate optic.

EXAMPLE 15-5

A zone plate designed for 1 Å x rays has a focal length of 20 cm. The source is placed 50 cm away. What is the radius of the first zone? Where is the image?

As for the pinhole in example 15-3, $R_1 = \sqrt{\lambda L_f} \approx 4.5\,\mu m$.
The simple thin-lens equation works for computing the first-order focus,

$$\frac{1}{L_f} \equiv \frac{1}{z} + \frac{1}{L} \Rightarrow L = \frac{1}{\dfrac{1}{L_f} - \dfrac{1}{z}} \approx \frac{1}{\dfrac{1}{20\,\text{cm}} - \dfrac{1}{50\,\text{cm}}} \approx 33\,\text{cm}.$$

While zone plates can behave much like thin lenses, focusing the radiation emitted from a source, they are fundamentally diffractive optics and so have some notable differences. One is strong achromatic aberration; that is, different wavelengths focus to different spots.

EXAMPLE 15-6

A zone plate designed for 1 Å x rays has a focal length of 20 cm at that wavelength. It is used with 1.05 Å x rays. The source is placed 50 cm away. Where is the image?

Since the zone plate was made for 1 Å x rays, it has a central blocked radius of $R_1 \approx 4.5\,\mu m$. This is a physical parameter of the lens, and is not changed by changing the source wavelength. The zone plate has a different focal length at the new wavelength,

$$L_f' = \frac{R_1^2}{\lambda'} \approx \frac{(4.5 \times 10^{-6}\ \text{m})^2}{1.05 \times 10^{-10}\ \text{m}} \approx 19\,\text{cm}.$$

Using this new focal length, we have

$$L = \frac{1}{\dfrac{1}{19\,\text{cm}} - \dfrac{1}{50\,\text{cm}}} = 30.5\,\text{cm}.$$

The spot moved about 2.5 cm (compared with the result in the previous example) when the wavelength was increased 5%.

Because the optics are diffractive, there are multiple diffraction orders, which result in multiple focal spots for a single wavelength of radiation. The radius of a Fresnel zone, from equation 15-21, depends on both the focal distance L_f and the zone number m. If a different focal length is chosen, there will be a different Fresnel zone radius, or a different number of Fresnel zones in the physical radius of the annulus. A focal length can be chosen that is an integral fraction of the original,

$$L_{f,\,N_{order}} = \frac{L_f}{N_{order}}, \tag{15-28}$$

where N_{order} is the order of the focus. The number of Fresnel zones within the original radius R_1, from equation 15-20, is then

$$\frac{R_1^2}{\lambda L_{f,\,N_{order}}} = \frac{\left(\sqrt{\lambda L_f}\right)^2}{\lambda\left\{\dfrac{L_f}{N_{order}}\right\}} = N_{order}. \tag{15-29}$$

Each of the annuli, which originally contained one Fresnel zone, now contains N_{order} zones.

The development of the third-order focus is shown in table 15-1. The table shows the number of zones, m, in each open annulus of the original zone plate if the plate is used for a higher-order focus with focal length L_f/N_{order}. For even orders, like $N_{order} = 2$, each open annulus has an even number of zones, which cancel exactly, so there is no intensity in the second-order focus. The physical radius of the start of the first blocked annulus is $R_a = R_1 = \sqrt{m\lambda L_f} = \sqrt{(1)\lambda L_f}$ when used in first order, $N_{order} = 1$. Used in third order, R_a does not change, but the focal length and hence the radius of the first Fresnel zone changes. Now, the focal length is $L_f/3$, and $R_a = R_3 = \sqrt{3(\lambda L_f)/3}$. The first blocked annulus has three zones. The first open annulus also has three zones, two of which cancel, so that the amplitude is decreased by a factor of 3. In general, the intensity is thus

$$I_{N_{order}} = \begin{cases} 0 & \text{even } N_{order} \\[2mm] \dfrac{I_{first}}{N_{order}^2} & \text{odd } N_{order} \end{cases} \tag{15-30}$$

EXAMPLE 15-7

A zone plate designed for 1 Å x rays has a focal length of 20 cm. The source is placed 50 cm away. Where is the third-order image?

The new focal length is $L_{f_3} \approx \dfrac{20\,\text{cm}}{3} \approx 6.6\,\text{cm}$. The image is now at

$$L_3 = \cfrac{1}{\cfrac{1}{6.6\,\text{cm}} - \cfrac{1}{50\,\text{cm}}} = 7.6$$

The third-order image is much closer to the zone plate than the first-order image.

TABLE 15-1. A zone plate with every other Fresnel zone blocked

	First order			Second order			Third order		
	m	Amplitude, E/E_A", per zone	E/E_A", for N_open zones	m	E/E_A", per zone	E/E_A", for N_open zones	m	E/E_A", per zone	E/E_A", for N_open zones
								9		
					6					
Blocked	3	Blocked			Blocked			8	Blocked	
					5			7		
								6	1	
					4	−1				
Open	2	1	N_open				0	5	−1	N_open/3
					3	1		4	1	
								3		
					2					
Blocked	1	Blocked			Blocked			2	Blocked	
					1			1		

Note: The first-order focal length is L_f, and the annuli radii are chosen so that each annulus is one Fresnel zone for this distance. The physical radii of the annuli cannot change when used in a different order m. When used in second order, each open annulus contains two Fresnel zones, which cancel. Used in third order, each open annulus has three open zones, two of which cancel. The focal length for the third-order focus is $L_f/3$.

Generally, when a zone plate is used as a lens to focus the beam, apertures are required to block the undesired higher-order beams, as shown in Figure 15-9.

Zone plates can be used for imaging, especially in microscopy, or for creating a small spot on a sample. The best possible diffraction-limited spot size from the zone plate is a result of the finite diameter of the zone plate, in the same manner as for a visible light lens, or for the refractive lens of section 12.1. The calculation is basically the same as when considering the effect of the finite crystal size, discussed in section 13.12.3, except using cylindrical coordinates, which yields a first-order Bessel function instead of a sinc function. For a lens or any optic of total radius R_{NZ}, at a focal distance L_f, the diffraction-limited spot size is

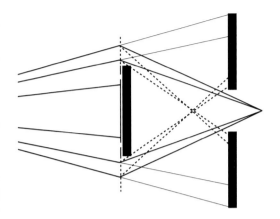

FIGURE 15-9. Zone plate focus, showing the on-axis beam stop to block the undiffracted beam, and the aperture to block the higher-order diffracted beams. The solid line passing through the final aperture is the desired first-order image. The dotted line is the third-order image, which forms closer to the zone plate. The diverging rays from the third-order image would distort the desired image unless blocked by the second aperture. From A. Michette, Zone Plates, chap. 40, *Handbook of Optics*, 3rd ed., vol. 5, McGraw-Hill, 2010.

$$R_{spot} \approx 1.22 \frac{L_f \lambda}{2R_{N_Z}}. \tag{15-31}$$

The factor 1.22 comes from the location of the first zero of the Bessel function, $J_1(1.22\pi) = 0$. The spot size is often expressed in terms of the width of the outermost zone, using equation 15-26, as

$$R_{spot} \approx 1.22 \frac{L_f \lambda}{2R_{N_Z}} = 1.22 \frac{L_f \lambda}{2\sqrt{N_Z L_f \lambda}} = 1.22 \frac{R_1}{2\sqrt{N_Z}} = 1.22\, w_{N_Z}. \tag{15-32}$$

Because the focal length decreases with focal order, as in equation 15-28, the spot size decreases as well,

$$R_{spot\, N_{order}} \approx 1.22 \frac{\left(\dfrac{L_f}{N_{order}}\right)\lambda}{2R_{N_Z}} = \frac{R_{spot1}}{N_{order}} = 1.22 \frac{w_{N_Z}}{N_{order}}. \tag{15-33}$$

EXAMPLE 15-8

A zone plate designed for 1 Å x rays has a focal length of 20 cm. The manufacturing process allows for a minimum feature size (blocked zone width) of 0.5 μm. What is the spot size for the first-order focus?

The spot size is

$$R_{spot\,N_{order}} \approx 1.22\frac{w_{N_Z}}{N_{order}} \approx 1.22\frac{0.5\,\mu m}{1} \approx 0.6\,\mu m.$$

Zone plates can currently be fabricated by electron-beam lithography with feature sizes, and hence spatial resolution, on the order of about 10 nm, considerably smaller than in the example (which was chosen to have fewer zones to make it easier to visualize).

The spot size R_{spot} is in the dimension perpendicular to the beam axis. There is also a depth of field, a spot length along the axis, which also can be computed analogously to that for a thin lens. For a lens the depth of field for a parallel input beam is

$$\Delta z \approx 2\left(\frac{L_f}{2R_{N_Z}}\right)^2 \lambda. \tag{15-34}$$

Substituting for the zone plate radius gives the depth of field,

$$\Delta z \approx 2\left(\frac{L_f}{2\sqrt{N_Z\lambda L_f}}\right)^2 \lambda = \frac{L_f}{N_Z}. \tag{15-35}$$

EXAMPLE 15-9

A zone plate designed for 1 Å x rays has a focal length of 20 cm. The manufacturing process allows for a minimum feature size (blocked zone width) of 0.5 μm. What is the depth of field for the first-order focus?

From example 15-4, the number of zones is $N_Z = 2N_{open} = 20$. The depth of field is thus $\Delta z \approx \dfrac{20\,cm}{20} \approx 1\,cm$. The length of the spot along the axis is orders of magnitude larger than the radial width, which is 0.6 μm.

Calculating the depth of field allows for an easy calculation of the wavelength bandwidth of the zone plate. As seen in example 15-6, the focal length of the zone plate changes with wavelength. If the chromatic aberration is to be small, that means the shift in focal position due to the change in wavelength should be less than the depth of field,

$$\left|\frac{\partial L_f}{\partial \lambda}\right|\Delta\lambda < \Delta z_{depth\,of\,field} \Rightarrow \left|\frac{\partial\left(\dfrac{R_1^2}{\lambda}\right)}{\partial \lambda}\right|\Delta\lambda < \frac{L_f}{N_Z} \Rightarrow \left|\frac{-R_1^2}{\lambda^2}\right|\Delta\lambda$$

$$= \frac{L_f}{\lambda}\Delta\lambda < \frac{L_f}{N_Z} \Rightarrow \frac{\Delta\lambda}{\lambda} < \frac{1}{N_Z}. \tag{15-36}$$

This wavelength restriction is algebraically equivalent to an alternative derivation. Zones were defined so that the phase due to the extra path length in going from the edge of radius R_m, compared with traveling on axis, is $m(\lambda/2)$. For the path touching the outer edge of the zone plate, the phase difference is $N_Z(\lambda/2)$. The wavelength bandwidth is defined to be the greatest change in wavelength for which this phase is changed by only π,

$$\frac{2\pi}{\lambda}\left(N_Z\frac{\lambda+\Delta\lambda}{2}-N_Z\frac{\lambda}{2}\right)<\pi\Rightarrow\frac{\Delta\lambda}{\lambda}<\frac{1}{N_Z}. \tag{15-37}$$

EXAMPLE 15-10

A zone plate designed for 1 Å x rays has a focal length of 20 cm. The manufacturing process allows for a minimum feature size (blocked zone width) of 0.5 μm. What is the wavelength bandwidth for the first-order focus?

From example 15-4, the number of zones is $N_Z = 2N_{open} = 20$. The bandwidth is thus $\frac{\Delta\lambda}{\lambda}<\frac{1}{N_Z}\Rightarrow\Delta\lambda=\frac{\lambda}{N_Z}\approx\frac{1\,\text{Å}}{20}\approx 0.05\,\text{Å}$. This much change in wavelength should change the focal length by 1 cm, the depth of field in example 15-9. This result was verified in example 15-6.

This section on zone plates has concentrated on *amplitude* zone plates, that is, zone plates which block the unwanted zones. Alternatively, instead of opaque regions, *phase* zone plates have transparent regions which add a phase change to allow the even zones to be in phase with the odd zones. In this case, less intensity is lost to absorption, and the overall focal spot efficiency is increased to a maximum of about 25%, compared with 10% efficiency for amplitude zone plates. Zone plates can be designed with a variety of intensity and phase-modulation patterns to maximize the desired performance. They are extensively used on synchrotron beamlines for microscopy and EUV projection lithography. Two common geometries are used for microscopy. In the first, used in scanning microscopy, the sample is translated through the focal spot, and the intensity is recorded at each position, to make a raster-scan image. A true-imaging system is shown in Figure 15-10.

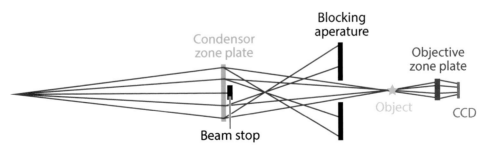

FIGURE 15-10. Sketch of a zone plate microscope. The vertical scale is very much exaggerated.

15.3 Crystal optics and multilayers

15.3.1 MONOCHROMATORS

Crystal monochromators are one of the most common types of x-ray optics (after pinholes and apertures). They are typically used before the sample to produce the monochromatic beam assumed for the diffraction measurements discussed in chapters 13 and 14, and are described by the same formalism as the diffraction measurements, especially Bragg's law, equation 13-45, $\lambda = 2d \sin \theta_B$. When used at relatively low photon energies, the crystals are usually employed in Bragg geometry, as in Figure 14-26, but can also be used in Laue geometry, as in Figure 14-25.

Monochromator crystals can also be used to polarize a conventional x-ray beam. If the crystal is chosen so that the Bragg angle is near 45°, as shown in Figure 15-11, then the diffracted beam is perpendicular to the incident beam,

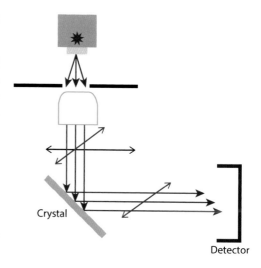

FIGURE 15-11. Setup for polarization by diffraction at 90°. After diffraction by the crystal, the beam is polarized in the out-of-plane direction. In this setup, a polycapillary optic is shown to collimate the incident beam. Collimating optics generally result in higher intensity off monochromators, regardless of the Bragg angle employed.

and so the direction of propagation is parallel to one of the two possible initial polarizations. Since polarization along the direction of propagation is not possible, the diffracted beam is linearly polarized, in the direction perpendicular to both the initial and final directions. Using a polarized beam to excite x-ray fluorescence can result in reduced background scatter if the detector is placed in the direction of polarization, for example, out of the page in Figure 15-11. The scatter cannot propagate in the direction of polarization. (This is the origin of the $\sin\chi$ term in equation 11-39.) Polarized beams are also employed to investigate the magnetic properties of materials. A particular difficulty in polarizing high-energy x-ray beams using this technique of 90° Bragg diffraction is that the small wavelengths require small d spacings and hence high-order diffraction (high (hkl) values), which generally results in low intensity.

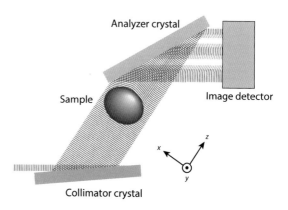

FIGURE 15-12. Diffraction-enhanced imaging using a crystal monochromator (labeled as a collimator crystal) and an analyzer crystal. From A. Momose, *Japanese Journal of Applied Physics* 44: 6355–67. Copyright 2005 The Japan Society of Applied Physics.

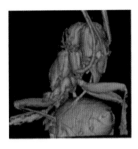

FIGURE 15-13. 3D phase map of an ant produced using the diffraction-enhanced imaging method. From A. Momose, *Japanese Journal of Applied Physics* 44: 6355–67. Copyright 2005 The Japan Society of Applied Physics.

In addition to being used to select an incident energy or polarization, monochromator crystals can also be employed after the sample to analyze the photon energies present in an x-ray beam, for example, in spectroscopy or fluorescence. In the x-ray fluorescence examples of section 4.7, it was generally assumed that the analysis was performed with an energy-sensitive detector. However, such detectors have somewhat limited energy resolution, often a few hundred electronvolts. For better energy resolution, the fluorescence emission can be analyzed by first collimating the emission by passing it through a small aperture or collimating optic, then observing the intensity as a monochromator crystal is rotated in the beam, in a manner similar to that of the θ-2θ diffractometer of Figure 13-11. Because Bragg's law refers to wavelength, this technique is generally known as *wavelength-dispersive x-ray fluorescence.*

Crystals are employed both as monochromators and as analyzers in diffraction-enhanced imaging, as shown in Figure 15-12. With no sample in place, the rays diffracted by the first monochromator crystal are diffracted by the second, analyzer, crystal. Refraction at the edges of interfaces within the sample changes the angle of the rays, so they are no longer diffracted by the analyzer crystal, resulting in dark bands and enhanced edge contrast. Imaging can also be performed in "dark-field" mode, in which the analyzer crystal is rotated slightly away from the Bragg angle, so that only rays refracted by the sample will be diffracted onto the detector. Quantitative measurements of the phase change imparted to the beam by the sample can be obtained by combining measurements at multiple analyzer crystal positions, as shown in Figure 15-13.

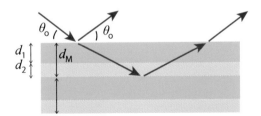

FIGURE 15-14. Bragg diffraction from multilayers with spacing d_M, where d_1 is the thickness of the first layer, for example, the molybdenum layer, and d_2 is the thickness of the other, for example, the silicon in a Mo/Si multilayer. The refraction of the ray at the surface is exaggerated (and there would also be refraction at the interface between the layers).

15.3.2 MULTILAYER OPTICS

Bragg's law cannot be satisfied for wavelengths greater than twice the plane spacing, so crystal optics are limited to low-wavelength, high-energy x rays. Multilayers, as shown in Figure 15-14, are artificial crystals of alternating layers of materials. The spacing d_M, which is the thickness of a layer pair,

$$d_M = d_1 + d_2, \tag{15-38}$$

replaces d in Bragg's law and can be much larger than crystalline plane spacings. Multilayers are therefore effective for lower-energy x rays. Because the index of refraction decrement δ for soft x rays is larger than for hard x rays, it is necessary to include the refraction correction which is generally ignored in employing Bragg's law. The Bragg angle θ_B inside the material is related to the angle in vacuum by Snell's law, equation 11-56, so that (assuming β is small),

$$\cos \theta_o = (1-\delta) \cos \theta_B, \tag{15-39}$$

where δ is the average index of refraction,

$$\delta = \frac{\delta_1 d_1 + \delta_2 d_2}{d_M}, \tag{15-40}$$

and δ_1 and δ_2 are the indexes of refraction for the individual materials in the multilayer. For hard x rays the difference between θ_B and θ_o is generally below the resolution of the measurement. In addition to correcting for the angle, matching the phase for the two reflected rays in Figure 15-14 should take into account the refractive index in the material. Using the artificial periodicity d_M, we then have for Bragg's law,

$$\lambda = 2d_M (1-\delta) \sin \theta_B. \tag{15-41}$$

Further, taking into account refraction, the interference condition becomes

$$\lambda = 2d_M (1-\delta) \sqrt{1 - \cos^2 \theta_B} = 2d_M (1-\delta) \sqrt{1 - \left(\frac{\cos \theta_o}{1-\delta} \right)^2}$$

$$= 2d_M \sqrt{(1-\delta)^2 - \cos^2 \theta_o} \approx 2d_M \sqrt{\sin^2 \theta_o - 2\delta}. \tag{15-42}$$

Solving for the angle in vacuum gives

$$\sin \theta_o \approx \frac{\lambda}{2d_M} \sqrt{1 + 2\delta \left(\left(\frac{2d_M}{\lambda} \right)^2 \right)}. \tag{15-43}$$

EXAMPLE 15-11

A multilayer diffractive optic is composed of many pairs of layers of Mo and Si. The Mo layers are 20 Å thick, and the silicon layers are 40 Å thick, so that each layer is 60 Å. The plasma energies of Mo and Si are approximately 60.3 eV and 30 eV, respectively. The multilayer is used with 0.2 keV radiation. What are the original Bragg angle ignoring the index, the real Bragg angle, and incident angle for this multilayer?

First, the wavelength is

$$\lambda = \frac{hc}{U} \approx \frac{12.4 \times 10^{-10} \text{ m} \cdot \text{keV}}{0.2 \text{ keV}} \approx 6.2 \times 10^{-9} \text{ m}.$$

Ignoring the index of refraction, we would have had a Bragg angle of

$$\theta'_B = \sin^{-1}\left(\frac{\lambda}{2d_M}\right) \approx \sin^{-1}\left(\frac{62\,\text{Å}}{2(60\,\text{Å})}\right) \approx 31.1°$$

The individual index decrements are

$$\delta_{Mo} = \frac{1}{2}\frac{U_{p,\,Mo}^2}{U^2} \approx \frac{1}{2}\left(\frac{60.3\,\text{eV}}{0.2\times10^3\text{eV}}\right)^2 \approx 0.045 \quad\text{and}$$

$$\delta_{Si} = \frac{1}{2}\frac{U_{p,\,Si}^2}{U^2} \approx \frac{1}{2}\left(\frac{30\,\text{eV}}{0.2\times10^3\text{eV}}\right)^2 \approx 0.011,$$

so that the average index is

$$\delta = \frac{\delta_{Mo}d_{Mo} + \delta_{Si}d_{Si}}{d_M} \approx \frac{(0.045)(20)+(0.011)(40)}{60} \approx 0.023.$$

The real Bragg angle is thus

$$\theta_B = \sin^{-1}\left(\frac{\lambda}{2d_M\,(1-\delta)}\right) \approx \sin^{-1}\left(\frac{62\,\text{Å}}{2\,(60\,\text{Å})(1-0.023)}\right) \approx 31.94°.$$

The incident angle can be found from Snell's law,

$$\theta_o = \cos^{-1}((1-\delta)\cos\theta_B) = \cos^{-1}((1-0.023)\cos(31.9°)) = 33.94°,$$

which shows that the incident angle needs to be 2° more than the Bragg angle. The incident angle can also be approximated by equation 15-43,

$$\theta \approx \sin^{-1}\left(\frac{\lambda}{2d_M}\sqrt{1+2\delta\left(\frac{2d_M}{\lambda}\right)^2}\right)$$

$$\approx \sin^{-1}\left(\frac{62\,\text{Å}}{2\,(60\text{Å})}\sqrt{1+2\,(0.023)\left(\frac{2(60\,\text{Å})}{62\,\text{Å}}\right)^2}\right) \approx 33.98°.$$

The approximation, which ignores terms in δ^2, is off by a few hundredths of a degree.

The reflectivity of the whole multilayer can be computed by first calculating the reflection coefficients for each layer, by using equation 11-77. The multilayer can then be represented using a matrix technique relating the fields in each layer, as sketched in Figure 15-15. The field headed down into the $(m+1)$th layer from the mth is

$$E_{d,m} = \boldsymbol{t}_{m-1,m}\,E_{d,m-1}\,e^{i\phi_m} + \boldsymbol{r}_{m,m-1}\,E_{up,m}\,e^{i2\phi_m}, \qquad (15\text{-}44)$$

where the exponent combines the phase change and absorption per layer,

$$\phi_m = \frac{4\pi}{\lambda}n_m d_m \sin\theta_m, \qquad (15\text{-}45)$$

where θ_m is the angle in the medium of the mth layer, d_m is the thickness of the individual layer, n_m is the complex index of refraction, and $r_{i,j}$ and $t_{i,j}$ are the reflection and transmission coefficients between the ith and jth layers, respectively, from equation 11-77. The zeroth layer is air, and, if there are N layers, the $(N+1)$th layer is the substrate. The real roughness of the surfaces can be taken into account in computing the reflection coefficients, as was done in equation 11-89. Similarly, the field heading up from the mth layer is

$$E_{up,m} = \boldsymbol{r}_{m,m+1}\,E_{d,m} + \boldsymbol{t}_{m+1,m}\,E_{up,m+1}\,e^{i\phi_{m+1}}. \tag{15-46}$$

These $2N$-coupled equations can be rearranged into matrix format,

$$\begin{pmatrix} E_d \\ E_{up} \end{pmatrix}_{m-1} = \frac{1}{\boldsymbol{t}_{m-1,m}}\,H_m \begin{pmatrix} E_d \\ E_{up} \end{pmatrix}_m, \tag{15-47}$$

where H_m is the transfer matrix. The reflection is, of course, sharply peaked when the real part of the phase given in equation 15-45 is an integral multiple of 2π, as shown in Figure 15-16. The calculation is similar to that for a chain of atoms, discussed in section 13.4, except in this case the number of multilayers is relatively small.

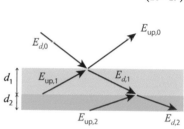

FIGURE 15-15. First pair of the multilayer stack showing the fields within each layer.

To develop strong interference effects the number of layers which are deposited should be larger than $1/\boldsymbol{r}$, where $\boldsymbol{r}$ is the magnitude of a typical reflection coefficient. The number of multilayers which can be seen by the x-ray beam can be smaller than the number physically present because of the strong absorption at low energies. The number of layers the electric field penetrates is approximately

$$N_{max} = \frac{\sin^2\theta_B}{2\pi\beta}, \tag{15-48}$$

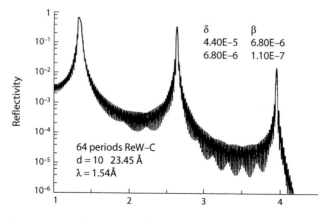

FIGURE 15-16. Computation of the reflectivity from a RhWC multilayer. E. Spiller, *Revue de Physique Appliquée* (Paris) 23, no. 10 (1988): 1687–1700.

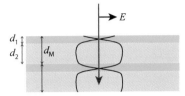

FIGURE 15-17. The field is a node within the higher-Z material, minimizing absorption.

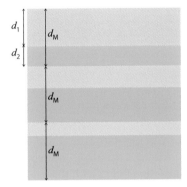

FIGURE 15-18. Quasiperiodic multilayer. The more absorptive material is thinner near the top, while the total period d_M remains constant.

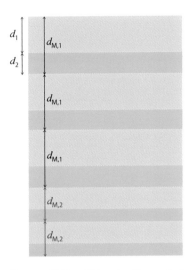

FIGURE 15-19. This depth-graded multilayer will, for a single incidence angle, reflect softer x rays from the top, and harder x rays from the bottom. A real multilayer will have a gradually decreasing period.

where β is the imaginary part of the index of refraction. To combat absorption, the multilayer is usually constructed so that the thinner layer is the one with the higher Z, in which the absorption is larger. The field is also designed to have a node in the high-Z layer, as shown in Figure 15-17. The multilayer can also be designed to have thinner layers of the absorptive material near the top, as shown for the quasiperiodic multilayer in Figure 15-18.

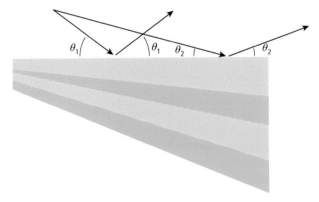

FIGURE 15-20. This laterally graded multilayer will, for a single energy, reflect a range of incident angles to collect more intensity from a point source.

The period can also be changed with depth. Especially for hard x rays, for which the penetration depth is large, a gradually changing period creates a broader spectral range for reflection and so higher intensity from a broadband beam. A depth-graded multilayer is shown in Figure 15-19. Alternatively, laterally graded multilayers will accept a broad range of incident angles, as shown in Figure 15-20. Coating flat or curved mirrors with specially designed multilayer films greatly increases their reflectivity. For example, curved coated mirrors are commonly used for both synchrotron beam applications and for x-ray astronomy. Multilayers can also be used in Laue (transmission) geometry for higher-energy x rays.

15.3.3 CURVED CRYSTALS

Another alternative to producing a larger monochromatic intensity is to curve the crystal. For a regular flat crystal, the majority of the incident radiation hits at an angle outside the narrow acceptance bandwidth of the crystal and is lost. Instead, the crystal can be curved so that all the rays from the source are incident at the optic at the Bragg angle and then are focused to the focal point, as shown in Figure 15-21. This is accomplished by placing the source, detector, and crystal surface on the Rowland circle, as described in Figure 14-9 and section 14.3. The crystal planes must then be bent to a radius of $2R$, where R is the radius of the Rowland circle. This requires both bending the crystal and machining its surface to maintain contact with the Rowland circle. In practice the machining of the surface is difficult, and most optics are simply bent, as shown in Figure 15-22. If the crystal is curved in one dimension it produces a line focus, with a focal length

$$L_f = 2R \sin \theta. \tag{15-49}$$

Three-dimensional point-to-point focusing can be achieved if the crystal is also curved in the out-of-plane direction with a radius of curvature

$$R_2 = 2R \sin^2\theta, \tag{15-50}$$

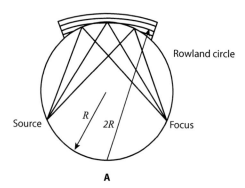

A

FIGURE 15-21. The Johansson geometry for a curved crystal. Z. W. Chen et al., *X-ray Optics and Instrumentation* 2008 (2008). doi: 10.1155/2008/318171.

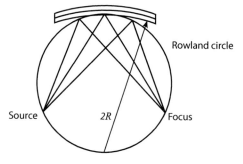

FIGURE 15-22. The Johansson geometry for a curved crystal. In this case, part of the crystal deviates from the Rowland circle, but no surface machining is required. Z. W. Chen et al., *X-ray Optics and Instrumentation* 2008 (2008). doi: 10.1155/2008/318171.

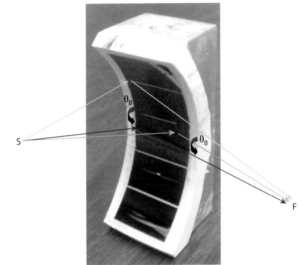

FIGURE 15-23. Photo of curved crystal optic. The doubly curved crystal optic is produced by pressing a thin silicon single crystal wafer between a curved substrate and a mold. This optic has five crystal segments. Optics used in commercial XRF systems often have a larger number of segments. Optic provided by XOS Inc. Photo courtesy of Robert Schmitz.

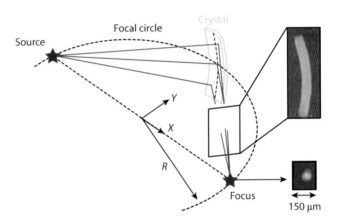

FIGURE 15-24. Three dimensional, point-to-point focusing from a doubly curved crystal optic
Z. W. Chen et al., *X-ray Optics and Instrumentation* 2008 (2008). doi: 10.1155/2008/318171.

to form a toroid (like the outer surface of a doughnut), as shown in Figures 15-23 and 15-24. Doubly curved crystal (DCC) optics collect and focus x rays emitted in a large solid angle by a laboratory point source, so they provide an increase in intensity compared with flat monochromators. DCC optics are increasingly becoming an established tool for x-ray applications, particularly in x-ray fluorescence. For x-ray fluorescence, the focused monochromatic excitation yields low background with high intensity for sensitive measurements in medical and material sciences and in manufacturing. Monochromatic beams also have potential application in medical imaging.

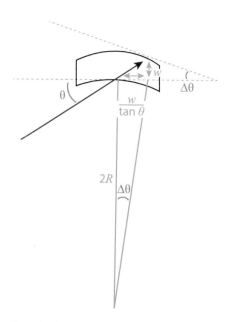

FIGURE 15-25. Geometry for calculating the crystal bandwidth for in-plane bending. Based on Bingolbali and MacDonald, *Nuclear Instruments and Methods in Physics Research Section B* 267 (2009): 832–41.

Curved crystals also have a larger acceptance bandwidth than flat crystals. Because of the oblique incidence, an x ray hits planes on the exit side of the crystal at a different angle than it hits planes at the surface. The change in angle, and hence crystal bandwidth due to in-plane bending, can be estimated using the geometry shown in Figure 15-25.

The resulting broadening is

$$\Delta\theta \approx \frac{w/\tan\theta_B}{2R}, \tag{15-51}$$

where $2R$ is the bending radius in the diffraction plane, w is the crystal thickness, and θ_B is the Bragg angle.

EXAMPLE 15-12

A curved crystal is used with 17.5 keV x rays and is produced from 40 μm (220) silicon wafers to have a focal length of 190 mm. What is the bending radius, and what is the approximate crystal bandwidth due to in-plane bending?

The lattice constant for silicon is $u \approx 5.4$ Å, so that $d = \dfrac{u}{\sqrt{2^2 + 2^2}} \approx 1.9$ Å. The wavelength is $\lambda = \dfrac{hc}{U} \approx 0.71$ Å.

The Bragg angle is $\theta = \sin^{-1}\left(\dfrac{\lambda}{2d}\right) \approx 11°$. $R = \dfrac{L_f}{2\sin\theta_B} \approx \dfrac{190\,\text{mm}}{2\sin(11°)} \approx 0.5\,\text{m}$.

$$\Delta\theta \approx \frac{w/\tan\theta_B}{2R} \approx \frac{0.04\,\text{mm}/\tan(11°)}{2(0.5\,\text{m})} \approx 0.2\,\text{mrad} = 0.012°.$$

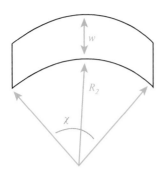

FIGURE 15-26. Geometry for calculating the crystal bandwidth for out-of-plane bending.

The geometry for broadening due to the out-of-plane bending is shown in Figure 15-26. The difference between the outer and the inner circumference gives a change in length, and hence a strain,

$$\frac{\Delta d}{d} = C_v \frac{\Delta L}{L} = C_v \frac{(R_2 + w)\chi - R_2\chi}{R_2\chi}$$

$$= C_v \frac{w}{R_2}, \tag{15-52}$$

where $\frac{\Delta d}{d}$ is the strain perpendicular to the planes, $\frac{\Delta L}{L}$ is the strain parallel to the planes, R_2 is the out-of-plane bending radius, Cv is Poisson's ratio, and χ is the angle subtended by the crystal. The resulting change in d spacing gives a change in Bragg angle, so that the broadening in crystal bandwidth due to the out-of-plane bending is

$$|\Delta\theta| = \frac{w}{R_2} C_v \tan \theta_B. \tag{15-53}$$

EXAMPLE 15-13

A curved crystal is used with 17.5 keV x rays and is produced from 40 μm (220) silicon wafers to have a focal length of 190 mm. The Poisson's ratio for silicon is approximately 0.28. What is the bandwidth due to the out-of-plane bending?

First, the radius of out-of-plane bending is $R_2 = R \sin^2 \theta_B \approx (0.5\,\text{m}) \sin^2 (11°) \approx 35\,\text{mm}$.

$$\Delta\theta \approx \frac{w}{R_2} C_v \tan \theta_B \approx \frac{0.04\,\text{mm}}{35\,\text{mm}} (0.28) \tan (11°) \approx 0.003°.$$

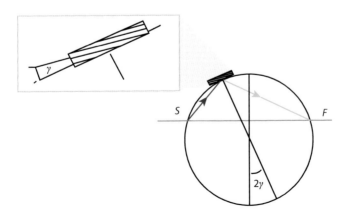

FIGURE 15-27. Asymmetric curved crystal, with different input and output focal lengths, in which case the crystal planes are no longer parallel to the surface. From Z. W. Chen et al., *X-ray Optics and Instrumentation* 2008 (2008). doi:10.1155/2008/318171.

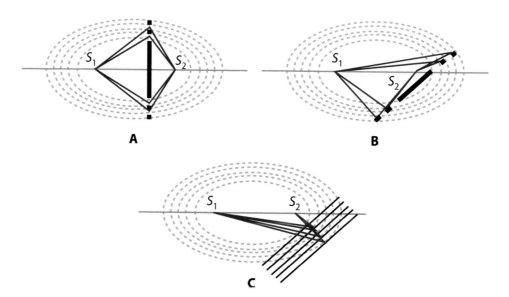

FIGURE 15-28. The three diffractive optics: A. Transmission zone plate; B. reflective zone plate; and C. curved crystal, or multilayer, can all be described in terms of an elliptical focus geometry. From A. Michette, Zone Plates, chap. 40, *Handbook of Optics*, 3rd ed., vol. 5, McGraw-Hill, 2010.

Curved crystals can also be used in asymmetric geometry, as shown in Figure 15-27. In that case, the planes are tilted relative to the crystal surface.

Zone plates, curved crystals, and mirrors can all be considered as different manifestations of an elliptical focusing geometry, as shown in Figure 15-28.

Problems

1. A zone plate with 40 open zones is designed for 10 nm radiation, with a focal length of 50 cm.
   a) What is the zone plate diameter?
   b) What is the focal spot size?
   c) Where is the third-order image of an object placed 40 cm from the zone plate?
2. A multilayer film is composed of alternating 6 nm layers of C, with a plasma energy of 29 eV and 6 nm layers of W, which has a plasma energy of 80 eV. The multilayer is to be used to diffract 0.3 keV x rays. What is the required incident angle?

Further reading

M. Bass, C. DeCusatis, J Enoch, V. Lakshminarayanan, G. Li, C. MacDonald, V. N. Mahajan, and E. Van Stryland, eds., **Handbook of Optics, 3rd** ed., vol. 5, McGraw-Hill, 2010.

R. Fitzgerald, Phase-sensitive x-ray imaging, *Physics Today* 53, no. 7 (2000): 23–27.

Y. Suzuki, A. Takeuchi, H. Takenaka, and I. Okada, Fabrication and performance test of Fresnel zone plate with 35 nm outermost zone width in hard x-ray region, *X-ray Optics and Instrumentation* (2008): doi:10.1155/2010/824387.

Z. W. Chen, W. M. Gibson, and H. Huang, High definition x-ray fluorescence: Principles and techniques, *X-ray Optics and Instrumentation* (2008): doi:10.1155/2008/318171.

SOLUTIONS TO END-OF-CHAPTER PROBLEMS

Chapter 1

1. Planck's constant is 6.6×10^{-34} J·s. 1 eV is the energy associated with an electron charge, $q_e \approx 1.6 \times 10^{-19}$, in a potential of 1 V. The speed of light is approximately 3×10^8 m/s. Verify equation 1-6.

$$h \approx (6.6 \times 10^{-34} \text{ J·s}) \left(\frac{1 \text{ eV}}{1.6 \times 10^{-19} \text{ J}} \right) \approx 4.13 \times 10^{-15} \text{ eV·s};$$

$$c \approx \left(3 \times 10^8 \frac{\text{m}}{\text{s}} \right) \left(10^{10} \frac{\text{Å}}{\text{m}} \right) \approx 3 \times 10^{18} \frac{\text{Å}}{\text{s}}.$$

$$\Rightarrow hc \approx \left(3 \times 10^{18} \frac{\text{Å}}{\text{s}} \right) (4.13 \times 10^{-15} \text{ eV·s}) \approx 1.24 \times 10^4 \text{ eV·Å}$$

$$\approx 12.4 \text{ keV·Å}.$$

2. What is the wavelength of 30 keV x rays?

$$\lambda = \frac{hc}{U} \approx \frac{12.4 \text{ keV·Å}}{30 \text{ keV}} \approx 0.4 \text{ Å}.$$

Chapter 2

1. Verify using equation 2-2 that after a time $t = t_{1/2}$ half of the original material remains.

$$N = N_o e^{-t \frac{\ln(2)}{t_{1/2}}} = N_o e^{-t_{1/2} \frac{\ln(2)}{t_{1/2}}} = N_o e^{-\ln(2)} = \frac{N_o}{2}.$$

2. How many atoms of ^{99m}Tc are required to produce an activity of 1 Ci? What mass of pure ^{99m}Tc is required?

$$\Gamma = N \frac{\ln(2)}{t_{1/2}} \Rightarrow N = \Gamma \frac{t_{1/2}}{\ln(2)} \approx 1 \text{Ci} \left(\frac{3.7 \times 10^{10} \frac{1}{\text{s}}}{\text{Ci}} \right) \left(\frac{6 \text{ hours}}{\ln(2)} \right) \left(\frac{3600 \text{ s}}{1 \text{ hr}} \right)$$

$$\approx 1.15 \times 10^{15} \text{ atoms}$$

Now we need the mass of that number of atoms

$$M = \frac{N}{N_A} M_M \approx \frac{1.15 \times 10^{15} \text{atoms}}{6 \times 10^{23} \frac{\text{atoms}}{\text{mole}}} \left(99 \frac{\text{g}}{\text{mole}} \right) \approx 1.9 \times 10^{-7} \text{g} \approx 0.2\,\mu\text{g}.$$

3. Assuming one photon per decay, what is the emission rate from a 69-year-old radioactive source that has a half-life of 10 years and an original strength of 2 mCi?

$$\Gamma = \Gamma_0\, e^{-\frac{\ln(2)}{t_{1/2}}t} \approx (2\,\text{mCi}) e^{-\frac{\ln(2)}{10}69} \approx (2\,\text{mCi}) e^{-\frac{\ln(2)}{10}69} \approx 17 \times 10^{-6}\,\text{Ci}$$

$$\approx (17 \times 10^{-6}\,\text{Ci}) \left(\frac{3.7 \times 10^{10} \frac{1}{\text{s}}}{1\,\text{Ci}} \right) \approx 6.2 \times 10^{5}\,\frac{1}{\text{s}}.$$

4. A sample of freshly produced ^{125}I, with a half-life of 60 days, was originally measured to have an activity of 1000 counts/s into a detector. It now gives 400 counts/s in the same geometry. How old is the sample?

There are a number of unknown factors: the fraction of the emitted counts that reach the detector, the number of counts per decay, and so on, but these are the same for both measurements and cancel out. We can simply set the ratio of activities equal to the ratio of measured counts to find the age:

$$\Gamma = \Gamma_0\, e^{-\frac{\ln(2)}{t_{1/2}}t} \Rightarrow t = -\frac{t_{1/2}}{\ln(2)} \ln\left(\frac{\Gamma}{\Gamma_0} \right) \approx -\frac{60\,\text{days}}{\ln(2)} \ln\left(\frac{400 \frac{\text{counts}}{\text{s}}}{1000 \frac{\text{counts}}{\text{s}}} \right) \approx 79\,\text{days}.$$

The age is slightly more than one half-life, since the count rate is down by slightly more than half.

5. To compute the activity, in millicuries, of a 100 kg person due to the normal concentration of 10^{-3} ppb (10^{-3} parts per billion, or 1 part per trillion, a concentration of 10^{-12} g/kg) of ^{14}C with a half-life of 5730 years, we will assume people are the same concentration as Lucite, $C_5H_8O_2$. a) Compute the number of moles of Lucite in 100 kg. b) Estimate the number of atoms of carbon in a 100 kg person c) Estimate the number of atoms of ^{14}C in a 100 kg person. d) Compute the activity of the person.

a) $M_M \approx 5\left(12\frac{\text{g}}{\text{mole}} \right) + 8\left(1\frac{\text{g}}{\text{mole}} \right) + 2\left(16\frac{\text{g}}{\text{mole}} \right) \approx 100\frac{\text{g}}{\text{mole}}.$

$$N_{moles} \approx \frac{100 \times 10^3 \text{g}}{100\frac{\text{g}}{\text{mole}}} \approx 10^3 \text{moles}.$$

b) $N_{atoms,\,C} \approx 10^3\,\text{moles}\left(6 \times 10^{23} \frac{\text{molecules}}{\text{mole}} \right) \left(\frac{5\,\text{atoms}}{\text{molecule}} \right)$

$$\approx 3 \times 10^{27}\,\text{carbon atoms}.$$

c) $N_o \approx (10^{-3}\,\text{ppb})(3 \times 10^{27}) \approx (10^{-3} \times 10^{-9})(3 \times 10^{27}) \approx 3 \times 10^{15}$ atoms of ^{14}C.

d) $\Gamma = N_o \dfrac{\ln(2)}{t_{1/2}} \approx 3 \times 10^{15} \dfrac{\ln(2)}{5730\,\text{yr}} \dfrac{1\,\text{yr}}{365\,\text{days}} \dfrac{1\,\text{day}}{24\,\text{hrs}} \dfrac{1\,\text{hr}}{3600\,\text{s}} \approx 1.1 \times 10^4 \dfrac{1}{\text{s}}$

$\approx \left(1.1 \times 10^4 \dfrac{1}{\text{s}}\right)\left(\dfrac{1\,\text{Ci}}{3.7 \times 10^{10}\,\dfrac{1}{\text{s}}}\right) \approx 3 \times 10^{-7}\,\text{Ci} \approx 0.3\,\mu\text{Ci}.$

6. An 8 mm diameter cylindrical pinhole is placed 100 mm from a source. What is the sensitivity of the pinhole?

$$Sen = \dfrac{\pi\left(\dfrac{W_a}{2}\right)^2}{4\pi(z)^2} \approx \dfrac{(4\,\text{mm})^2}{4(100\,\text{mm})^2} \approx 4 \times 10^{-4}.$$

7. If the cylindrical hole in Figure 2-5 is 1 mm in diameter, and is placed 200 mm from the tumor, what fraction of the emitted radiation is intercepted by the pinhole? What activity, in millicuries, is required to produce 1000 photons/s through the pinhole (assuming one photon per decay)?

$$Sen = \dfrac{\pi\left(\dfrac{W_a}{2}\right)^2}{4\pi z^2} \approx \dfrac{(1\,\text{mm})^2}{16(200\,\text{mm})^2} \approx 1.6 \times 10^{-6}.$$

$$\Gamma \approx \dfrac{1000\,\dfrac{\text{photons}}{\text{s}}}{1.6 \times 10^{-6}}\left(\dfrac{1\,\text{Ci}}{3.7 \times 10^{10}\,\dfrac{\text{photons}}{\text{s}}}\right) \approx 0.017\,\text{Ci} \approx 17\,\text{mCi}.$$

8. A 10 keV x-ray photon is captured by a NaI(Tl) scintillator. How many blue-violet photons are emitted?

$$N \approx \dfrac{\left(\dfrac{10\,\text{keV}}{\text{x-ray photon}}\right)(0.13)}{\left(\dfrac{3\,\text{eV}}{\text{blue-violet photon}}\right)} \approx 433\,\dfrac{\text{blue-violet photons}}{\text{x ray}}.$$

If 1% of the blue-violet photons land on a PMT tube, and each creates a single photoelectron, how many electrons are there at the entrance to the PMT? If the PMT has a gain of 10^6, how many electrons exit the PMT? If these are collected with 2% efficiency onto a 1 pF capacitor, what is the total charge on the capacitor? What is the height of the resultant voltage peak? What is the conversion factor (ratio of the electronic signal in volts to photon energy in kiloelectron-volts)?

$$N_{elec,\,enter} \approx (433\,\text{blue-violet photons})(0.01)\left(\dfrac{1\,\text{electron}}{\text{photon}}\right) \approx 4.3.$$

$$N_{elec,\ exit} \approx 4.3\left(10^6\right) \approx 4.3 \times 10^6 \text{ electrons.}$$

$$q \approx 2\% \left(4.3 \times 10^6 \text{ electrons}\right)\left(\frac{1.6 \times 10^{-19} \text{ Coul}}{\text{electron}}\right) \approx 1.38 \times 10^{-14} \text{ Coul.}$$

$$\Phi = \frac{q}{C} \approx \frac{1.4 \times 10^{-14} \text{ Coul}}{1 \times 10^{-12} \text{ farad}} \approx 0.014 \text{ V.}$$

The conversion factor is 1.4 mV/keV.

9. A radioactive source produces a count rate of 10 counts/s. How many seconds must the count be recorded to keep the quantum noise to less than 1%?

$$\frac{\Delta N}{N} \approx \frac{\sqrt{N}}{N} \approx \frac{1}{\sqrt{N}} \approx 0.01 \Rightarrow N \approx \left(\frac{1}{0.01}\right)^2 \approx 10^4 \Rightarrow t = \frac{10^4}{10\frac{1}{s}} \approx 1000 \text{ s.}$$

10. A pinhole with a sensitivity of 10^{-6} is to be used to image a tumor with three hot spots, with activity of 1, 2, and 3 μCi, respectively. How long a count time is required to distinguish the intensities?

We can consider the signal to be $N_3 - N_2$. The noise associated with N_3 is $\sqrt{N_3}$, and with N_2 is $\sqrt{N_2}$. Noise adds in quadrature, so the noise is $\sqrt{\left(\sqrt{N_3}\right)^2 + \left(\sqrt{N_2}\right)^2} = \sqrt{N_3 + N_2}$. We need the noise to be less than the signal. If we want it to be less by a factor of f, then we need $f\sqrt{N_3 + N_2} < (N_3 - N_2)$. (The largest noise is associated with 3 mCi; the minimum separation, with $3 - 2$ μCi.)

$$N_3 \approx (10^{-6})\, 3 \times 10^{-6} \text{ Ci} \frac{3.7 \times 10^{10}\frac{1}{s}}{\text{Ci}} t \approx \left(0.11\frac{1}{s}\right) t;$$

$$N_2 \approx 2 \times 10^{-6} \text{ Ci} \frac{3.7 \times 10^{10}\frac{1}{s}}{\text{Ci}} t \approx \left(0.074\frac{1}{s}\right) t.$$

$$\Delta N = N_3 - N_2 \approx \left(0.037\frac{1}{s}\right) t.$$

$$\Delta N = f\sqrt{N_3 + N_2} \Rightarrow \left(0.037\frac{1}{s}\right) t \approx f\sqrt{\left(0.11\frac{1}{s}\right) t + \left(0.074\frac{1}{s}\right) t}$$

$$\Rightarrow t \approx \frac{0.184\frac{1}{s}}{\left(0.037\frac{1}{s}\right)^2} f^2 \approx (135 \text{ s})\, f^2.$$

If we say the signal must be larger than the noise ($f = 1$), then we need 1354 seconds. If we want the signal to be 3 times the noise, we need 1200 seconds, or 20 minutes.

11. An area detector placed in one location records an intensity of (04020) µCi on its five pixels. A second detector, placed at 90°, records $\begin{pmatrix} 0 \\ 3 \\ 0 \\ 3 \\ 0 \end{pmatrix}$. You don't know how many sources there are. Find at least three possible distributions of integer source values (three possible 5×5 arrays) that fit the data. *Hint:* The first row and column of the 5×5 array must be all zeros for the first pixel of the detector to read zero in the two orientations. Possible solutions are

$$\begin{pmatrix} 0 & 0 & 0 & 0 & 0 \\ 0 & 3 & 0 & 0 & 0 \\ 0 & 0 & 0 & 0 & 0 \\ 0 & 1 & 0 & 2 & 0 \\ 0 & 0 & 0 & 0 & 0 \end{pmatrix}, \begin{pmatrix} 0 & 0 & 0 & 0 & 0 \\ 0 & 2 & 0 & 1 & 0 \\ 0 & 0 & 0 & 0 & 0 \\ 0 & 2 & 0 & 1 & 0 \\ 0 & 0 & 0 & 0 & 0 \end{pmatrix}, \text{ or } \begin{pmatrix} 0 & 0 & 0 & 0 & 0 \\ 0 & 1 & 0 & 2 & 0 \\ 0 & 0 & 0 & 0 & 0 \\ 0 & 3 & 0 & 0 & 0 \\ 0 & 0 & 0 & 0 & 0 \end{pmatrix}.$$

Chapter 3

1. Verify the value of U_R in equation 3-3.

$$U_R = \frac{M_e q_e^4}{8\varepsilon_o^2 h^2} \approx \frac{(9.11 \times 10^{-31} \text{ kg})(1.6 \times 10^{-19} \text{ Coul})^4}{8 \left(8.854 \times 10^{-12} \frac{\text{Coul}^2}{\text{N} \cdot \text{m}^2} \right)^2 (6.626 \times 10^{-34} \text{ J} \cdot \text{s})^2}$$

$$\approx \left(2.18 \times 10^{-18} \frac{\text{kg} \cdot \text{m}^2}{\text{s}^2} \right) \frac{1 \text{ eV}}{1.6 \times 10^{-19} \text{ J}} \approx 13.6 \text{ eV}.$$

2. Show that the ratio of radiation intensity at two high temperatures is approximately $\frac{I(T_1)}{I(T_2)} \approx e^{\frac{U}{k_B}\left(\frac{1}{T_2} - \frac{1}{T_1} \right)}$.

$$I_\nu d\nu = \frac{8\pi}{(hc)^2} \frac{(h\nu)^3}{\left(e^{\left(\frac{h\nu}{k_B T}\right)} - 1 \right)} d\nu \Rightarrow I_U dU = \frac{8\pi}{(hc)^2} \frac{(U)^3}{\left(e^{\left(\frac{U}{k_B T}\right)} - 1 \right)} dU$$

$$\Rightarrow \frac{I(T_1)}{I(T_2)} = \frac{e^{\frac{U}{k_B T_2}} - 1}{e^{\frac{U}{k_B T_1}} - 1} \approx \frac{e^{\frac{U}{k_B T_2}}}{e^{\frac{U}{k_B T_1}}} \approx e^{\frac{U}{k_B}\left(\frac{1}{T_2} - \frac{1}{T_1} \right)},$$

and then estimate the ratio of intensity of 1 keV radiation at 500,000 K to that at a temperature of "only" 100,000 K:

$$\frac{1}{T_2} - \frac{1}{T_1} \approx \frac{1}{10^5 K} - \frac{1}{5 \times 10^5 \, K} \approx 8 \times 10^{-6} \, \frac{1}{K};$$

$$\frac{U}{k_B} \approx \frac{1000 \, eV}{8.6 \times 10^{-5} \, \frac{eV}{K}} \approx 1.16 \times 10^7 \, K$$

$$\Rightarrow \frac{I(T_1)}{I(T_2)} \approx e^{\frac{U}{k_B}\left(\frac{1}{T_2} - \frac{1}{T_1}\right)} \approx e^{\left(8 \times 10^{-6} \frac{1}{K}\right)(1.16 \times 10^7 K)} \approx e^{93} \approx 10^{40}.$$

Increasing the temperature by a factor of 5 increased the radiation at that frequency by 40 orders of magnitude!

3. Compute the energy of the (visible light) hydrogen Balmer α line, associated with the transition of an electron from the $m = 3$ to the $m = 2$ level in hydrogen.

$$U_2 = -U_R \frac{Z^2}{m^2} \approx -(13.6 \, eV)\frac{1^2}{2^2} \approx -3.4 \, eV$$

$$U_3 \approx -(13.6 \, eV)\frac{1^2}{3^2} \approx -1.5 \, eV$$

$$U = U_3 - U_2 \approx -1.5. \, eV - (-3.4 \, eV) \approx 1.9 \, eV.$$

4. a) Estimate the velocity of nickel atoms at 1000 K by setting $k_B T \approx M_{Nickel} v^2$, then use that result to estimate the distance a hot nickel atom could travel during a 1 ps laser pulse.

$$v \approx \sqrt{\frac{k_B T}{M_{atom}}} \approx \sqrt{\frac{\left(8.6 \times 10^{-5} \frac{eV}{K}\right)(1000 \, K)\left(1.6 \times 10^{-19} \frac{J}{eV}\right)}{\left(59 \times 10^{-3} \frac{kg}{mole}\right)\left(\frac{1 \, mole}{6 \times 10^{23} \, atoms}\right)}} \approx 377 \, \frac{m}{s}.$$

$$x = vt \approx \left(377 \, \frac{m}{s}\right)(10^{-12} \, s) \approx 4 \, Å, \quad \text{no appreciable movement.}$$

b) Calculate the approximate power density due to a 2 mJ laser pulse of 1 ps duration focused to a diameter of 0.5 μm that is absorbed within a depth of 50 nm of a nickel surface.

We can ignore the tiny movement of the atoms and use a simple calculation for volume,

$$V \approx \pi \left(\frac{5 \times 10^{-7} m}{2}\right)^2 (50 \times 10^{-9} m) \approx 9.8 \times 10^{-21} \, m^3.$$

$$P \approx \frac{2 \times 10^{-3} \, J}{10^{-12} \, s} \approx 2 \times 10^9 \, W.$$

$$\frac{P}{V} \approx \frac{2 \times 10^9 \, W}{10^{-20} \, m^3} \approx 2 \times 10^{29} \, \frac{W}{m^3}.$$

A modest laser pulse can create enormous power densities.

5. A gas discharge in a vacuum is used to create a plasma. The chamber is filled with He, then evacuated to a density of 10^{-9} kg/m³. The spark has a temperature of about 1 million K. Calculate the electron density, the plasma frequency, and the Debye length.

$$k_B T \approx \left(8.6 \times 10^{-5} \, \frac{eV}{K} \right)(10^6 K) \approx 86 \, eV, \text{ so the He is fully ionized.}$$

$$\rho_e = ZN_A \frac{\rho}{M} \approx \left(2 \, \frac{electrons}{atom} \right)\left(6.02 \times 10^{23} \, \frac{atoms}{mole} \right)\left(\frac{10^{-9} \, \frac{kg}{m^3}}{4 \times 10^{-3} \, \frac{kg}{mole}} \right)$$

$$\approx 3 \times 10^{17} \, \frac{electrons}{m^3}.$$

$$\omega_p = \sqrt{\frac{\rho_e q_e^2}{\varepsilon_0 M_e}} \approx \sqrt{\frac{\left(3 \times 10^{17} \, \frac{electrons}{m^3} \right)\left(1.6 \times 10^{-19} \, \frac{Coul}{electron} \right)^2}{\left(8.85 \times 10^{-12} \, \frac{Coul^2}{N \cdot m^2} \right)(9.1 \times 10^{-31} \, kg)}}$$

$$\approx 3.1 \times 10^{10} \, \frac{rad}{s}. \quad (\approx 5 \, GHz, \text{ microwave})$$

$$\Lambda_D = \sqrt{\frac{\varepsilon_0 k_B T}{\rho_e q_e^2}} \approx \sqrt{\frac{\left(8.85 \times 10^{-12} \, \frac{Coul^2}{N \cdot m^2} \right)\left(1.38 \times 10^{-23} \, \frac{J/electron}{K} \right)(10^6 \, K)}{\left(3 \times 10^{17} \, \frac{electrons}{m^3} \right)\left(1.6 \times 10^{-19} \, \frac{Coul}{electron} \right)^2}}$$

$$\approx 0.13 \, mm.$$

6. A plasma has a plasma frequency of 1 GHz and a Debye length of 60 μm. The volume is reduced by a factor of 4, increasing the electron density by the same factor. The temperature is unchanged.

a) What is the new plasma frequency?

$$\frac{\omega_p}{2\pi} = \frac{1}{2\pi}\sqrt{\frac{q_e^2 \rho_e}{\varepsilon_0 M_e}} = \frac{1}{2\pi}\sqrt{\frac{q_e^2 \, 4\rho_{eo}}{\varepsilon_0 M_e}} \approx \frac{2\omega_{po}}{2\pi} \approx 2 \, GHz.$$

b) What is the new screening length?

$$\Lambda_D = \sqrt{\frac{\varepsilon_0 k_B T}{\rho_e q_e^2}} = \sqrt{\frac{\varepsilon_0 k_B T}{4\rho_{eo} q_e^2}} \approx \frac{\Lambda_{Do}}{2} \approx 30 \, \mu m.$$

7. Show that the energy corresponding to the plasma frequency of fully ionized SiO_2, with a density of 2.2 g/cm³, is 30 eV.

$$\rho_e = ZN_A \frac{\rho}{M} \approx \left((14 + 2 \times 8) \frac{\text{electrons}}{\text{molecule}} \right) \left(6.02 \times 10^{23} \frac{\text{molecules}}{\text{mole}} \right)$$

$$\times \left(\frac{2.2 \times 10^3 \frac{\text{kg}}{\text{m}^3}}{(28 + 2 \times 16) \times 10^{-3} \frac{\text{kg}}{\text{mole}}} \right) \approx 6.6 \times 10^{29} \frac{\text{electrons}}{\text{m}^3}.$$

$$\omega_p = \sqrt{\frac{\rho_e q_e^2}{\varepsilon_0 M_e}} \approx \sqrt{\frac{\left(6.6 \times 10^{29} \frac{\text{electrons}}{\text{m}^3} \right) \left(1.6 \times 10^{-19} \frac{\text{Coul}}{\text{electron}} \right)^2}{\left(8.85 \times 10^{-12} \frac{\text{Coul}^2}{\text{N} \cdot \text{m}^2} \right) (9 \times 10^{-31} \text{ kg})}}$$

$$\approx 4.6 \times 10^{16} \frac{\text{rad}}{\text{s}}.$$

$$E_p = \hbar \omega_p \approx (1.05 \times 10^{-34} \text{ J} \cdot \text{s}) \left(4.6 \times 10^{16} \frac{\text{rad}}{\text{s}} \right) \left(\frac{1.6 \times 10^{-19} \text{ J}}{\text{eV}} \right) \approx 30 \text{ eV}.$$

Chapter 4

1. X rays are produced by an x-ray tube with a molybdenum anode, with $Z = 42$.

a) What is the approximate energy of the K, L, and M levels, assuming screening constants of 3.2, 12, and 21, respectively?

$$U_K = -U_R' \frac{(Z - \sigma)^2}{n^2} \approx -(13.6 \text{ eV}) \frac{(42 - 3.2)^2}{1^2} \approx -20500 \text{ eV} \approx -20.5 \text{ keV}.$$

$$U_L \approx -(13.6 \text{ eV}) \frac{(42 - 12)^2}{2^2} \approx -3.1 \text{ keV}.$$

$$U_M \approx -(13.6 \text{ eV}) \frac{(42 - 21)^2}{3^2} \approx -0.7 \text{ keV}.$$

b) What are the resulting characteristic energies of the Kα, Kβ, and Lα lines?

For Kα, $U = U_L - U_K \approx -3.1 + 20.5 \text{ keV} \approx 17.4 \text{ keV}.$

For Kβ, $U = U_M - U_K \approx -0.7 + 20.5 \text{ keV} \approx 19.8 \text{ keV}.$

For Lα, $U = U_M - U_L \approx -0.7 + 3.1 \text{ keV} \approx 2.4 \text{ keV}.$

This approximation is reasonable for estimating line energies. The actual measured energies are 20.0 keV for the K level, and 19.6 keV, 17.5 keV, and 2.3 keV, for the Kβ, Kα, and Lα lines, respectively.

2. The energy levels of silver, Ag, are listed in the table with their quantum numbers.

	E, keV	n	ℓ	j
N_3	−0.058	4	1	3/2
N_2	−0.06	4	1	1/2
N_1	−0.1	4	0	1/2
M_5	−0.368	3	2	5/2
M_4	−0.37	3	2	3/2
M_3	−0.57	3	1	3/2
M_2	−0.60	3	1	1/2
M_1	−0.7	3	0	1/2
L_3	−3.4	2	1	3/2
L_2	−3.5	2	1	1/2
L_1	−3.8	2	0	1/2
K_1	−25.5	1	0	1/2

a) What are the energies of all allowed Kα lines?

Kα means an electron transition from an L to a K shell. The only K shell is K_1, which has $\ell = 0$. The $\Delta\ell = \pm1$ rule means that only transitions from L_2 and L_3 are possible:

L_3 to K_1: 3.4 keV − (−25.5 keV) ≈ 22.1 keV;

L_2 to K_1: −3.5 keV − (−25.5 keV) ≈ 22 keV.

b) What is the minimum x-ray tube voltage required for a silver anode to produce Kα radiation?

The electron beam must have enough energy to ionize the K electron, so 25.5 kV.

c) What are the energies of all allowed Kβ lines?

Similarly, Kβ means an electron transition from an M to a K shell, and the $\Delta\ell = \pm1$ rule means that only transitions from M_2 and M_3 are possible:

M_3 to K_1: −0.57 keV − (−25.5 keV) ≈ 24.93 keV;

M_2 to K_1: −0.60 keV − (−25.5 keV) ≈ 24.9 keV.

d) If core electrons are knocked out of silver atoms, what fraction of the emission is x-ray photons (as opposed to Auger electrons; see section 4.5)?

$$\text{Probability for Ag} \approx \left| \frac{Z^4}{A + Z^4} \right| \approx \left| \frac{47^4}{10^6 + 47^4} \right| \approx 83\%.$$

e) X-ray emission from a tungsten anode tube is used to excite Ag fluorescence to determine the concentration of Ag in a sample (see section 4.7). As the tube voltage is reduced from 50 kV to 20 kV the Ag fluorescence count rate drops abruptly to zero. Why?

As in part b, the voltage must be above 25.5 kV to knock out the core electron.

3. What is the relative emission rate of the two lines in the Mo Kβ doublet?

The Kβ doublet originates from electrons falling from the M level to the K level. Since the K, 1s, level has $\ell = 0$, the doublet is from the $2p_{3/2}$ and $2p_{1/2}$ levels, which have $\ell = 1$, and $J = 3/2$ or $1/2$, so

$$\frac{\Gamma_{3/2}}{\Gamma_{1/2}} = \frac{2\left(\frac{3}{2}\right)+1}{2\left(\frac{1}{2}\right)+1} = \frac{4}{2} = 2.$$

4. A Mo anode is run with a tube voltage of 25 kV and a current of 1 mA.

a) Assuming that the efficiency coefficient, C_{char}, of Kα of characteristic line emission is 2×10^{-4} x rays per electron, and the exponent O_{char} is 1.6, what is the emission rate of Kα x rays in photons per second? The binding energy $q_e\Phi_b$ is 20 keV.

$$\Gamma = C_{char}\frac{J}{q_e}\left(\frac{\Phi - \Phi_o}{\Phi_o}\right)^{1.6} \approx \left(2\times10^{-4}\frac{photons}{electron}\right)\left(\frac{1\times10^{-3}\dfrac{Coul}{s}}{1.6\times10^{-19}\dfrac{Coul}{electron}}\right)$$

$$\times\left(\frac{25\,kV - 20\,kV}{20\,kV}\right)^{1.6} \approx 1.4\times10^{11}\frac{photons}{s}.$$

b) Assuming the x rays are emitted isotropically, what is the count rate of Kα x rays for a 10 mm square detector 150 mm from the source?

$$\Gamma_D = \Gamma\frac{W_D^2}{4\pi z^2} \approx \left(1.4\times10^{11}\frac{photons}{s}\right)\frac{(10\,mm)^2}{4\pi(150\,mm)^2} \approx 4.8\times10^7\frac{photons}{s}.$$

c) By what factor does the count rate increase if the tube voltage is doubled to 50 kV, still at 1 mA?

The count rate is increased by a factor of

$$Ratio = \left(\frac{\Phi_2 - \Phi_o}{\Phi_1 - \Phi_o}\right)^{1.6} \approx \left(\frac{50\,kV - 20\,kV}{25\,kV - 20\,kV}\right)^{1.6} \approx 18,$$

an order of magnitude more than the increase in voltage.

5. What is the x-ray emission probability (as opposed to Auger emission) for an innermost core hole vacancy in carbon? In tungsten?

$$\text{Probability for C} \approx \left|\frac{Z^4}{C_A + Z^4}\right| = \left|\frac{6^4}{10^6 + 6^4}\right| \approx 0.1\%.$$

$$\text{Probability for W} \approx \left|\frac{74^4}{10^6 + 74^4}\right| \approx 97\%.$$

6. Doppler broadening and other factors tend to create Gaussian line widths, or lines which are mixes of Gaussian and Lorentzian shapes. Find the FWHM Δv_{FWHM} when the frequency distribution is Gaussian, $I(v) = F_o e^{-\frac{(v-v_o)^2}{2\sigma^2}}$.

$$I(v) = F_o e^{-\frac{(v-v_o)^2}{2\sigma^2}} = \frac{F_o}{2} \Rightarrow e^{-\frac{(v-v_o)^2}{2\sigma^2}} = \frac{1}{2} \Rightarrow -\frac{(v-v_o)^2}{2\sigma^2} = -\ln(2)$$

$$\Rightarrow v = v_o \pm \sqrt{2\sigma^2 \ln 2} \Rightarrow FWHM = v_+ - v_- = 2\sqrt{2\ln 2}\,\sigma.$$

7. Find the Fourier transform of the time dependence of the quantum state $F(v) = \Im\{\psi(t)\} = \int_0^\infty \psi(t)e^{-2\pi i vt}\,dt$ for the decaying exponential of equation 4-14, and show that FF^* is proportional to the Lorentzian distribution of equation 4-15. (The constant of proportionality is resolved by setting the intensity at the peak, $I(v)|_{v=v_o} = CF(v)F^*(v)|_{v=v_o}$, equal to the value at the peak, $I(v_o)$.)

$$F(v) = \int_0^\infty \psi_o e^{-\frac{t}{\tau}} e^{2\pi i v_o t} e^{-2\pi i vt}\,dt = \int_0^\infty \psi_o e^{\left(2\pi i(v_o - v) - \frac{1}{\tau}\right)t}\,dt$$

$$= \psi_o \frac{e^{\left(2\pi i(v_o - v) - \frac{1}{\tau}\right)t}\Big|_0^\infty}{2\pi i(v_o - v) - \frac{1}{\tau}} = \psi_o \frac{-1}{2\pi i(v_o - v) - \frac{1}{\tau}}.$$

$$FF^* = \psi_o^2 \left(\frac{-1}{2\pi i(v_o - v) - \frac{1}{\tau}}\right)\left(\frac{-1}{-2\pi i(v_o - v) - \frac{1}{\tau}}\right)$$

$$= \psi_o^2 \left(\frac{1}{4\pi^2(v_o - v)^2 + \frac{1}{\tau^2}}\right) = \psi_o^2 \left(\frac{\tau^2}{4\pi^2\tau^2(v_o - v)^2 + 1}\right)$$

$$= I(v_o)\frac{1}{(2\pi\tau(v_o - v))^2 + 1} \quad \text{if} \quad I(v_o) = \tau^2\psi_o^2.$$

8. An energy level has a lifetime with a FWHM of 1 fs. What is the energy width of the emission line created when an electron falls from this level? From example 4-5,

$$\Delta v_{FWHM} = \frac{1}{\pi\tau}, \text{ so}$$

$$\Delta U_{FWHM} = h\Delta v_{FWHM} = \frac{h}{\pi\tau} \approx \frac{4.14 \times 10^{-15} \text{ eV} \times \text{s}}{\pi 10^{-15}\text{s}} \approx 1.3 \text{ eV}.$$

9. A fluorescence system has an MDL of 500 ppm for copper with a measurement time of 100 s. The background is 50 counts/s/bin for 20 eV wide bins. The peak is 300 eV wide. What is the total count rate (fluorescence plus background) for a 10% copper sample?

$$N_{bins} \approx \left(\frac{300\,\text{eV}}{20\,\text{eV/bin}} \right) \approx 15.$$

$$\Gamma_{Bkgnd} \approx \left(50\,\frac{\text{counts/s}}{\text{bin}} \right)(15\,\text{bins}) \approx 750\,\frac{\text{counts}}{\text{s}}.$$

$$N_{Bkgnd} \approx \left(750\,\frac{\text{counts}}{\text{s}} \right)(100\,\text{s}) \approx 7.5 \times 10^4.$$

$$MDL \approx 3X\,\frac{\sqrt{N_{Bkgnd}}}{N_s} \Rightarrow N_s \approx 3X\,\frac{\sqrt{N_{Bkgnd}}}{MDL} \approx 3(0.1)\,\frac{\sqrt{7.5 \times 10^4}}{500/10^6} \approx 1.6 \times 10^5.$$

$$\Gamma_s \approx \frac{1.6 \times 10^5\,\text{counts}}{100\text{s}} \approx 1.6 \times 10^3\,\frac{\text{counts}}{\text{s}}.$$

$$\Gamma = \Gamma_s + \Gamma_{Bkgnd} \approx 1.6 \times 10^3\,\frac{\text{counts}}{\text{s}} + 750\,\frac{\text{counts}}{\text{s}} \approx 2.4 \times 10^3\,\frac{\text{counts}}{\text{s}}.$$

10. An artificial noise-free fluorescence peak for a 10% copper sample is shown in Figure A-1. The data plotted in the figure are summed over 100 eV energy bins. Answer the questions by estimating from the graph.

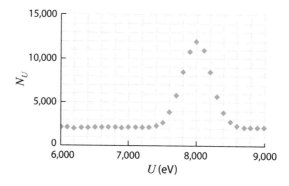

FIGURE A-1. Example fluorescence peak for a 10% Ni sample.

a) What is the background level per bin?
From the graph, it is about 2000 per bin outside the peak.

b) What is the FWHM of the signal peak?
The maximum number of counts per bin is 12000. The maximum signal counts per bin is thus approximately $12000 - 2000 = 10000$. Half the maximum is 5000. The graph in the figure includes background, so the curve at half maximum should be at about 7000 counts/bin. The graph goes through 7000 counts/bin at about 7750 eV, so the FWHM is roughly 500 eV.

c) What is the background?
The answer of course depends on what is taken as the peak width. Using a standard width of the FWHM, there are 5 bins in the peak, so the background is

$$N_{Bkgnd} \approx (5)(2000) \approx 10000\,\text{counts}.$$

d) What is the signal?

The total is the sum of the 5 points in the peak, roughly $N \approx (8500 + 11{,}000 + 12{,}000 + 11{,}000 + 8{,}500) \approx 51000$.

Then, the signal is $N_s = N - N_{Bkgnd} \approx 51000 - 10000 = 41000$.

e) What is the expected noise?

$$\boldsymbol{N} = \sqrt{N} \approx \sqrt{51000} \approx 226.$$

f) What is the expected signal-to-noise ratio?

$$\frac{N_s}{\boldsymbol{N}} \approx \frac{41000}{226} \approx 182.$$

g) With noise, what would be the MDL, in parts per million?

$$MDL \approx 3 \frac{\sqrt{N_{Bkgnd}}}{N_S} X \approx 3 \frac{\sqrt{10000}}{41000} (0.1) \approx 7.3 \times 10^{-4} \approx 730 \text{ ppm.}$$

Chapter 5

1. Consider a Mo anode tube source with a tube voltage of 40 kV, current of 1 mA, and 2×10^{-4} x rays per electron when the tube voltage is twice the binding energy. The Kα energy is 17.5 keV, the linewidth is 7 eV, the source has a diameter of 50 μm and is 100 mm away from a 5 mm diameter detector. Considering only the characteristic Kα emission, give the emission rate, power, intensity at 100 mm, angular intensity, photon intensity at 100 mm, angular photon intensity, brightness and brilliance, the power, and count rate into the detector. *Hint*: Recall equation 4-9.

Emission rate:

$$\Gamma = C_{char} \frac{J}{q_e} \left(\frac{\Phi - \Phi_o}{\Phi_o} \right)^{1.6} \approx \left(2 \times 10^{-4} \frac{\text{photons}}{\text{electron}} \right) \left(\frac{1 \times 10^{-3} \dfrac{\text{Coul}}{\text{s}}}{1.6 \times 10^{-19} \dfrac{\text{Coul}}{\text{electron}}} \right)$$

$$\times \left(\frac{40 \text{ kV} - 20 \text{ kV}}{20 \text{ kV}} \right)^{1.6} \approx 1.2 \times 10^{12} \frac{\text{photons}}{\text{s}}.$$

Power:

$$P = \Gamma \bar{U} \approx \left(1.2 \times 10^{12} \frac{\text{photons}}{\text{s}} \right) \left(17.5 \times 10^3 \frac{\text{eV}}{\text{photon}} \right) \left(1.6 \times 10^{-19} \frac{\text{J}}{\text{eV}} \right)$$

$$\approx 3.5 \times 10^{-3} \text{ W} \approx 3.5 \text{ mW.}$$

Intensity: $I = \dfrac{P}{4\pi z^2} \approx \dfrac{3.5 \times 10^{-3} \text{ W}}{4\pi (100 \text{ mm})^2} \approx 2.8 \times 10^{-8} \dfrac{\text{W}}{\text{mm}^2} \approx 28 \dfrac{\text{mW}}{\text{m}^2}.$

Angular intensity: $I_\Omega = \dfrac{P}{\Omega_s} \approx \dfrac{3.5 \times 10^{-3}\,\text{W}}{4\pi} \approx 2.8 \times 10^{-4}\,\dfrac{\text{W}}{\text{Sr}}$.

or

$$I_\chi \approx \dfrac{3.5 \times 10^{-3}\,\text{W}}{2\pi\left(1 - \cos\dfrac{2\pi}{2}\right)\left(\dfrac{10^3\,\text{mrad}}{\text{rad}}\right)^2} \approx 2.8 \times 10^{-10}\,\dfrac{\text{W}}{\text{mrad}^2}.$$

Photon intensity:

$$\Psi = \dfrac{\Gamma}{4\pi z^2} \approx \dfrac{1.25 \times 10^{12}\,\text{photons/s}}{4\pi\,(100\,\text{mm})^2} \approx 1 \times 10^7\,\dfrac{\text{photons}}{\text{mm}^2 \cdot \text{s}} \approx 10^{13}\,\dfrac{\text{photons}}{\text{m}^2 \cdot \text{s}}.$$

Photon angular intensity:

$$\Psi_\Omega = \dfrac{\Gamma}{\Omega_s} \approx \dfrac{10^{13}\,\text{photons/s}}{4\pi} \approx 10^{11}\,\dfrac{\text{photons/s}}{\text{Sr}} \quad \text{or} \quad \Psi_\chi \approx 10^5\,\dfrac{\text{photons/s}}{\text{mrad}^2}.$$

To determine brightness, we need the fraction of photons within 0.1 % of the central energy. The bandwidth is $\delta U = 0.1\% U = 10^{-3}(17.5\,\text{keV}) = 17.5\,\text{eV}$. The energy spread was given as $\Delta U = 7\,\text{eV}$, so the fraction of the intensity in that range is

$$fraction_{Bndwth} = 1, \quad \text{and} \quad \Psi_{\chi Bndwth} \approx 10^5\,\dfrac{\text{photons}}{\text{s} \cdot \text{mrad}^2 \cdot 0.1\%\ \text{bandwidth}}.$$

Brilliance: $b_A \approx \dfrac{10^5\,\dfrac{\text{photons}}{\text{s} \cdot \text{mrad}^2 \cdot 0.1\% \text{bandwidth}}}{\pi\left(\dfrac{50 \times 10^{-3}\,\text{mm}}{2}\right)^2}$

$$\approx 5 \times 10^7\,\dfrac{\text{photons}}{\text{s} \cdot \text{mrad}^2 \cdot \text{mm}^2 \cdot 0.1\% \text{bandwidth}}.$$

Power into the detector: $P_D = I A_D \approx \left(2.8 \times 10^{-8}\,\dfrac{\text{W}}{\text{mm}^2}\right)\left(\pi\,\dfrac{5\,\text{mm}}{2}\right)^2 \approx 0.5\,\mu\text{W}$.

Count rate into detector:

$$\Gamma_D = \Psi A_D \approx \left(10^7\,\dfrac{\text{photons}}{\text{mm}^2 \cdot \text{s}}\right)\left(\pi\,\dfrac{5\,\text{mm}}{2}\right)^2 \approx 2 \times 10^8\,\dfrac{\text{photons}}{\text{s}}.$$

2. A synchrotron x-ray beam has a height of 2 mm at a distance of 20 m from the source, and is restricted by apertures to a width of 5 mm in the horizontal direction. Give the horizontal and vertical global divergence of the beam.

$$\Theta_{vert} = \arctan\left(\dfrac{W_{vert}}{z}\right) \approx \dfrac{W_{vert}}{z} \approx \dfrac{2 \times 10^{-3}\,\text{m}}{(20\,\text{m})} \approx 10^{-4}\,\text{rad} \approx 100\,\mu\text{rad}.$$

$$\Theta_{horiz} \approx \dfrac{W_{horiz}}{z} \approx \dfrac{5 \times 10^{-3}\,\text{m}}{(20\,\text{m})} \approx 2.5 \times 10^{-4}\,\text{rad} \approx 250\,\mu\text{rad}.$$

3. What is the local divergence of the source in problem 1, as viewed from the detector?

$$\xi \approx \frac{W_s}{z} \approx \frac{50 \times 10^{-3}\,\text{mm}}{(100\,\text{mm})} \approx 5 \times 10^{-4}\,\text{rad} \approx 0.03°.$$

4. What is the minimum image resolution for a 100 mm thick object placed 500 mm from the source in problem 1, but with the detector moved back to 600 mm?

$$\text{At } z = 500\,\text{mm}, \ \xi \approx \frac{W_s}{z} \approx \frac{50 \times 10^{-3}\,\text{mm}}{(500\,\text{mm})} \approx 100 \times 10^{-6}\,\text{rad} \approx 100\,\mu\text{rad}.$$

$$\Delta y = \xi L \approx (100 \times 10^{-6})(100\,\text{mm}) \approx 0.01\,\text{mm} \approx 10\,\mu\text{m}.$$

5. Show that the time average of the product of sine and cosine, $\langle \sin(\omega t)\cos(\omega t) \rangle$, is zero.

$$\langle \sin(\omega t)\cos(\omega t) \rangle = \frac{1}{T} \int_0^T \sin \omega t \cos \omega t \, dt = \frac{1}{2\omega T} \sin^2(\omega t) \Big|_0^T$$

$$= \frac{1}{2\omega T} \sin^2\left(\frac{2\pi}{T} t\right) \Big|_0^T = 0.$$

6. Prove equation 5-54.

Using $\langle \cos^2(\kappa r_1 - \omega t) \rangle = \langle \cos^2(\kappa r_2 - \omega t) \rangle = \frac{1}{2}, \ \frac{1}{2} c\varepsilon_o E_o^2 = I_1$, and $\Delta r = r_2 - r_1$,

$$I = c\varepsilon_o \langle (E_o \cos(\kappa r_1 - \omega t) + E_o \cos(\kappa r_2 - \omega t + \phi_s))^2 \rangle$$

$$= c\varepsilon_o \left(\begin{array}{l} E_o^2 \langle \cos^2(\kappa r_1 - \omega t) \rangle + 2E_o^2 \langle \cos(\kappa r_1 - \omega t)\cos(\kappa r_2 - \omega t + \phi_s) \rangle \\ + E_o^2 \langle \cos^2(\kappa r_2 - \omega t + \phi_s) \rangle \end{array} \right)$$

$$= I_1 + 4I_1 \langle \cos(\kappa r_1 - \omega t)\cos(\kappa r_2 - \omega t + \phi_s) \rangle + I_1$$

$$= 2I_1 + 2I_1 \langle \cos(\kappa r_1 - \omega t)\cos(\kappa r_2 - \omega t + \phi_s) - \sin(\kappa r_1 - \omega t)\sin(\kappa r_2 - \omega t + \phi_s) \rangle$$

$$\quad + 2I_1 \langle \cos(\kappa r_1 - \omega t)\cos(\kappa r_2 - \omega t + \phi_s) + \sin(\kappa r_1 - \omega t)\sin(\kappa r_2 - \omega t + \phi_s) \rangle$$

$$= 2I_1 + 2I_1 \left\langle \begin{array}{l} \cos[(\kappa r_2 - \omega t + \phi_s) + (\kappa r_1 - \omega t)] + \cos[((\kappa r_2 - \omega t + \phi_s) \\ - (\kappa r_1 - \omega t))] \end{array} \right\rangle$$

$$= 2I_1 + 2I_1 \langle \cos(\kappa \Delta r + \phi_s) \rangle + 2I_1 \langle \cos(\kappa(r_2 + r_1) - 2\omega t + \phi_s) \rangle$$

$$= 2I_1 + 2I_1 \langle \cos(\phi_{tot}) \rangle + 0.$$

7. What is the spacing of the interference fringes on a screen 10 m from slits separated by 0.5 μm and irradiated with 10 keV radiation?

$$y_m = m \frac{\lambda L}{d} \Rightarrow y_{m+1} - y_m = \frac{\lambda L}{d} \approx \frac{(1.24 \times 10^{-10}\,\text{m})(10\,\text{m})}{0.5 \times 10^{-6}\,\text{m}}$$

$$\approx 2.5 \times 10^{-3}\,\text{m} \approx 2.5\,\text{mm}.$$

8. How far do the fringes in problem 7 shift if the source, 10 cm from the slits, is moved 10 μm?

$$y_{shift} = \frac{wL}{z} \approx \frac{(10 \times 10^{-6} \text{ m})(10 \text{ m})}{10 \times 10^{-2} \text{ m}} \approx 10^{-3} \text{ m} \approx 1 \text{ mm}.$$

This is almost half the fringe separation.

9. What is the maximum source size that can be employed to see the interference pattern in problem 7?

$$W_s < \frac{\lambda z}{d} \approx \frac{(1.24 \times 10^{-10} \text{ m})(10 \times 10^{-2} \text{ m})}{(0.5 \times 10^{-6} \text{ m})} \approx 2.5 \times 10^{-5} \text{ mm} \approx 25 \text{ μm},$$

which, as expected, is slightly more than twice the shift in problem 8.

10. What is the transverse coherence length for the source in problem 1, at the detector distance of 100 mm?

$$\lambda \approx \frac{12.4 \times 10^{-10} \text{ m}}{17.5} \approx 0.71 \times 10^{-10} \text{ m} \Rightarrow Y_c = \frac{\lambda z}{W_s} = \frac{\lambda}{\xi} \approx \frac{0.71 \times 10^{-10} \text{ m}}{(5 \times 10^{-4} \text{ rad})}$$

$$\approx 1.4 \times 10^{-7} \text{ m} \approx 0.14 \text{ μm}.$$

11. What is the longitudinal coherence length of the source in problem 1?

$$L_c = c\tau = c\frac{1}{\Delta f} = \frac{hc}{\Delta U} \approx \frac{12.4 \times 10^3 \text{ eV Å}}{7 \text{ eV}} \approx 1.8 \times 10^3 \text{ Å} \approx 0.18 \text{ μm}.$$

Chapter 6

1. Show that equation 6-10 is the same as equation 9-9 when $v \ll c$, and find the direction of polarization of the electric field if the velocity and acceleration are in the z direction, and the observer is in the y direction

$$\vec{E} = \frac{q_e}{4\pi\varepsilon_o c} \left. \frac{\hat{r} \times \left\{ \left(\hat{r} - \dfrac{\vec{v}}{c} \right) \times \dfrac{\vec{a}}{c} \right\}}{r\left(1 - \dfrac{\vec{v} \cdot \hat{r}}{c} \right)^3} \right|_{v/c \to 0} \approx \frac{q_e}{4\pi\varepsilon_o c} \frac{\hat{r} \times \left\{ (\hat{r} - 0) \times \dfrac{\vec{a}}{c} \right\}}{r(1-0)^3}$$

$$= \frac{q_e}{4\pi\varepsilon_o cr} \hat{r} \times \left(\hat{r} \times \frac{\vec{a}}{c} \right). \quad \text{q.e.d.}$$

If $\vec{v} = v\hat{z}$ and $\vec{a} = a\hat{z}$, then writing $\hat{r} = r\hat{y}$ gives

$$\left(\hat{r} \times \frac{\vec{a}}{c} \right) = \frac{a}{c}\hat{x}, \quad \text{so} \quad \hat{r} \times \left(\hat{r} \times \frac{\vec{a}}{c} \right) = \frac{a}{c}\hat{y} \times \hat{x} = -\frac{a}{c}\hat{z}.$$

The beam is polarized in the z direction, the direction of the acceleration.

2. X rays are produced by an x-ray tube with a voltage of 30 kV, a current of 1 mA, a source size of 0.5 mm × 0.5 mm, and a thick molybdenum anode.

a) What is the minimum x-ray wavelength produced?

$$\lambda_{min} = \frac{hc}{U_{max}} \approx \frac{12.4\,\text{keV}\,\text{Å}}{30\,\text{keV}} \approx 0.4\,\text{Å}.$$

b) What is the bremsstrahlung power?

$$P = \eta J\Phi = C_\eta Z J\Phi^2 \approx \left(\frac{1.1\times10^{-9}}{V}\right)(42)(10^{-3}\,\text{A})(30\times10^3\,\text{V})^2 \approx 42\,\text{mW}.$$

c) Assuming the bremsstrahlung has a typical triangular frequency spectrum, with its peak at half the maximum energy, what is the emission rate within an energy window of width 0.5 keV centered on the Kα line?

From equation 6-30, the intercept at the peak is

$$\Gamma_P = \frac{6P}{U_{max}\,(U_{max}+U_P\,)} \approx \frac{6(0.04\,\text{W})}{(30\,\text{keV})(45\,\text{keV})}\left(\frac{1\,\text{keV}}{10^3\,\text{eV}}\right)\left(\frac{1\,\text{eV}}{1.6\times10^{-19}\,\text{J}}\right)$$

$$\approx 1.2\times10^{12}\,\frac{\text{photons}}{\text{s}\cdot\text{keV}}.$$

(Note in this case, since the average energy is clearly 15 keV, the total rate of photon emission can be found from the power divided by the average energy,

$$\Gamma = \frac{P}{\bar{U}} \approx \frac{(0.04\,\text{W})}{(15\,\text{keV})}\left(\frac{1\,\text{keV}}{10^3\,\text{eV}}\right)\left(\frac{1\,\text{eV}}{1.6\times10^{-19}\,\text{J}}\right) \approx 1.73\times10^{13}\,\frac{\text{photons}}{\text{s}},$$

and then the intercept can be found from the total emission rate,

$$\Gamma = \frac{1}{2}\Gamma_P U_{max} \Rightarrow \Gamma_P = \frac{2\Gamma}{U_{max}} \approx \frac{2\left(1.73\times10^{13}\,\frac{\text{photons}}{\text{s}}\right)}{30\,\text{keV}} \approx 1.2\times10^{12}\,\frac{\text{photons}}{\text{s}\cdot\text{keV}}.\right)$$

The energy of the Kα line is 17.5 keV, which is larger than the energy at which the spectrum peaks, so the emission rate per kiloelectronvolt at the energy of the Kα line is

$$\Gamma_U = \Gamma_P\frac{U_{max}-U}{U_{max}-U_P} \approx \left(1.2\times10^{12}\,\frac{\text{photons}}{\text{s}\cdot\text{keV}}\right)\frac{30-17.5}{30-15} \approx 9.6\times10^{11}\,\frac{\text{photons}}{\text{s}\cdot\text{keV}}.$$

The amount in the bin is then

$$\Gamma_{0.5\,\text{keV bin}} = \Gamma_U\,\Delta U \approx \left(9.6\times10^{11}\,\frac{\text{photons}}{\text{s}\cdot\text{keV}}\right)(0.5\,\text{keV}) \approx 4.8\times10^{11}\,\frac{\text{photons}}{\text{s}}.$$

d) What is the brilliance of the bremsstrahlung emission in that energy window?

In this case the energy window width is 0.1% of 17.5 keV, which is 17.5 eV. The emission rate into that bin is

$$\Gamma_{0.5\,keV\,bin} \approx \Gamma_{0.5\,keV\,bin} \frac{17.5\,eV}{500\,eV} \approx \left(4.8\times10^{11}\,\frac{photons}{s}\right)\frac{17.5\,eV}{500\,eV}$$

$$\approx 1.7\times10^{10}\,\frac{photons}{s\cdot(0.1\%\,bw)}.$$

The brightness is then

$$\Psi_{\chi Bndwth} \approx \frac{1.7\times10^{10}\,\dfrac{photons}{s\cdot(0.1\%\,bw)}}{4\pi\left(\dfrac{10^3\,mrad}{rad}\right)^2} \approx 1.3\times10^3\,\frac{photons}{mrad^2\cdot s\cdot(0.1\%\,bw)}.$$

The brilliance is

$$b_A = \frac{\Psi_{\chi Bndwth}}{A} \approx \frac{1.3\times10^3\,\dfrac{photons}{mrad^2\cdot s\cdot(0.1\%\,bw)}}{(0.5\,mm)^2}$$

$$\approx 5.4\times10^3\,\frac{photons}{mrad^2\cdot s\cdot(0.1\%\,bw)\cdot mm^2}.$$

This is several orders of magnitude less than the brilliance of a characteristic line.

e) Sketch the x-ray energy spectrum, including both the bremsstrahlung and the characteristic lines (assume doublets are not resolved) measured with a detector with 0.5 keV energy bins. Assume the Kβ line has one-fifth the emission rate of the Kα line, and use $C_{char} = 3\times10^{-4}$. The energies and the heights of the bremsstrahlung and the characteristic lines should be correct.

The bremsstrahlung is linear, rising from 0 at $U=0$ to a peak at 15 keV and then falling back to zero at 30 keV. The peak emission rate (at 15 keV) into a 0.5 keV bin is

$$\Gamma_{brem,\,peak} = \Gamma_P\,\Delta U \approx \left(1.2\times10^{12}\,\frac{photons}{s\cdot keV}\right)(0.5\,keV) \approx 5.8\times10^{11}\,\frac{photons}{s}.$$

We need the characteristic flux, as in chapter 5:

$$\Gamma = C_{char}\frac{J}{q_e}\left(\frac{\Phi-\Phi_b}{\Phi_b}\right)^{1.6} \approx \left(3\times10^{-4}\,\frac{photons}{electron}\right)\left(\frac{1\times10^{-3}\,\dfrac{Coul}{s}}{1.6\times10^{-19}\,\dfrac{Coul}{electron}}\right)$$

$$\times\left(\frac{30\,kV-20\,kV}{20\,kV}\right)^{1.6} \approx 6\times10^{11}\,\frac{photons}{s}.$$

This flux is slightly higher than the bremsstrahlung peak. The sketch is shown in Figure A-2.

f) Repeat part e if the voltage is increased to 60 kV (and the bremsstrahlung peak is at half the new maximum).

Doubling the voltage quadruples the bremsstrahlung power, but the emission rate in the peak bin is unchanged (the number of bins doubles, and the average photon energy in each bin doubles, but the number of photons per bin remains the same). The characteristic emission goes up by a factor of 10:

$$\Gamma = C_{char}\frac{J}{q_e}\left(\frac{\Phi - \Phi_b}{\Phi_b}\right)^{1.6} \approx \left(3\times10^{-4}\frac{photons}{electron}\right)\left(\frac{1\times10^{-3}\frac{Coul}{s}}{1.6\times10^{-19}\frac{Coul}{electron}}\right)$$

$$\times\left(\frac{60\,kV - 20\,kV}{20\,kV}\right)^{1.6} \approx 6\times10^{12}\frac{photons}{s}.$$

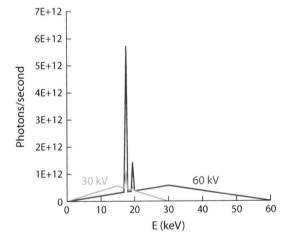

FIGURE A-2. Solution to problem 6-2.

3. An x-ray tube has a thick tungsten, W, anode, a tube voltage of 100 kV, a current of 20 mA, and an electron spot size of 0.2 mm × 2 mm, viewed at an angle of 10°. Assume the emission rate peaks at 0 keV, as in Figure 6-10.

a) What is the bremsstrahlung emission rate within an energy window of width 1 keV centered on 50 keV?

In this case the power is

$$P = \eta J\Phi = C_\eta ZJ\Phi^2 \approx \left(\frac{1.1\times10^{-9}}{V}\right)(74)(20\times10^{-3}A)(100\times10^3V)^2 \approx 16\,W.$$

The intercept at the peak is

$$\Gamma_P = \frac{6P}{U_{max}(U_{max}+U_P)} \approx \frac{6(16\,\mathrm{W})}{(100\,\mathrm{keV})(100\,\mathrm{keV})}\left(\frac{1\,\mathrm{keV}}{10^3\,\mathrm{eV}}\right)\left(\frac{1\,\mathrm{eV}}{1.6\times10^{-19}\,\mathrm{J}}\right)$$

$$\approx 6.1\times10^{13}\,\frac{\mathrm{photons}}{\mathrm{s\cdot keV}}.$$

The emission rate at 50 keV is

$$\Gamma_U = \Gamma_P\frac{U_{max}-U}{U_{max}-0} \approx \left(6.1\times10^{13}\,\frac{\mathrm{photons}}{\mathrm{s\cdot keV}}\right)\frac{100-50}{100} \approx 3\times10^{13}\,\frac{\mathrm{photons}}{\mathrm{s\cdot keV}}.$$

Since the bin is 1 keV wide, $\Gamma \approx 3\times10^{13}\,\dfrac{\mathrm{photons}}{\mathrm{s}}$.

b) What is the brilliance of the bremsstrahlung emission at 50 keV? The bandwidth is 50 eV, so brightness

$$\Psi\chi_{Bndwth} = \frac{3\times10^{13}\,\dfrac{\mathrm{photons}}{\mathrm{s\cdot keV}}}{4\pi\left(\dfrac{10^3\,\mathrm{mrad}}{\mathrm{rad}}\right)^2}(50\times10^{-3}\,\mathrm{keV}) \approx 1.2\times10^5\,\frac{\mathrm{photons}}{\mathrm{mrad}^2\cdot\mathrm{s}\cdot(0.1\%\,\mathrm{bw})}.$$

The source area depends on the observation angle, so the brilliance is

$$b_A = \frac{\Psi\chi_{Bndwth}}{A} \approx \frac{1.2\times10^5\,\dfrac{\mathrm{photons}}{\mathrm{mrad}^2\cdot\mathrm{s}\cdot(0.1\%\,\mathrm{bw})}}{(0.2\,\mathrm{mm})(2\,\mathrm{mm})\sin(10°)}$$

$$\approx 1.8\times10^6\,\frac{\mathrm{photons}}{\mathrm{mrad}^2\cdot\mathrm{s}\cdot(0.1\%\,\mathrm{bw})\,\mathrm{mm}^2}.$$

4. A plot of the spectrum from a silver, Ag, anode tube with a tube voltage of 40 kV and an energy bin of 0.5 keV is shown in Figure A-3. (The K, L, and M energies of silver are approximately 25.5, 3.5, and 0.7 keV, respectively)

a) Give the numerical values of U_A, U_B, and U_C (located at the labeled features on the plot).

$$U_A = U_{K\alpha} = U_L - U_K \approx 22\,\mathrm{keV}.$$
$$U_B = U_{K\beta} = U_M - U_K \approx 24.8\,\mathrm{keV}.$$
$$U_C = q_e\Phi \approx 40\,\mathrm{keV}.$$

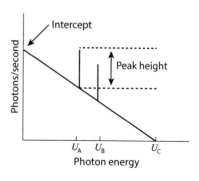

FIGURE A-3. Flux spectrum for problem 6-4.

b) Fill in the table with the approximate factor (for example, if the tube voltage doubles, U_C goes up by a factor of 2, so a 2 has been filled in for that item):

By what factor does the item below change if:	The tube voltage doubles?	The tube current doubles?	The bin size doubles?	The atomic number doubles?
U_A	$a=1$	$b=1$	$c=1$	$d=4$
U_c	2	$e=1$	$f=1$	$g=1$
y axis intercept	$h=1$	$i=2$	$j=2$	$k=2$
The peak height for the peak at U_A	$l=8.3$	$m=2$	$n=1$	$o=0$

a–c) The Kα energy does not change with voltage, current or bin size, so the factor is 1.

d) $U_A = U_{K\alpha} \approx U'_R \left[\dfrac{(Z-\varsigma_1)^2}{1^2} - \dfrac{(Z-\varsigma_2)^2}{2^2} \right]$. Doubling Z increases U_A by a

factor of ~2^2. (The actual Kα energy for Pu, with $Z=94$, is 104 keV, ~4.7 times the Kα energy of silver.)

e–g) Neither current, atomic number, nor bin size changes the maximum kinetic energy of the incoming electrons, so the energy U_C is unchanged, and the factor is one.

h–k) $\Gamma_P \Delta U = \dfrac{6P}{U_{max}(U_{max}+U_P)} \Delta U = \dfrac{6(C_\eta \Phi Z)\Phi J}{q_e \Phi(q_e \Phi)} \Delta U = \dfrac{6C_\eta Z J}{q_e^2} \Delta U,$

so no change with Φ (the power quadruples, but the average photon energy doubles, so the total photon emission rate only doubles. That is spread over twice as many bins, so the intercept does not change with voltage). The intercept doubles with J, ΔU, and Z.

l–n) The peak height is $\Gamma = C_{char} \dfrac{J}{q_e} \left(\dfrac{\Phi - \Phi_o}{\Phi_o} \right)^{1.6}$, so doubling Φ changes the

peak height by a factor of $\left(\dfrac{80-25.5}{40-25.5} \right)^{1.6} \approx 8.3$. It doubles with J and is

unchanged by increasing ΔU.

o) From part d, the ionization energy quadruples to ~100 keV, and the voltage is no longer high enough to excite the Kα peak.

Chapter 7

1. What is the electron velocity if $\gamma = 3$?

$$\gamma = \frac{1}{\sqrt{1-\frac{v^2}{c^2}}} \Rightarrow \frac{1}{\gamma^2} = 1 - \frac{v^2}{c^2} \Rightarrow v = c\sqrt{1-\frac{1}{\gamma^2}} \approx c\sqrt{1-\frac{1}{3^2}}$$

$$\approx 0.943\,c \approx 2.8 \times 10^8\ \frac{m}{s}.$$

2. Consider a 2000 MeV beam and a magnet with a field of 0.5 T. Find the orbit radius, and the relativistic factor γ.

The relativistic factor is $\gamma = \dfrac{U}{U_e} = \dfrac{2000.5\,\text{MeV}}{0.51\,\text{MeV}} \approx 3900,$ so that the radius is

$$R = \gamma \frac{M_e c}{q_e B} \approx (3900)\,\frac{(9.1 \times 10^{-31}\ \text{kg})\left(3 \times 10^8\ \dfrac{m}{s}\right)}{(1.6 \times 10^{-19}\ \text{Coul})(0.5\ \text{T})} \approx 13\ \text{m}.$$

3. Show that equation 7-28 implies both sides of equation 7-29.

$$P_\chi = \frac{q_e}{6\pi\varepsilon_o R}\gamma^4 J = \frac{q_e}{3\pi 2\varepsilon_o R}\frac{q_e hc}{q_e hc}\gamma^4 J = \frac{hc\alpha}{3\pi q_e}\gamma^4\frac{J}{R}.$$

$$\frac{hc\alpha}{3\pi q_e}\gamma^4\frac{J}{R} = \frac{hc\alpha}{3\pi q_e}\left(\frac{U}{U_{eo}}\right)^4\frac{J}{R}$$

$$\approx \frac{12.4 \times 10^3\ \text{eV} \cdot \text{Å}\left(1.6 \times 10^{-19}\ \dfrac{\text{J}}{\text{eV}}\right)\dfrac{10^{-10}\ \text{m}}{\text{Å}}\dfrac{1}{137}}{3\pi\,(1.6 \times 10^{-19}\ \text{Coul})\dfrac{\text{A}}{\text{Coul/s}}\left(0.51\,\text{MeV}\dfrac{10^{-3}\ \text{GeV}}{\text{MeV}}\right)^4}\frac{10^{-3}\ \text{rad}}{\text{mrad}}\frac{JU^4}{R}$$

$$\approx \left(14.1\frac{\text{W} \cdot \text{m}}{\text{A (mrad)}\,\text{GeV}^4}\right)\frac{JU^4}{R}.$$

4. Repeat examples 7-3 and 7-4 for a 2000 MeV beam and a magnet with a field of 0.5 T and a current of 200 mA (with an observation angle of $\theta = 0$). We already have $\gamma = 3900$ and $R = 13$ m from problem 2.

$$P_\chi \approx 14.1\frac{\text{W} \cdot \text{m}}{\text{A (mrad)(GeV)}^4}\frac{(0.2\ \text{A})(2\ \text{GeV})^4}{13\ \text{m}} \approx 3.4\ \frac{\text{W}}{\text{mrad}}.$$

$$P_{\text{detector}} = P_\chi \Delta\chi \approx \left(3.4\,\frac{\text{W}}{\text{mrad}}\right)(1\ \text{mrad}) \approx 3.4\ \text{W}. \quad \Delta\theta \approx \frac{2}{\gamma} \approx 0.5\ \text{mrad}.$$

$$I_\chi \approx \frac{3.4 \dfrac{\text{W}}{\text{mrad}}}{0.5\,\text{mrad}} \approx 6.6\,\frac{\text{W}}{\text{mrad}^2}.$$

$$U_{char} \approx \frac{3}{4\pi} \frac{12.4\,\text{keV Å}}{\left(13\,\text{m}\,\dfrac{10^{10}\,\text{Å}}{\text{m}}\right)}(3900)^3 \approx 1.3\,\text{keV}.$$

$$P_{\chi U} \approx \frac{P_\chi}{2U_{char}} \approx 1.3\,\frac{\text{W}}{\text{keV}}.$$

$$\Psi_{\chi Bndwth} \approx \frac{\dfrac{1}{2}\dfrac{P_\chi}{U_{char}}\dfrac{\Delta U_{bw}}{U_{char}}\dfrac{1}{\Delta\theta}}{} \approx \frac{(6.6\,\text{W}/\text{mrad}^2)}{2\left(1300\,\dfrac{\text{eV}}{\text{ph}}\right)\left(\dfrac{1.6\times10^{-19}\,\text{J}}{\text{eV}}\right)}10^{-3}$$

$$\approx 1.6\times10^{13}\,\frac{\text{photons}}{\text{s}\cdot\text{mrad}^2\,(0.1\%\,\text{bw})}.$$

5. Use the Web resource of Figure 7-13 to check your answers for problem 4. The web result is shown in Figure A-4. The peak energy and brightness agree with the calculation of problem 4

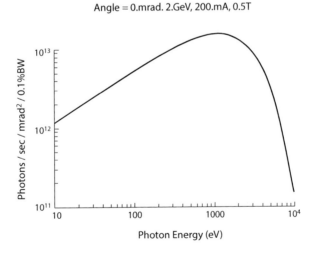

Angle = 0.mrad. 2.GeV, 200.mA, 0.5T

FIGURE A-4. Solution to problem 7-6.

6. For the beam in problem 5, compute the brilliance, source divergence, and transverse coherence length if the electron bunch size is 50 μm, and the experiment is performed at the characteristic energy at a distance of 20 m.

$$b_A = \frac{\Psi_{\chi Bndwth}}{A} \approx \frac{1.5 \times 10^{13} \dfrac{\text{photons}}{\text{s} \cdot \text{mrad}^2 \cdot (0.1\% \, \text{bw})}}{\pi \left(\dfrac{50 \times 10^{-3} \, \text{mm}}{2} \right)^2}$$

$$\approx 7.8 \times 10^{15} \frac{\text{photons}}{\text{s} \cdot \text{mrad}^2 \cdot \text{mm}^2 \cdot (0.1\% \, \text{bw})}.$$

$$\xi = \frac{W_s}{L} \approx \frac{50 \times 10^{-6} \, \text{m}}{20 \, \text{m}} \approx 2.5 \times 10^{-6};$$

$$\lambda = \frac{hc}{U_{char}} \approx \frac{12.4 \, \text{keV Å}}{1.3 \, \text{keV}} \approx 9 \, \text{Å};$$

$$Y_c = \frac{\lambda}{\xi} \approx \frac{9 \times 10^{-10} \, \text{m}}{2.5 \times 10^{-6}} \approx 0.4 \, \text{mm}.$$

7. An insertion device used with a beam of 2000 MeV electrons at 200 mA is 10 m long. Each magnet has a magnetic field of 0.5 T. Calculate the deflection parameter K_{insert}, the cone angle, and the effective radius when the device has a 1 m period and then when it has a 5 mm period. Note that one is a wiggler, and one is an undulator.

These are the same energy parameters as for problem 4, so $\gamma = 3900$.

$$K_{insert} = \frac{q_e B_o d}{2\pi M_e c} \approx \frac{(1.6 \times 10^{-19} \, \text{Coul})(0.5 \, \text{T})}{2\pi \, (9.1 \times 10^{-31} \, \text{kg}) \left(3 \times 10^8 \, \dfrac{\text{m}}{\text{s}} \right)} d \approx \frac{46.7}{\text{m}} d.$$

For $d = 1$ m, $K_{insert} = 47$, and the device is a wiggler. Then, $\Delta\theta = \dfrac{K_{insert}}{\gamma} = 12 \, \text{mrad}.$

$R = \dfrac{d}{2\pi} \dfrac{\gamma}{K_{insert}} \approx 13.4 \, \text{m}$, as for the bending magnet of problem 2. The number of

periods is $N = \dfrac{L}{d} \approx \dfrac{10 \, \text{m}}{1 \, \text{m}} = 10.$

For $d = 5$ mm, $K_{insert} = 0.2$, and the device is an undulator.

$$\Delta\theta = \frac{\sqrt{1 + \dfrac{K_{insert}^2}{2}}}{\gamma \sqrt{N}} \approx \frac{\sqrt{1 + \dfrac{0.2^2}{2}}}{3900 \sqrt{2000}} \approx 0.006 \, \text{mrad}.$$

$R_u = \dfrac{d}{2\pi} \dfrac{\gamma}{K_{insert}} \approx 13.4 \, \text{m}$, which is unchanged, as the d cancels the d in the

deflection parameter.

Chapter 8

1. What is the relative rate of stimulated to spontaneous emission at 100 eV for a plasma in equilibrium at 500,000 K?

$$\frac{\Gamma_{\downarrow,stim}}{\Gamma_{\downarrow,spon}} = \frac{BI_U}{A} \approx e^{-\frac{U}{k_BT}} \approx \exp\left(-\frac{100\,eV}{8.3\times10^{-5}\,\frac{eV}{K}\,5\times10^5\,K}\right) \approx 0.1.$$

About 10% of the emission is stimulated.

2. Consider an undulator with a small insertion parameter, to be used to build a free-electron laser. What is the resonant wavelength of an undulator with $d = 15$ mm and $\gamma = 3000$?

$$\lambda_1 = \frac{d}{2\gamma^2} \approx \frac{15\,mm}{2(3000)^2} \approx 8\,Å.$$

Chapter 9

1. Aluminum has atomic number 13 and density 2.7 g/cm³.

a) For photon energies of 10 and 20 keV, estimate the atomic cross section for photoelectric absorption.

The K edge energy of Al is quite low, about 1.6 keV, so both photon energies are enough to knock out K electrons. The cross sections are

$$\sigma_{10} \approx \sigma_o\left(\frac{U_o}{U}\right)^3 Z^4 \approx (38.8\,\text{barns})\left(\frac{1\,keV}{10\,keV}\right)^3 (13)^4 \approx 1.1\times10^3\,\text{barns.}$$

$$\sigma_{20} \approx \sigma_o\left(\frac{U_o}{U}\right)^3 Z^4 \approx (38.8\,\text{barns})\left(\frac{1\,keV}{20\,keV}\right)^3 (13)^4 \approx 139\,\text{barns.}$$

b) For a photon energy of 20 keV, estimate the absorption coefficient, the mass absorption coefficient, and the absorption length.

The mass absorption coefficient:

$$\frac{\mu_{ab}}{\rho} = \frac{N_A}{M}\sigma_{20} \approx \frac{\left(6.02\times10^{23}\,\frac{\text{atoms}}{\text{mole}}\right)}{27\,\frac{g}{\text{mole}}}(139\times10^{-24}\,cm^2) \approx 3\,\frac{cm^2}{g}.$$

The absorption coefficient: $\mu_{ab} = \rho\frac{\mu_{ab}}{\rho} \approx \left(2.7\,\frac{g}{cm^3}\right)\left(3\,\frac{cm^2}{g}\right) \approx 8.3\,\frac{1}{cm}.$

The absorption length: $\mu_{ab}^{-1} = \frac{cm}{8.3} \approx 0.12\,cm \approx 1.2\,mm.$

c) Estimate the transmission through a 60 μm foil at 20 keV

$$T = e^{-\mu_{ab}L} \approx e^{-\left(8.3\frac{1}{cm}10^{-4}\frac{cm}{\mu m}\right)(60\,\mu m)} \approx 95\%.$$

d) Estimate the imaginary parts of the index of refraction and of the atomic scattering factor at 20 keV.

$$\beta = \frac{\lambda}{4\pi}\mu_{ab} \approx \left(\frac{0.62\times10^{-8}\,cm}{4\pi}\right)(8.3\,cm^{-1}) \approx 4.1\times10^{-9}.$$

$$f_2 = \frac{\sigma_{ab}}{2r_e\lambda} \approx \frac{139\times10^{-24}\,cm^2}{2(2.8\times10^{-13}\,cm)(0.62\times10^{-8}\,cm)} \approx 0.04.$$

2. Lucite, $C_5H_8O_2$, has a density of 0.9 g/cm³. For a photon energy of 30 keV, estimate the atomic cross sections for photoelectric absorption for elemental C, H, and O, and use that information to estimate the absorption coefficient and hence the transmission through a 5 cm thick block.

$$\sigma_C \approx \sigma_o\left(\frac{E_o}{E}\right)^3 Z^4 \approx (38.8\,barns)\left(\frac{1\,keV}{30\,keV}\right)^3 (6)^4 \approx 1.86\,barns.$$

$$\sigma_H \approx \sigma_o\left(\frac{E_o}{E}\right)^3 Z^4 \approx (38.8\,barns)\left(\frac{1\,keV}{30\,keV}\right)^3 (1)^4 \approx 1.4\times10^{-3}\,barn.$$

$$\sigma_O \approx \sigma_o\left(\frac{E_o}{E}\right)^3 Z^4 \approx (38.8\,barns)\left(\frac{1\,keV}{30\,keV}\right)^3 (8)^4 \approx 5.89\,barns.$$

$$\rho_{molecule} = \frac{\rho N_A}{M_{molecule}} \approx \left(0.9\frac{g}{cm^3}\right)\frac{\left(6.02\times10^{23}\frac{molecules}{mole}\right)}{(5\times12+8\times1+2\times16)\frac{g}{mole}}$$

$$\approx 5.4\times10^{21}\frac{molecules}{cm^3}.$$

$$\mu_{ab} = \sum_i \rho_i\sigma_i = \rho_{molecule}\sum_i m_i\sigma_i \approx \left(5.4\times10^{21}\frac{molecules}{cm^3}\right)$$

$$\times\left[5\frac{atoms}{molecule}\left(1.86\frac{barns}{atom}\right)+2\frac{atoms}{molecule}\left(5.86\frac{barns}{atom}\right)\right]$$

$$\times\left(10^{-24}\frac{cm^2}{barn}\right) \approx 0.11\frac{1}{cm}.$$

$$T = e^{-\mu_{ab}L} \approx e^{-\left(0.11\frac{1}{cm}\right)(5\,cm)} \approx 57\%.$$

3. A system designer wants to build an x-ray fluorescence system to use filtered Kα x rays to excite the fluorescence. The table lists (make-believe) elements with their energies. Identify the best choice among the elements on the list (and give a reason) for each of the following.

Element	Kα, keV	Kβ, keV	Ionization energy, keV
A	5	7	10
B	20	23	24
C	22	25	26
D	32	35	40
E	60	70	80

a) The anode: The anode should be chosen so that a filter can be found with an ionization energy between the Kα and Kβ, so that the Kβ will be suppressed. Element C works, since it has a Kβ of 25 keV.

b) The filter: Element B should be used for the filter (the absorption edge at 24 keV means that it is effective at removing the K β.

c) The sample emitting the fluorescence: The ionization energy of the sample should be less than the Kα of element C, so element A works.

d) Identify a reasonable tube voltage.

To excite K emission from element C, the voltage should be greater than 26 kV. To avoid passage of high-energy bremsstrahlung through the filter, the voltage should not be too high. $V = 35$ kV is probably fine.

4. Use the absorption coefficient of Al computed in problem 1 to answer the following:

a) What is the subject contrast of a 1 mm thick void in a 1 cm thick slab of Al at 20 keV?

$$C = \frac{I_1 - I_2}{I_1} = 1 - e^{-\mu_{ab} \Delta z} \approx 1 - e^{-(8.3\,\text{cm}^{-1})(0.1\,\text{cm})} \approx 0.57.$$

b) What is the signal-to-noise ratio for detecting the void if the detected count rate through the material is 10,000 counts/s and the measurement is done for 1 s?

$$Signal \approx C N = C \Gamma t \approx (0.57)\left(10,000 \frac{\text{photons}}{\text{s}}\right)(1\,\text{s}) \approx 5658 \text{ photons}.$$

$$N = \sqrt{N} = \sqrt{\Gamma t} \approx \sqrt{\left(10,000 \frac{\text{photons}}{\text{s}}\right)(1\,\text{s})} \approx 100.$$

$$SNR \approx 57.$$

c) What is the dose, in gray, to the 1 cm × 1 cm × 1 cm Al block?

We need I_o. $I = I_o e^{-\mu_{ab}L} \Rightarrow I_o = Ie^{+\mu_{ab}L}$ and $I = \dfrac{\Gamma U}{A}$.

$$Dose = \frac{I_o(1 - e^{-\mu_{ab}L})t}{\rho d} = \frac{\Gamma e^{+\mu_{ab}L}U(1 - e^{-\mu_{ab}L})t}{A\rho d} = \frac{(\Gamma t)U(e^{+\mu_{ab}L} - 1)}{A\rho d}$$

$$= \frac{(10{,}000)(e^{8.3\,\mathrm{cm}^{-1}(5\,\mathrm{cm})} - 1)(20\,\mathrm{keV})}{(1\,\mathrm{cm})^2 \left(2.7\,\dfrac{\mathrm{g}}{\mathrm{cm}^3}\right)(1\,\mathrm{cm})} \left(\frac{1.6 \times 10^{-16}\,\mathrm{J}}{\mathrm{keV}}\right) \left(10^3\,\frac{\mathrm{g}}{\mathrm{kg}}\right)$$

$$\approx 5 \times 10^{-5}\,\frac{\mathrm{J}}{\mathrm{kg}} \approx 50\,\mu\mathrm{Gy}.$$

Chapter 10

1. Compute the scattered photon energy for 100 keV x rays scattered at an angle of
a) 0°, b) 90°, and c) 180°.

$$\lambda = \frac{hc}{U} \approx \frac{12.4\,\mathrm{keV} \cdot \text{Å}}{100\,\mathrm{keV}} \approx 0.124\,\text{Å}$$

$$\lambda' = \lambda + \frac{h}{M_e c}(1 - \cos\theta) \approx 0.124\,\text{Å} + (0.024\,\text{Å})(1 - \cos\theta).$$

a) For $\theta = 0°$, $1 - \cos(\theta) = 0$, so the final energy is 100 keV.

b) For $\theta = 90°$, $1 - \cos(\theta) = 1$,

$$\lambda' \approx 0.124\,\text{Å} + 0.024\,\text{Å} \approx 0.148\,\text{Å} \Rightarrow U = \frac{hc}{\lambda'} \approx 84\,\mathrm{keV}.$$

c) For $\theta = 180°$, $1 - \cos(\theta) = 2$,

$$\lambda' \approx 0.124\,\text{Å} + 0.0485\,\text{Å} \approx 0.173\,\text{Å} \Rightarrow U = \frac{hc}{\lambda'} \approx 72\,\mathrm{keV}.$$

2. What is the backscattered photon energy observed in the lab frame if the initial
visible light wavelength was $\lambda_o = 0.5\,\mu\mathrm{m}$, and the light was scattered off 25 MeV
electrons?

$$\gamma = \frac{U + M_e c^2}{M_e c^2} \approx \frac{25.5\,\mathrm{MeV}}{0.51\,\mathrm{MeV}} \approx 50 \Rightarrow \lambda'' \approx \frac{\lambda_o}{4\gamma^2} \approx \frac{0.5\,\mu\mathrm{m}}{4(50)^2} \approx 0.5\,\text{Å}$$

$$\Rightarrow U'' = \frac{hc}{\lambda''} \approx \frac{12.4\,\mathrm{keV}\,\text{Å}}{0.5\,\text{Å}} \approx 25\,\mathrm{keV}.$$

3. Prove equation 10-13.

$$SF = \frac{I_S}{I_S + I_P} \Rightarrow 1 - SF = \frac{I_S + I_P}{I_S + I_P} - \frac{I_S}{I_S + I_P} = \frac{I_P}{I_S + I_P} \Rightarrow \frac{SF}{1 - SF} = \frac{I_S}{I_P}. \quad \text{q.e.d.}$$

4. A 10 cm thick patient is imaged at a photon energy with an attenuation length
of 10 cm. At that energy, the subject contrast for a tumor in the patient is
$C = 0.1$. The image is taken in 10 s with an incident photon intensity of

2.72×10^6 photons/cm²/s, and a detector pixel size of $100\,\mu m \times 100\,\mu m$. The scatter-to-primary ratio measured after the patient is 9.

a) What is the scatter fraction?

$$SF = \frac{I_s}{I_s + I_p} = \frac{I_s/I_p}{I_s/I_p + 1} \approx \frac{9}{9+1} \approx 0.9$$

b) What is the image contrast?

$$\boldsymbol{C}_s = \boldsymbol{C}(1-SF) \approx 0.1(1-0.9) \approx 0.01.$$

c) What are the total and primary photon intensities at the detector (after absorption in the patient)?

$$I_p = I_o e^{-\mu_{tot}L} = I_o e^{-\frac{L}{\mu_{tot}^{-1}}} \approx \left(2.72 \times 10^6\, \frac{\text{photons}}{\text{cm}^2 \cdot \text{s}}\right) e^{-\frac{10\,\text{cm}}{10\,\text{cm}}} \approx 10^6\, \frac{\text{photons}}{\text{cm}^2 \cdot \text{s}}.$$

The scatter intensity is 9 times this value, $I_s \approx 9 \times 10^6\, \dfrac{\text{photons}}{\text{cm}^2 \cdot \text{s}}$, so the total intensity is $I_{tot} \approx 10^7\, \dfrac{\text{photons}}{\text{cm}^2 \cdot \text{s}}$.

d) What is the number of photons per pixel in a region of normal tissue outside the tumor?

$$N_{total} = IAt \approx \left(10^7\, \frac{\text{photons}}{\text{cm}^2 \cdot \text{s}}\right)\left(100\,\mu m\, \frac{10^{-4}\,\text{cm}}{\mu m}\right)^2 (10\,\text{s}) \approx 10^4;$$

$$N_p = (1-SF)(N_{total}) \approx (0.1)10^4 \approx 10^3.$$

e) What would the signal-to-noise ratio be if the scatter was removed before it hit the detector?

Since only the primary beam hits the detector: $S = \boldsymbol{C}\, N_p \approx (0.1)(10^3) \approx 10^2.$

The noise is $\boldsymbol{N} = \sqrt{N_p} \approx 32 \Rightarrow SNR = \dfrac{S}{\boldsymbol{N}} \approx \dfrac{100}{32} \approx 3.2.$

f) What is the signal-to-noise ratio with the scatter?

The signal is still only from the primary beam and so is unchanged. (Alternatively, $S = \boldsymbol{C}_s\, N_{tot} \approx (0.01)(10^4) \approx 10^2$.) However, both the primary and the scatter hit the detector and contribute to the noise:

$\boldsymbol{N} = \sqrt{N_{tot}} \approx \sqrt{10^4} \approx 100 \Rightarrow SNR = \dfrac{S}{\boldsymbol{N}} \approx 1.$ The scatter has rendered the tumor undetectable.

5. An image has an image contrast of 0.05 and a scatter fraction of 80%. A grid is to be designed to increase the contrast by a factor of 3. The primary transmission is 70%.

a) What is the subject contrast?

$$\boldsymbol{C}_s = \boldsymbol{C}(1-SF) \Rightarrow \boldsymbol{C} = \frac{\boldsymbol{C}_s}{1-SF} \approx \frac{0.05}{1-0.8} \approx 0.25.$$

b) What is the maximum scatter fraction that will give three times the original image contrast?

$$C_{s2} = 3C_s = C(1 - SF_2) \Rightarrow SF_2 = 1 - \frac{3C_s}{C} \approx 1 - \frac{0.15}{0.25} \approx 0.4.$$

c) Solve for the maximum allowable scatter transmission for the grid

The original scatter-to-primary ratio is $SPR = \dfrac{I_S}{I_P} = \dfrac{SF}{1 - SF} \approx \dfrac{0.8}{1 - 0.8} \approx 4.$ The

scatter-to-primary ratio after the grid will be $SPR_2 = \dfrac{SF_2}{1 - SF_2} \approx \dfrac{0.4}{1 - 0.4} \approx 0.67.$

$$SPR_2 = \frac{T_s I_s}{T_p I_p} = \frac{T_s}{T_p} SPR_1 \Rightarrow T_s = T_p \frac{SPR_2}{SPR_1} \approx (0.7)\frac{0.67}{4} \approx 0.12.$$

The grid must not transmit more than 12% of the scatter.

Chapter 11

1. Copper has a density of 9 g/cm³, atomic weight of 63 g/mole, and 29 electrons per atom. It is irradiated with a 10 keV x-ray beam.

a) Compute its plasma energy, in electronvolts.

$$\rho_e = Z\rho_{atom} = Z\frac{\rho N_A}{M_m} \approx (29)\left(9\,\frac{g}{cm^3}\right)\left(\frac{10^2 cm}{m}\right)^3 \frac{\left(6.02 \times 10^{23}\,\dfrac{molecules}{mole}\right)}{63\,\dfrac{g}{mole}}$$

$$\approx 2.5 \times 10^{30}\,\frac{1}{m^3}.$$

$$\omega_p = \sqrt{\frac{\rho_e q_e^2}{M_e \varepsilon_o}} = \sqrt{\frac{\left(2.5 \times 10^{30}\,\dfrac{1}{m^3}\right)(1.6 \times 10^{-19}\,C)^2}{(9.1 \times 10^{-31}\,kg)\left(8.85 \times 10^{-12}\,\dfrac{C^2}{N \cdot m^2}\right)}} \approx 8.9 \times 10^{16}\,\frac{rad}{s}.$$

$$U_p = \hbar\omega_p \approx \left(\frac{4.1 \times 10^{-15}\,eV \cdot s}{2\pi}\right)\left(8.9 \times 10^{16}\,\frac{1}{s}\right) \approx 58.7\,eV.$$

b) Compute the index of refraction decrement δ.

$$\delta = \frac{1}{2}\frac{\omega_p^2}{\omega^2} = \frac{1}{2}\frac{U_p^2}{U^2} \approx \frac{1}{2}\left(\frac{58.3\,eV}{10 \times 10^3 eV}\right)^2 \approx 1.7 \times 10^{-5}.$$

c) Compute the real part of the atomic scattering factor.

$$f_1 = \frac{2\pi}{r_e \lambda^2 \rho_{atom}}\delta. \text{ Here } \lambda = \frac{hc}{U} \approx \frac{12.4\,keV \cdot \text{Å}}{10\,keV} \approx 1.24\,\text{Å. and}$$

$$\rho_{atom} = \frac{\rho N_A}{M_m} \approx 8.6 \times 10^{28}\,\frac{1}{m^3}. \text{ As usual, we obtain } f_1 = Z = 29.$$

d) Compute the phase velocity.

$$v_p = \frac{c}{n} \approx c(1+\delta) \approx 1.00002 \, c.$$

e) Compute the cross section for scattering for a single free electron.

$$\sigma_s = \frac{8}{3}\pi r_e^2 \approx \frac{8}{3}\pi(2.82 \times 10^{-13} \text{ cm})^2 \approx 6.6 \times 10^{-25} \text{ cm}^2 \approx 0.66 \text{ barn}.$$

f) Compute the scattering coefficient μ.

$$\mu_s = \rho_{atom}\sigma_s f_1 = \rho_e \sigma_s \approx \left(2.5 \times 10^{24} \frac{1}{\text{cm}^3}\right)(0.67 \times 10^{-24} \text{ cm}^2) \approx 1.66 \text{ cm}^{-1}.$$

g) Compute the fraction loss to scattering if the beam passes through 1 mm of copper.

$$Loss = I_o - I; \textit{ fractional loss} = \frac{I_o - I}{I_o} = \frac{I_o - I_o e^{-\mu_s L}}{I_o}$$

$$= 1 - e^{-\mu_s L} \approx 1 - e^{-\left(1.6\frac{1}{\text{cm}}\right)(0.1\,\text{cm})} \approx 1 - 0.85 \approx 0.15.$$

h) Compute the critical angle θ_c.

$$\theta_c = \sqrt{2\delta} \approx \sqrt{2(1.7 \times 10^{-5})} \approx \frac{U_p}{U} = 5.9 \text{ mrad}.$$

i) Compute the reflectivity for x rays incident at $1.1\theta_c$ and $0.9\theta_c$ with and without including β and Debye-Waller-like surface roughness with a height of 5 Å.

In the absence of roughness $\boldsymbol{R} = \boldsymbol{R_o}$.

$$\theta_{A,\,B} = \sqrt{\mp\frac{\Delta}{2} + \frac{1}{2}\sqrt{\Delta^2 + 4\beta^2}}; \quad \Delta = \theta_c^2 - \theta^2.$$

$$\boldsymbol{R_o} \approx \frac{(\theta - \theta_A)^2 + \theta_B^2}{(\theta + \theta_A)^2 + \theta_B^2}.$$

For the case in which we include absorption we need β for copper. The cross section for absorption is

$$\sigma_{10} \approx \sigma_o \left(\frac{U_o}{U}\right)^3 Z^4 \approx (38.8 \text{ barns})\left(\frac{1\,\text{keV}}{10\,\text{keV}}\right)^3 (29)^4 \approx 27 \times 10^3 \text{ barns}.$$

$$\mu_{ab} = \rho_{atom}\sigma_{ab} \approx \left(8.6 \times 10^{22} \frac{\text{atoms}}{\text{cm}^3}\right)\left(27 \times 10^3 \frac{\text{barns}}{\text{atom}}\right)\left(10^{-24} \frac{\text{cm}^2}{\text{barn}}\right)$$

$$\approx 2.4 \times 10^3 \frac{1}{\text{cm}}.$$

So $\beta = \dfrac{\lambda \mu_{ab}}{4\pi} \approx 2.3 \times 10^{-6}$. We also need the Debye-Waller roughness factor,

$$\Delta R_{DW} \approx R_o \frac{16\pi^2}{\lambda^2} z_{rms}^2 \, \theta^2 \approx 2.6 \times 10^3 \, R_o \theta^2.$$ The results are given in the table.

θ_c	θ	Δ	β	θ_A	θ_B	R_o	ΔR_{DW}	Net R
5.80E–03	5.28E–03	6.54E–06	0	0.0E+00	2.6E–03	1.00	0.07	0.93
	5.28E–03	6.54E–06	2.30E–06	8.6E–04	2.7E–03	0.59	0.04	0.55
	6.45E–03	–7.2E–06	0	2.7E–03	0.0E+00	0.17	0.02	0.15
	6.45E–03	–7.2E–06	2.30E–06	2.8E–03	8.3E–04	0.16	0.02	0.14

For the smaller angle, the reflectivity including β has dropped 40% compared with the case without absorption.

Chapter 12

1. a) Compute the focal length for a copper lens with a negative 1 cm radius of curvature at 10 keV (use the parameters for the problems from chapter 11).

$$L_f = \frac{R/2}{n-1} \approx \frac{R}{-2\delta} \approx \frac{-1\,\text{cm}}{2(-1.7 \times 10^{-5})} \approx 290\,\text{m}.$$

b) Compute the focal length for a compound optic with a radius of -0.1 cm, and 100 lenses.

$$L'_f = \frac{R}{N2\delta} \approx \frac{-0.1\,\text{cm}}{2(100)(-1.7 \times 10^{-5})} \approx 0.29\,\text{m}.$$

c) Estimate the smallest possible focal spot from this optic. If it were diffraction limited it would be quite small:

$$R_{spot} \approx \frac{\lambda L_f}{2R} \approx \frac{(1.24 \times 10^{-10}\,\text{m})(0.29\,\text{m})}{2(10^{-3}\,\text{m})} \approx 0.02\,\mu\text{m}.$$

2. An elliptical mirror is made from copper to be used at 10 keV. If the beam to be focused is 2 mm thick, and the source is 3 m away, how long must the mirror be?

$$L \approx \frac{y}{\theta_c} \approx \frac{2 \times 10^{-3}\,\text{m}}{(5.9 \times 10^{-3})} \approx 0.34\,\text{m}.$$ The distance to the mirror is irrelevant.

3. What is the expected focal spot size at 10 keV from a polycapillary optic with a focal length of 20 mm and a channel size of 2 μm.

$$\theta_c \approx \frac{U_p}{U} \approx \frac{30\,\text{eV}}{10\,\text{keV}} \approx 3\,\text{mrad}.$$

$$w_{spot} \approx 1.3 L_f \theta_c + w \approx 1.3(20 \times 10^{-3}\,\text{m})(3 \times 10^{-3}) + 2\,\mu\text{m} \approx 80\,\mu\text{m}.$$

4. What is the focal length of a micropore optic with a radius of curvature of 1 m?

$$L_f = \frac{R}{2} = 0.5 \text{ m.}$$

Chapter 13

1. A face-centered cubic (fcc) crystal has 4 atoms per unit cell: at a corner and at the center of the three nearest faces: (0, 0, 0), (0, 1/2, 1/2), (1/2, 0, 1/2), and (1/2, 1/2, 0). Show that the structure factor for an fcc crystal is

$$\boldsymbol{F} = f \left(1 + (-1)^{h+k} + (-1)^{h+l} + (-1)^{k+l}\right), \text{ when } h, k, l \text{ are all even or all odd.}$$

The lattice is cubic, so, as in example 13-5, $\vec{u}_1 = u\hat{x}, \vec{u}_2 = u\hat{y}, \vec{u}_3 = u\hat{z}$, and $\vec{G} = \frac{2\pi}{u}(h\hat{x} + k\hat{y} + l\hat{z})$. The four atom locations are as follows: for the one at the corner, $\vec{r}_1 = (0\vec{u}_1 + 0\vec{u}_2 + 0\vec{u}_3) = 0$, and for the three faces,

$$\vec{r}_2 = \left(0\vec{u}_1 + \frac{1}{2}\vec{u}_2 + \frac{1}{2}\vec{u}_3\right) = \frac{u}{2}(\hat{y} + \hat{z}), \ \vec{r}_3 = \frac{u}{2}(\hat{x} + \hat{z}), \text{ and } \vec{r}_4 = \frac{u}{2}(\hat{x} + \hat{y}). \text{ Thus}$$

$$\vec{G} \cdot \vec{r}_1 = 0, \ \vec{G} \cdot \vec{r}_2 = \frac{2\pi}{u}(h\hat{x} + k\hat{y} + l\hat{z}) \cdot \frac{u}{2}(\hat{y} + \hat{z}) = \pi(k+l), \ \vec{G} \cdot \vec{r}_3 = \pi(h+l), \text{ and}$$

$\vec{G} \cdot \vec{r}_4 = \pi(h+k)$. Since $e^{i\pi} = \cos(\pi) + i\sin(\pi) = -1$, the structure factor is

$$\boldsymbol{F}_{FCC}(h, k, l) = \sum_j f_j e^{i\vec{G} \cdot \vec{r}_j} = f\left(1 + e^{i\pi(h+k)} + e^{i\pi(h+l)} + e^{i\pi(k+l)}\right)$$

$$= f\left(1 + (-1)^{(h+k)} + (-1)^{(h+l)} + (-1)^{(k+l)}\right)$$

$$= \begin{cases} 4f & \text{if } h, k, l \text{ are all even or all odd} \\ 0 & \text{otherwise} \end{cases}.$$

2. A polycrystalline sample of an fcc crystal with a unit cell cube edge of 0.3 nm is irradiated with 10 keV photons.

   a) What is the reciprocal lattice constant (the length of the edge of the reciprocal lattice cube)?

   $$g = \frac{2\pi}{u} \approx \frac{2\pi}{3\text{Å}} \approx 2.1 \text{ Å}^{-1} \approx 21 \text{ nm}^{-1}.$$

   b) Find the diameters of all the diffraction rings on a 300 mm diameter detector placed 100 mm on the source side of the sample (the detector has a hole to allow the incident beam to pass through). $\lambda = \frac{hc}{U} \approx \frac{12.4 \text{ keV} \cdot \text{Å}}{10 \text{ keV}} \approx 1.24 \text{ Å}.$

   The calculations are shown in the table. Because the detector is on the source side of the sample, we will see rings only for $2\theta > 180°$. The ring radius will then be $R = L \tan(\pi - 2\theta)$. Sufficient angle for backscattering requires large (hkl), but if (hkl) is too large, it is not possible to satisfy Bragg's law with the given wavelength. Thus there are only two diffraction rings.

h	k	l	$h^2+k^2+l^2$	$F=1+(-1)^{h+k}$ $+(-1)^{h+l}+(-1)^{k+l}$	$d=\dfrac{u}{\sqrt{h^2+k^2+l^2}}$	$\sin\theta_B=\dfrac{\lambda}{2d}$	2θ	$L\,tan(\pi-2\theta)$
0	0	0	0	4	∞	0	0.0	0.0
1	0	0	1	F is zero; no ring				
1	1	0	2	0				
1	1	1	3	4	1.73	0.36	41.9	Forward directed (so does not hit detector)
2	0	0	4	4	1.50	0.41	48.8	Forward
2	1	0	5	0				
2	1	1	6	0				
2	2	1	9	0				
3	0	0	9	0				
3	0	1	10	0				
3	1	1	11	4	0.90	0.69	86.5	Forward
2	2	2	12	4	0.87	0.72	91.4	Forward
3	2	1	14	0				
4	0	0	16	4	0.75	0.83	111.5	253.7, misses detector
3	2	2	17	0				
4	0	1	17	0				
4	1	1	18	0				
3	3	0	18	0				
3	3	1	19	4	0.69	0.90	128.5	125.7
4	2	0	20	4	0.67	0.92	135.1	99.6
4	2	1	21	0				
3	3	2	22	0				
4	2	2	24	4	0.61	1.01	Not	allowed

c) The grain size is 50 nm. What is the width of the rings due to grain-size broadening?

For the smallest ring:

$$\frac{\Delta\theta}{\tan\theta} = \frac{d}{W_{grain}} \Rightarrow \Delta 2\theta = 2\tan\theta\,\frac{d}{W_{grain}} \approx 2\tan\left(\frac{135°}{2}\right)\frac{0.67\,\text{Å}}{500\,\text{Å}}$$

$$\approx 6.5\,\text{mrad} \approx 0.37°.$$

For the other ring:

$$\Delta 2\theta_2 = 2\tan\left(\frac{128°}{2}\right)\frac{0.69\,\text{Å}}{500\,\text{Å}} = 5.7\,\text{mrad} = 0.33°.$$

d) What is the largest sample size that will give a ring width small enough to observe the grain-size broadening in part c if the source is 200 mm from the sample?

Following equation 13-79, but in the case of backscatter,

$$2\theta = \arctan\left(\frac{r_{ring}}{L}\right) = \arctan\left(\tan\left(\pi - 2\theta_B \pm \frac{\Theta}{2}\right) \pm \frac{W}{2L}\right)$$

$$\approx 2\theta_B \pm \frac{\Theta}{2} \pm \frac{W}{2L}\cos^2(\pi - 2\theta_B),$$

where W is the sample size. The global divergence is $\Theta \approx \dfrac{W}{z}$. Assuming the sample size broadening must be less than the smallest grain size broadening from part c,

$$\Delta 2\theta_{size} \approx \frac{W_{sample}}{L}\cos^2(\pi - 2\theta_B) + \frac{W_{sample}}{z} < \Delta 2\theta_{grain}$$

$$\Rightarrow W_{sample} < \frac{\Delta 2\theta_{grain}}{\dfrac{1}{z} + \dfrac{\cos^2(2\theta_B)}{L}} \approx \frac{5.7\times 10^{-3}\,\text{rad}}{\dfrac{1}{200\,\text{mm}} + \dfrac{1}{100\,\text{mm}}\cos^2(128°)} \approx 0.6\,\text{mm}.$$

3. Silicon is diamond cubic. A diamond cubic crystal has eight atoms per unit cell in two groups of four: the first group includes a corner and the center of the three nearest faces: (0, 0, 0), (0, 1/2, 1/2), (1/2, 0, 1/2), and (1/2, 1/2, 0)

The second group is displaced 1/4, 1/4, 1/4 from those four: (1/4, 1/4, 1/4), (1/4, 3/4, 3/4), (3/4, 1/4, 3/4), and (3/4, 3/4, 1/4).

Show that the structure factor for the (100), (200) and (300) planes is zero.

Following the pattern of problem 13-1, there are eight terms in the structure factor, but they occur in two groups of four. Because the structure factor can be expressed as a product, both terms must be nonzero for diffraction to be observed.

$$\mathbf{F}_{dc} = f \left(\begin{array}{c} 1 + e^{i\pi(h+k)} + e^{i\pi(h+l)} + e^{i\pi(k+l)} + e^{i\frac{\pi}{2}i(h+k+l)} + e^{i\frac{\pi}{2}(3h+3k+l)} \\ + e^{i\frac{\pi}{2}(3h+k+3l)} + e^{i\frac{\pi}{2}(h+3k+3l)} \end{array} \right)$$

$$= f \left(1 + e^{i\frac{\pi}{2}(h+k+l)} \right) \left(1 + e^{i\pi(h+k)} + e^{i\pi(h+l)} + e^{i\pi(k+l)} \right)$$

$$= F_{fcc} \left(1 + e^{i\frac{\pi}{2}(h+k+l)} \right)$$

$$= \begin{cases} 8f & \text{if } (h,k,l \text{ are all even}) \text{ AND } h+k+l=4n \\ 4f(1\pm i) & \text{if } h,k,l \text{ all odd} \\ 0 & \text{else} \end{cases} .$$

The (100) and (300) planes are not all even or all odd. The (200) plane does not satisfy $h+k+l=4n$.

Chapter 14

1. A powder diffractometer uses 10 keV x rays. What is the radius of the Ewald sphere?

$R_{Ewald} = \dfrac{2\pi}{\lambda} \approx \dfrac{2\pi}{1.24 \,\text{Å}} \approx 5.1 \,\text{Å}^{-1}$. This can be compared with the reciprocal lattice length of $g \approx 2.1 \,\text{Å}^{-1}$ for a crystal with lattice constant of $u = 3 \,\text{Å}$ in problem 13-2a.

2. A face-centered cubic (fcc) single crystal with lattice constant $u = 3 \,\text{Å}$ is irradiated with a bremsstrahlung source producing x rays of energy ranging from 0 to 40 keV. The x-ray beam is incident in the $-y$ direction. The detector is 30 mm from the sample (on the side away from the source). Find the wavelength for the diffraction from the (111) plane and the (x, y) coordinate of the spots on the detector.

For the $(hkl) = (111)$ plane, the plane spacing is $d = \dfrac{u}{\sqrt{h^2 + k^2 + l^2}} \approx \dfrac{3\,\text{Å}}{\sqrt{3}} \approx 1.7 \,\text{Å}$.

Following example 14-1, using $\vec{\kappa}_f = -\dfrac{2\pi}{\lambda}\hat{y} + \dfrac{2\pi}{u}(h\hat{x} + k\hat{y} + l\hat{z})$, where

$\vec{\kappa}_f \cdot \vec{\kappa}_f = \left(\dfrac{2\pi}{\lambda}\right)^2$ yields $\lambda = \dfrac{2ku}{h^2 + k^2 + l^2}$. For the (111) plane

$\lambda = \dfrac{2ku}{h^2 + k^2 + l^2} \approx \dfrac{2(1)(3\,\text{Å})}{1^2 + 1^2 + 1^2} \approx 2 \,\text{Å}$. Then applying Bragg's law as usual gives

$\theta = \sin^{-1}\left(\dfrac{\lambda}{2d}\right) \approx \sin^{-1}\left(\dfrac{2\,\text{Å}}{2(1.7\,\text{Å})}\right) \approx 35°$. So the radius for forward scatter is

$R = L \tan(2\theta) \approx (30 \text{ mm}) \tan(70.5°) \approx 85 \text{ mm}$. In this case instead of a ring we will

get a single spot, so we also need the azimuthal angle,

$$\varphi = \tan^{-1}\left(\frac{k_z}{k_x}\right) = \tan^{-1}\left(\frac{l}{k}\right) \approx 45°.$$

The location on the detector is thus $(R \cos(\varphi), R \sin(\varphi)) \approx (60, 60)$ mm.

3. A bcc crystal in a θ-2θ diffractometer run with Cu Kα radiation has a peak for the (111) planes at $2\theta = 40°$.

a) What is the d spacing of the crystal?

$$\lambda \approx \frac{12.4 \text{ keV} \cdot \text{Å}}{8 \text{ keV}} \approx 1.55 \text{ Å}. \text{ Then the plane spacing from Bragg's law is}$$

$$d = \frac{\lambda}{2 \sin \theta} \approx \frac{1.55 \text{ Å}}{2 \sin(20°)} \approx 2.266 \text{ Å}.$$

b) The sample is strained by 1% (the lattice constant increases by 1%). What is the new peak angle if the strain is parallel to the (111) planes?
This changes the spacing in the plane but doesn't change the d spacing between the planes, or the diffraction angle.

c) If the strain is perpendicular to the planes?

$$d' = (1.01)d = 2.289 \text{ Å} \Rightarrow 2\theta' = 2\sin^{-1}\left(\frac{\lambda}{2d'}\right) = 2\sin^{-1}\left(\frac{1.55 \text{ Å}}{2(2.289 \text{ Å})}\right) = 39.6°.$$

The angle changed by 0.4°, so broadening effects from sample and beam sizes need to be small to measure the strain.

4. Compute the Darwin width for diffraction from a (111) silicon crystal at 8 keV. The structure factor is given in problem 13-3.

$$\boldsymbol{F}(h, k, l) = \boldsymbol{F}_{FCC}\left(1 + e^{i\frac{\pi}{2}(h+k+l)}\right) = 4f_o(1 - i), \text{ so that } \sqrt{|\boldsymbol{F}|^2} = (4f_o)\sqrt{2}.$$

Silicon has a lattice constant of 0.54 nm, so the plane spacing is

$$d = \frac{u}{\sqrt{h^2 + k^2 + l^2}} \approx \frac{5.4 \text{ Å}}{4} \approx 3.1 \text{ Å},$$

and the Bragg angle is

$$\theta = \sin^{-1}\left(\frac{\lambda}{2d}\right) \approx \sin^{-1}\left(\frac{1.55 \text{ Å}}{2(3.1 \text{ Å})}\right) \approx 14°.$$

The factor proportional to the momentum transfer is $\dfrac{\sin(\theta)}{\lambda} \approx 0.16 \text{ Å}^{-1}$, and, from Figure 13-4, the scattering factor is about 11. The Darwin width is then

$$\Delta\theta = 2r_e\lambda^2\left(\frac{1+\cos^2\theta}{2}\right)\frac{\sqrt{|\mathbf{F}|^2}}{\pi V_c \sin 2\theta}$$

$$\approx (2.8\times10^{-15}\,\text{m})(1.55\times10^{-10}\,\text{m})^2\left(\frac{1+\cos^2(14°)}{2}\right)\frac{4(11)\sqrt{2}}{\pi(5.4\times10^{-10}\,\text{m})^3\sin(28°)}$$

$$\approx 34\,\mu\text{rad} \approx (34\times10^{-6})\left(\frac{180\,\text{deg}}{\pi}\right)\left(\frac{60\,\text{min}}{\text{deg}}\right)\left(\frac{60\,\text{sec}}{\text{min}}\right) \approx 7\,\text{arcsec.}$$

Chapter 15

1. A zone plate with 40 open zones is designed for 10 nm radiation, with a focal length of 50 cm.

   a) What is the zone plate diameter?

   The number of zones in the zone plate area is $N_Z = 2N_{open} = 2(40) = 80$. The radius of the outermost zone is then

   $$R_{N_Z} = \sqrt{N_Z\lambda L_f} \approx \sqrt{80(10\times10^{-9}\,\text{m})(50\times10^{-2}\,\text{m})} \approx 630\,\mu\text{m,}$$

   so the diameter is 1.3 mm.

   b) What is the focal spot size?

   $$R_{spot} \approx 1.22 w_{N_Z} \approx 1.22\left(R_{N_Z} - R_{N_Z-1}\right) \approx 1.22\frac{\sqrt{\lambda L_f}}{2\sqrt{N_Z}}$$

   $$\approx \frac{1.22\sqrt{(10\times10^{-9}\,\text{m})(50\times10^{-2}\,\text{m})}}{2\sqrt{80}} \approx 4.8\,\mu\text{m.}$$

   c) Where is the third-order image of an object placed 40 cm from the zone plate?

   $$L_{f3} = \frac{L_f}{3}.$$ The focal length can be used in the thin-lens equation,

   $$L = \frac{1}{\dfrac{1}{L_{f3}} - \dfrac{1}{z}} \approx \frac{1}{\dfrac{3}{50\,\text{cm}} - \dfrac{1}{40\,\text{cm}}} \approx 29\,\text{cm.}$$

   The image is 29 cm from the zone plate, 69 cm from the object.

2. A multilayer film is composed of alternating 6 nm layers of C, with a plasma energy of 29 eV, and 6 nm layers of W, which has a plasma energy of 80 eV. It is to be used to diffract 0.3 keV x rays. What is the required incident angle?

First, the wavelength is

$$\lambda = \frac{hc}{U} \approx \frac{12.4 \times 10^{-10} \, \text{m} \cdot \text{keV}}{0.3 \, \text{keV}} \approx 4.1 \times 10^{-9} \, \text{m}.$$

If we ignored the index of refraction, the Bragg angle would have been

$$\theta'_B = \sin^{-1}\left(\frac{\lambda}{2d_M}\right) \approx \sin^{-1}\left(\frac{4.1 \, \text{nm}}{2(12 \, \text{nm})}\right) \approx 9.9°.$$

The individual index decrements are

$$\delta_C = \frac{1}{2}\frac{U_{p,\,Mo}}{U^2} \approx \frac{1}{2}\left(\frac{29 \, \text{eV}}{0.3 \times 10^3 \text{eV}}\right)^2 \approx 0.005$$

$$\text{and} \quad \delta_W = \frac{1}{2}\frac{U^2_{p,\,Mo}}{U^2} \approx \frac{1}{2}\left(\frac{80 \, \text{eV}}{0.3 \times 10^3 \text{eV}}\right)^2 \approx 0.036,$$

so that the average index is

$$\delta = \frac{d_C\delta_C + d_W\delta_W}{d_M} \approx \frac{6(0.005) + 6(0.036)}{12} \approx 0.028.$$

The real Bragg angle is thus

$$\theta_B = \sin^{-1}\left(\frac{\lambda}{2d_M(1-\delta)}\right) \approx \sin^{-1}\left(\frac{41 \, \text{Å}}{2(120 \, \text{Å})(1-0.028)}\right) \approx 10.2°.$$

The incident angle can be found from Snell's law,

$$\theta_o = \cos^{-1}((1-\delta)\cos\theta_B) \approx \cos^{-1}((1-0.028)\cos(10°)) \approx 17°.$$

For these soft x rays, the incident angle needs to be nearly twice that calculated from the usual Bragg's law.